MATHEMATICAL STATISTICS FOR APPLIED ECONOMETRICS

MATHEMATICAL STATISTICS FOR APPLIED ECONOMETRICS

CHARLES B. MOSS
University of Florida
Gainesville, USA

CRC Press is an imprint of the
Taylor & Francis Group, an **informa** business
A CHAPMAN & HALL BOOK

First published in paperback 2024

First published 2015 by CRC Press

Published 2019 by CRC Press
2385 NW Executive Center Drive, Suite 320, Boca Raton FL 33431

and by CRC Press
4 Park Square, Milton Park, Abingdon, Oxon, OX14 4RN

CRC Press is an imprint of Taylor & Francis Group, LLC

Publisher's Note
The publisher has gone to great lengths to ensure the quality of this reprint but points out that some imperfections in the original copies may be apparent.

Library of Congress Cataloging-in-Publication Data

Moss, Charles B. (Charles Britt)
Mathematical statistics for applied econometrics / Charles B. Moss.
pages cm
Includes bibliographical references and index.
ISBN 978-1-4665-9409-8 (hardcover : alk. paper) 1. Econometrics. 2. Economics--Mathematical models. I. Title.

HB141.M67 2015
330.01'5195--dc23 2014028241

ISBN: 978-1-4665-9409-8 (hbk)
ISBN: 978-1-03-292102-0 (pbk)
ISBN: 978-0-429-16889-5 (ebk)

DOI: 10.1201/b17503

Visit the Taylor & Francis Web site at
http://www.taylorandfrancis.com

and the CRC Press Web site at
http://www.crcpress.com

Dedication

This book is dedicated to the memory of Henri Theil.

Over the course of my training in econometrics, I had several motivated instructors. However, I really did not understand econometrics until I collaborated with Hans.

Contents

III Econometric Applications 215

List of Figures

List of Tables

Preface

This book is drawn from the notes that I developed to teach AEB 6571 – Econometric Methods I in the Food and Resource Economics Department at the University of Florida from 2002 through 2010. The goal of this course was to cover the basics of statistical inference in support of a subsequent course on classical introductory econometrics. One of the challenges in teaching a course like this is the previous courses that students have taken in statistics and econometrics. Specifically, it is my experience that most introductory courses take on a cookbook flavor. If you have this set of data and want to analyze that concept – apply this technique. The difficulty in this course is to motivate the why. The course is loosely based on two courses that I took at Purdue University (Econ 670 and Econ 671).

While I was finishing this book, I discovered a book titled *The Lady Tasting Tea: How Statistics Revolutionized Science in the Twentieth Century* by David Salsburg. I would recommend any instructor assign this book as a companion text. It includes numerous pithy stories about the formal development of statistics that add to numerical discussion in this textbook. One of the important concepts introduced in *The Lady Tasting Tea* is the debate over the meaning of probability. The book also provides interesting insight into statisticians as real people. For example, William Sealy Gosset was a statistician who developed the Student's t distribution under the name Student while working in his day job with Guiness.

Another feature of the book is the introduction of symbolic programs Maxima and *Mathematica*™ in Appendix A. These programs can be used to reduce the cost of the mathematical and numerical complexity of some of the formulations in the textbook. In addition, I typically like to teach this course as a "numbers" course. Over the years I have used two programs in the classroom – Gauss™ by Aptech and R, an open-source product of the R-Project. In general, I prefer the numerical precision in Gauss. However, to use Gauss efficiently you need several libraries (i.e., CO – Constrained Optimization). In addition, Gauss is proprietary. I can typically make the code available to students through Gauss-Lite based on my license. The alternative is R, which is open-source, but has a little less precision. The difficulties in precision are elevated in the solve() command for the inverse. In this textbook, I have given the R code for a couple of applications.

Of course, writing a book is seldom a solitary enterprise. It behooves me to recognize several individuals who contributed to the textbook in a variety of ways. First, I would like to thank professors who taught my econometric courses over the years, including Paul Beaumont of Florida State University, who taught Econ 670 and Econ 671 at Purdue University; James Binkley, who taught Ag Econ 650 at Purdue; and Wade Brorson of Oklahoma State University, who taught Ag Econ 651 at Purdue University during my time there. I don't think that any of these professors would have pegged me to write this book. In addition, I would like to thank Scott Shonkwiler for our collaboration in my early years at the University of Florida. This collaboration included our work on the inverse hyperbolic sine transformation to normality. I also would like to thank the students who suffered through AEB 6571 – Econometrics Methods I at the University of Florida. Several, including Cody Dahl, Grigorios Livanis, Diwash Neupane, Matthew Salois, and Dong Hee Suh, have provided useful comments during the writing process. And in a strange way, I would like to thank Thomas Spreen, who assigned me to teach this course when he was the Food and Resource Economics Department's graduate coordinator. I can honestly say that this is not a course that I would have volunteered to teach. However, I benefitted significantly from the effort. My econometric skills have become sharper because of the assignment.

Finally, for the convenience of the readers and instructors, most of my notes for AEB 6571 are available online at http://ricardo.ifas.ufl.edu/aeb6571.econometrics/. The datasets and programs used in this book are available at http://www.charlesbmoss.com:8080/MathStat/.

1

Defining Mathematical Statistics

CONTENTS

At the start of a course in mathematical statistics students usually ask three questions. Two of these questions are typically what is this course going to be about and how is this different from the two or three other statistics courses that most students have already taken before mathematical statistics? The third question is how does the study of mathematical statistics contribute to my study of economics and econometrics? The simplest answer to the first question is that we are going to develop statistical reasoning using mathematical techniques. It is my experience that most students approach statistics as a toolbox, memorizing many of the statistical estimators and tests (see box titled **Mathematical Statistics – Savage**). This course develops the student's understanding of the reasons behind these tools. Ultimately, mathematical statistics form the basis of econometric procedures used to analyze economic data and provide a firm basis for understanding decision making under risk and uncertainty.

Mathematical statistics share the same linkage to statistics that mathematical economics has to economics. In mathematical economics, we develop the consequences of economic choice of such primal economic concepts as consumer demand and producer supply. Focusing on demand, we conceptualize how a concave set of ordinal preferences implies that consumers will choose a unique bundle of goods given any set of prices and level of income. By extension, we can follow the logic to infer that these conditions will lead to demand curves that are downward sloping and quasi-convex in price space by Roy's identity. Thus, any violation of these conditions (i.e., downward sloping and quasi-convexity) implies that the demand curve is not consistent with a unique point on an ordinal utility map. Hence, the development of logical connections using mathematical precision gives us a logical structure for our analysis.

Mathematical Statistics – Savage

In the present century there has been and continues to be extraordinary interest in mathematical treatment of problems of inductive inference. For reasons I cannot and need not analyze here, this activity has been strikingly concentrated in the English-speaking world. It is known under several names, most of which stress some aspects of the subject that seemed of overwhelming importance at the moment when the name was coined. "Mathematical statistics," one of its earliest names, is still the most popular. In this name, "mathematical" seems to be intended to connote rational, theoretical, or perhaps mathematically advanced, to distinguish the subject from those problems of gathering and condensing numerical data that can be considered apart from the problem of inductive inference, the mathematical treatment of which is generally trivial. The name "statistical inference" recognizes that the subject is concerned with inductive inference. The name "statistical decision" reflects the idea that inductive inference is not always, if ever, concerned with what to believe in the face of inconclusive evidence, but that at least sometimes it is concerned with what action to decide upon under such circumstances [41, p. 2].

The linkage between mathematical statistics and statistics is similar. The theory of statistical inference is based on primal concepts such as estimation, sample design, and hypothesis testing. Mathematical statistics allow for the rigorous development of this statistical reasoning. Conceptually, we will define what is meant by a random variable, how the characteristics of this random variable are linked with a distribution, and how knowledge of these distributions can be used to design estimators and hypothesis tests that are meaningful (see box titled **the Role of Foundations – Savage**).

The Role of Foundations – Savage

It is often argued academically that no science can be more secure than its foundations, and that, if there is controversy about the foundations, there must be even greater controversy about the higher parts of the science. As a matter of fact, the foundations are the most controversial parts of many, if not all sciences. Physics and pure mathematics are excellent examples of this phenomenon. As for statistics, the foundations include, on any interpretation of which I have ever heard, the foundations of probability, as controversial a subject as one could name. As in other sciences, controversies over the foundations of statistics reflect themselves to some

> extent in everyday practice, but not nearly so catastrophically as one might imagine. I believe that here as elsewhere, catastrophe can be avoided, primarily because in practical situations common sense generally saves all but the most pedantic of us from flagrant error [41, p. 1].

However, in our development of these mathematical meanings in statistics, we will be forced to consider the uses of these procedures in economics. For much of the twentieth century, economics attempted to define itself as the science that studies the allocation of limited resources to meet unlimited and competing human wants and desires [4]. However, the definition of economics as a science may raise objections inside and outside the discipline. Key to the definition of a field of study as a science is the ability or willingness of its students and practitioners to allow its tenets to be empirically validated. In essence, it must be possible to reject a cherished hypothesis based on empirical observation. It is not obvious that economists have been willing to follow through with this threat. For example, remember our cherished notion that demand curves must be downward sloping and quasi-convex in price space. Many practitioners have estimated results where these basic relationships are violated. However, we do not reject our so-called "Law of Demand." Instead we expend significant efforts to explain why the formulation yielding this result is inadequate. In fact, there are several possible reasons to suspect that the empirical or econometric results are indeed inadequate, many of which we develop in this book. My point is that despite the desire of economists to be classified as a scientists, economists are frequently reticent to put theory to an empirical test in the same way as a biologist or physicist. Because of this failure, economics largely deserves the suspicion of these white coated practitioners of more basic sciences.

1.1 Mathematical Statistics and Econometrics

The study of mathematical statistics by economists typically falls under a broad sub-discipline called econometrics. Econometrics is typically defined as the use of statistics and mathematics along with economic theory to describe economic relationships (see the boxes titled **Tinbergen on Econometrics** and **Klein on Econometrics**). The real issue is what do we mean by describe? There are two dominant ideas in econometrics. The first involves the scientific concept of using statistical techniques (or more precisely, statistical inference) to test implications of economic theory. Hence, in a traditional scientific paradigm, we expose what we think we know to experience (see the box titled **Popper on Scientific Discovery**). The second use of econometrics involves the estimation of parameters to be used in policy analysis. For example, economists working with a state legislature may be interested in estimating the effect of a sales tax holiday for school supplies on the government's sales tax revenue. As a result, they may be more interested in imposing economically justified restrictions that add additional information to their data rather than testing these hypotheses. The two uses of econometrics could then be summarized as scientific uses versus the uses of planners.

Tinbergen on Econometrics

Econometrics is the name for a field of science in which mathematical-economic and mathematical-statistical research are applied in combination. Econometrics, therefore, forms a borderland between two branches of science, with the advantages and disadvantages thereof; advantages, because new combinations are introduced which often open up new perspectives; disadvantages, because the work in this field requires skill in two domains, which either takes up too much time or leads to insufficient training of its students in one of the two respects [51, p. 3].

Klein on Econometrics

The purely theoretical approach to econometrics may be envisioned as the development of that body of knowledge which tells us how to go about measuring economic relationships. This theory is often developed on a fairly abstract or general basis, so that the results may be applied to any one of a variety of concrete problems that may arise. The empirical work in econometrics deals with actual data and sets out to make numerical estimates of economic relationships. The empirical procedures are direct applications of the methods of theoretical econometrics [24, p. 1].

Popper on Scientific Discovery

A scientist, whether theorist or experimenter, puts forward statements, or systems of statements, and tests them step by step. In the field of the empirical sciences, more particularly, he constructs hypotheses, or systems of theories, and tests them against experience by observation and experiment.
I suggest that it is the task of the logic of scientific discovery, or logic of knowledge, to give a logical analysis of this procedure; that is to analyse the method of empirical sciences [38, p. 3].

1.1.1 Econometrics and Scientific Discovery

The most prominent supporters of the traditional scientific paradigm to econometrics are Theil, Kmenta, and Spanos. According to Theil,

> Econometrics is concerned with the empirical determination of economic laws. The word "empirical" indicates that the data used for this determination have been obtained from observation, which may be either controlled experimentation designed by the econometrician interested, or "passive" observation. The latter type is as prevalent among economists as it is among meterologists [49, p.1].

Kamenta [26] divides statistical applications in economics into descriptive statistics and statistical inference. Kmenta contends that most statistical applications in economics involve applications of statistical inference, that is, the use of statistical data to draw conclusions or test hypotheses about economic behavior. Spanos states that "econometrics is concerned with the systematic study of economic phenomena using observed data" [45, p. 3].

How it all began – Haavelmo

The status of general economics was more or less as follows. There were lots of deep thoughts, but a lack of quantitative results. Even in simple cases where it can be said that some economic magnitude is influenced by only one causal factor, the question of how strong is the influence still remains. It is usually not of very great practical or even scientific interest to know whether the influence is positive or negative, if one does not know anything about the strength. But much worse is the situation when an economic magnitude to be studied is determined by many different factors at the same time, some factors working in one direction, others in the opposite directions. One could write long papers about so-called tendencies explaining how this factor might work, how that factor might work and so on. But what is the answer to the question of the total net effect of all the factors? This question cannot be answered without measures of the strength with which the various factors work in their directions. The fathers of modern econometrics, led by the giant brains of Ragnar Frisch and Jan Tinbergen, had the vision that it would be possible to get out of this situation for the science of economics. Their program was to use available statistical material in order to extract information about how an economy works. Only in this way could one get beyond the state of affairs where talk of tendencies was about all one could have as a result from even the greatest brains in the science of economics [15].

Nature of Econometrics – Judge et al.

If the goal is to select the best decision from the economic choice set, it is usually not enough just to know that certain economic variables are related. To be really useful we must usually also know the direction of the relation and in many cases the magnitudes involved. Toward this end, econometrics, using economic theory, mathematical economics, and statistical inference as an analytical foundation and economic data as the information base, provides an inferential basis for:
(1) Modifying, refining, or possibly refuting conclusions contained in economic theory and/or what represents current knowledge about economic processes and institutions.
(2) Attaching signs, numbers, and reliability statements to the coefficient of variables in economic relationships so that this information can be used as a basis for decision making and choice [23, p. 1].

A quick survey of a couple of important economics journals provides a look at how econometrics is used in the development of economic theory. Ashraf and Galor [2] examine the effect of genetic diversity on economic growth. Specifically, they hypothesize that increased genetic diversity initially increases economic growth as individuals from diverse cultures allow the economy to quickly adopt a wide array of technological innovations. However, this rate of increase starts to decline such that the effect of diversity reaches a maximum as the increased diversity starts to impose higher transaction costs on the economy. Thus, Ashraf and Galor hypothesize that the effect of diversity on population growth is "hump shaped." To test this hypothesis, they estimate two empirical relationships. The first relationship examines the effect of genetic diversity on each country's population density.

$$\ln(P_i) = \beta_0+\beta_1 G_i+\beta_2 G_i^2+\beta_3 \ln(T_i)+\beta_4 \ln(X_{1i})+\beta_5 \ln(X_{2i})+\beta_6 \ln(X_{3i})+\epsilon_i \tag{1.1}$$

where $\ln(P_i)$ is the natural logarithm of the population density for country i, G_i is a measure of genetic diversity in country i, T_i is the time in years since the establishment of agriculture in country i, X_{1i} is the percentage of arable land in country i, X_{2i} is the absolute latitude of country i, X_{3i} is a variable capturing the suitability of land in country i for agriculture, and ϵ_i is the residual. The second equation then estimates the effect of the same factors on each country's income per capita.

$$\ln(y_i) = \gamma_0+\gamma_1 \hat{G}_i+\gamma_2 \hat{G}_i^2+\gamma_3 \ln(T_i)+\gamma_4 \ln(X_{1i})+\gamma_5 \ln(X_{2i})+\gamma_6 \ln(X_{3i})+\nu_i \tag{1.2}$$

where y_i represents the income per capita and $\hat{G}_i$ is the estimated level of genetic diversity. Ashraf and Galor use the estimated genetic diversity to adjust for the relationship between genetic diversity and the path of development

TABLE 1.1
Estimated Effect of Genetic Diversity on Economic Development

Variable	Population Density	Income per Capita
Genetic Diversity (G_i)	225.440***	203.443**
	(73.781)[a]	(83.368)
Genetic Diversity Squared (G_i^2)	-3161.158**	-142.663**
	(56.155)	(59.037)
Emergence of Agriculture ($\ln(T_i)$)	1.214***	-0.151
	(0.373)	(0.197)
Percent of Arable Land ($\ln(X_{1i})$)	0.516***	-0.112
	(0.165)	(0.103)
Absolute Latitude ($\ln(X_{2i})$)	-0.162	0.163
	(0.130)	(0.117)
Land Suitability ($\ln(X_{3i})$)	0.571*	-0.192**
	(0.294)	(0.096)
R^2	0.89	0.57

[a] Numbers in parenthesis denote standard errors. *** denotes statistical significance at the 0.01 level of confidence, ** denotes statistical significance at the 0.05 level of confidence, and * denotes statistical significance at the 0.10 level of confidence.
Source: Ashraf and Galor [2]

from Africa to other regions of the world (i.e., the "Out of Africa" hypothesis). The statistical results of these estimations presented in Table 1.1 support the theoretical arguments of Ashraf and Galor.

In the same journal, Naidu and Yuchtman [35] examine whether the "Master and Servant Act" used to enforce labor contracts in Britain in the nineteenth century affected wages. At the beginning of the twenty-first century a variety of labor contracts exist in the United States. Most hourly employees have an implicit or continuing consent contract which is not formally binding on either the employer or the employee. By contrast, university faculty typically sign annual employment contracts for the upcoming academic year. Technically, this contract binds the employer to continue to pay the faculty member the contracted amount throughout the academic year unless the faculty member violates the terms of this contract. However, while the faculty member is bound by the contract, sufficient latitude is typically provided for the employee to be released from the contract before the end of the academic year without penalty (or by forfeiting the remaining payments under the contract). Naidu and Yuchtman note that labor laws in Britain (the Master and Servant Act of 1823) increased the enforcement of these labor contracts by providing both civil and criminal penalties for employee breach of contract. Under this act employees who attempted to leave a job for a better opportunity could be forced back into the original job under the terms of the contract.

TABLE 1.2
Estimates of the Effect of Master and Servant Prosecutions on Wages

Variable	Parameter
Fraction of Textiles × ln(Cotton Price)	159.3***
	(42.02)[a]
Iron County × ln(Iron Price)	51.98**
	(19.48)
Coal County × ln(Coal Price)	41.25***
	(10.11)
ln(Population)	79.13**
	(35.09)

[a] Numbers in parenthesis denote standard errors. *** denotes statistical significance at the 0.01 level of confidence, and ** denotes statistical significance at the 0.05 level of confidence.
Source: Naidu and Yuchtman [35]

Naidu and Yuchtman develop an economic model which indicates that the enforcement of this law will reduce the average wage rate. Hence, they start their analysis by examining factors that determine the number of prosecutions under the Master and Servant laws for counties in Britain before 1875.

$$\begin{aligned} Z_{it} = \alpha_0 + \alpha_1 S_i \times X_{1,t} + \alpha_2 I_{2,i} \times \ln\left(X_{2,t}\right) + \alpha_3 I_{3,i} \ln\left(X_{3,t}\right) \\ + \alpha_4 \ln\left(p_{i,t}\right) + \epsilon_{it} \end{aligned} \tag{1.3}$$

where Z_{it} is the number of prosecutions under the Master and Servant Act in county i in year t, S_i is the share of textile production in county i in 1851, $X_{1,t}$ is the cotton price at time t, $I_{2,i}$ is a dummy variable that is 1 if the county produces iron and 0 otherwise, $X_{2,t}$ is the iron price at time t, $I_{3,i}$ is a dummy variable that is 1 if the county produces coal and 0 otherwise, $X_{3,t}$ is the price of coal, $p_{i,t}$ is the population of county i at time t, and ϵ_{it} is the residual. The results for this formulation are presented in Table 1.2. Next, Naidu and Yuchtman estimate the effect of these prosecutions on the wage rate.

$$\begin{aligned} w_{it} = \beta_0 + \beta_1 I_{4,t} \times \ln\left(\bar{Z}_i\right) + \beta_2 X_{5,it} + \beta_3 X_{6,it} + \beta_4 \ln\left(X_{7,it}\right) \\ + \beta_5 \ln\left(p_{it}\right) + \beta_6 X_{8,it} + \nu_{it} \end{aligned} \tag{1.4}$$

where w_{it} is the average wage rate in county i at time t, I_t is a dummy variable that is 1 if $t > 1875$ (or after the repeal of the Master and Servant Act) and 0 otherwise, $X_{5,it}$ is the population density of county i at time t, $X_{6,it}$ is the proportion of the population living in urban areas in county i at time t, $X_{7,it}$ is the average income in county i at time t, $X_{8,it}$ is the level of union membership in county i at time t, and ν_{it} is the residual. The results presented in Table 1.3 provide weak support (i.e., at the 0.10 level of significance) that prosecutions under the Master and Servant Act reduced wages. Specifically, the positive

TABLE 1.3
Effect of Master and Servant Prosecutions on the Wage Rate

Variable	Parameter
Post-1875 × ln(Average Prosecutions)	0.0122*
	(0.0061)
Population Density	-0.0570
	(0.0583)
Proportion Urban	-0.0488
	(0.0461)
ln(Income)	0.0291
	(0.0312)
ln(Population)	0.0944**
	(0.0389)
Union Membership	0.0881
	(0.0955)

[a] Numbers in parenthesis denote standard errors. ** denotes statistical significance at the 0.05 level of confidence and * denotes statistical significance at the 0.10 level of confidence.
Source: Naidu and Yuchtman [35]

coefficient on the post-1875 variable indicates that wages were 0.0122 shillings per hour higher after the Master and Servant Act was repealed in 1875.

As a final example, consider the research of Kling et al. [25], who examine the role of information in the purchase of Medicare drug plans. In the Medicare Part D prescription drug insurance program consumers choose from a menu of drug plans. These different plans offer a variety of terms, including the price of the coverage, the level of deductability (i.e., the lower limit required for the insurance to start paying benefits), and the amount of co-payment (e.g., the share of the price of the drug that must be paid by the senior). Ultimately consumers make a variety of choices. These differences may be driven in part by differences between household circumstances. For example, some seniors may be in better health than others. Alternatively, some households may be in better financial condition. Finally, the households probably have different attitudes toward risk. Under typical assumptions regarding consumer behavior, the ability to choose maximizes the benefits from Medicare Part D to seniors. However, the conjecture that consumer choice maximizes the benefit from the Medicare drug plans depends on the consumer's ability to understand the benefits provided by each plan. This concept is particularly important given the complexity of most insurance packages. Kling et al. analyze the possibility of comparison friction. Comparison friction is a bias from switching to a possibly better product because the two products are difficult to compare. To analyze the significance of comparison friction Kling et al. construct a sample of seniors who purchase Medicare Part D coverage. Splitting this sample into a control group and an intervention (or treatment) group, the intervention group was then provided personalized information about how each alternative would affect the household. The control group was then given access to a webpage which could be used to construct the same information. The researchers then observed which households switched their coverage. The sample was then used to estimate

$$\begin{aligned} D_i = \alpha_0 + \alpha_1 Z_i + \alpha_2 X_{1i} + \alpha_3 X_{2i} + \alpha_4 X_{3i} + \alpha_5 X_{4i} + \alpha_6 X_{5i} \\ \alpha_7 X_{6i} + \alpha_8 X_{7i} + \alpha_9 X_{8i} + \alpha_{10} X_{9i} + \alpha_{11} X_{10i} + \epsilon_i \end{aligned} \tag{1.5}$$

where D_i is one if the household switches its plan and zero otherwise, Z_i is the intervention variable equal to one if the household was provided individual information, X_{1i} is a dummy variable which is one if the head of household is female, X_{2i} is one if the head of household is married, X_{3i} is one if the individual finished high school, X_{4i} is one if the participant finished college, X_{5i} is one if the individual completed post-graduate studies, X_{6i} is one if the participant is over 70 years old, X_{7i} is one if the participant is over 75 years old, X_{8i} is one if the individual has over four medications, X_{9i} is one if the participant has over seven medications, and X_{10i} is one if the household is poor.

Table 1.4 presents the empirical results of this model. In general these results confirm a comparison friction since seniors who are given more information about alternatives are more likely to switch (i.e., the estimated intervention parameter is statistically significant at the 0.10 level). However, the empirical results indicate that other factors matter. For example, married couples are more likely to switch. In addition, individuals who take over seven medications are more likely to switch. Interestingly, individual levels of education (i.e., the high school graduate, college graduate, and post-college graduate variables) are not individually significant. However, further testing would be required to determine whether education was statistically informative. Specifically, we would have to design a statistical test that simultaneously restricted all three parameters to be zero at the same time. As constructed, we can only compare each individual effect with the dropped category (probably that the participant did not complete high school).

In each of these examples, data is used to test a hypothesis about individual behavior. In the first study (Ashraf and Galor [2]), the implications of individual actions on the aggregate economy (i.e., nations) are examined. Specifically, does greater diversity lead to economic growth? In the second study, Naidu and Yuchtman [35] reduced the level of analysis to the region, asking whether the Master and Servant Act affected wages at the parish (or county) level. In both scenarios the formulation does not model the actions themselves (i.e., whether genetic diversity improves the ability to carry out a variety of activities through a more diverse skill set or whether the presence of labor restrictions limited factor mobility) but the consequences of those actions. The last example (Kling et al. [25]) focuses more directly on individual behavior. However, in all three cases an economic theory is faced with observations.

On a somewhat related matter, econometrics positions economics as a positive science. Econometrics is interested in what happens as opposed to what should happen (i.e., a positive instead of a normative science; see box **The**

TABLE 1.4
Effect of Information on Comparison Friction

Variable	Parameter
Intervention	0.098*
	(0.041)
Female	−0.023
	(0.045)
Married	0.107*
	(0.045)
High School Graduate	−0.044
	(0.093)
College Graduate	0.048
	(0.048)
Post-college Graduate	−0.084
	(0.062)
Age 70+	−0.039
	(0.060)
Age 75+	0.079
	(0.048)
4+ Medications	−0.054
	(0.050)
7+ Mediations	0.116*
	(0.052)
Poor	0.097*
	(0.045)

[a] Numbers in parenthesis denote standard errors. * denotes statistical significance at the 0.10 level of confidence.
Source: Kling et al. [25]

Methodology of Positive Economics – Friedman). In the forgoing discussion we were not interested in whether increased diversity should improve economic growth, but rather whether it could be empirically established that increased diversity was associated with higher economic growth.

The Methodology of Positive Economics – Friedman

... the problem how to decide whether a suggested hypothesis or theory should be be tentatively accepted as part of the "body of systematized knowledge concerning what is." But the confusion [John Neville] Keynes laments is still so rife and so much a hindrance of the recognition that economics can be, and in part is, a positive science that it seems to preface the main body of the paper with a few remarks about the relation between positive and normative economics.

... Self-proclaimed "experts" speak with many voices and can hardly all be regarded as disinterested; in any event, on questions that matter so much, "expert" opinion could hardly be accepted soley on faith even if the "experts" were nearly unanimous and clearly disinterested The conclusions of positive economics seem to be, and are, immediately relevant to important normative problems, to questions of what ought to be done and how any given goal can be attained. Laymen and experts alike are inevitably tempted to shape positive conclusions to fit strongly held normative preconceptions and to reject positive conclusions if their normative implications – or what are said to be their normative implications – are unpalatable.

Positive economics is in principle independent of any particular ethical position or normative judgments. As Keynes says, it deals with "what is," not with "what ought to be." Its task is to provide a system of generalizations that can be used to make correct predictions about the consequences of any change in circumstances. Its performance is to be judged by the precision, scope, and conformity with experience of the predictions it yields [13, pp. 3–5].

1.1.2 Econometrics and Planning

While the interaction between governments and their economies is a subject beyond the scope of the current book, certain features of this interaction are important when considering the development of econometrics and the role of mathematical statistics within that development. For modern students of economics, the history of economics starts with the classical economics of Adam Smith [44]. At the risk of oversimplication, Smith's insight was that markets allowed individuals to make choices that maximized their well-being. Aggregated over all individuals, these decisions acted like an invisible hand that allocated resources toward the production of goods that maximized the overall well-being of the economy. This result must be viewed within the context of the economic thought that the classical model replaced – mercantilism [43]. Historically the mercantile system grew out of the cities. Each city limited the trade in raw materials and finished goods in its region to provide economic benefits to the city's craftsmen and merchants. For example, by prohibiting the export of wool (or by imposing significant taxes on those exports) the resulting lower price would benefit local weavers. Smith's treatise demonstrated that these limitations reduced society's well-being.

The *laissez-faire* of classical economics provided little role for econometrics as a policy tool. However, the onset of the Great Depression provided a significantly greater potential role for econometrics (see box **Government and**

Economic Life – Staley). Starting with the Herbert Hoover administration, the U.S. government increased its efforts to stimulate the economy. This shift to the managed economy associated with the administration of Franklin Roosevelt significantly increased the use of econometrics in economic policy. During this period, the National Income and Product Accounts (NIPA) were implemented to estimate changes in aggregate income and the effect of a variety of economic policies on the aggregate economy. Hence, this time period represents the growth of econometrics as a planning tool which estimates the effect of economic policies such as changes in the level of money supply or increases in the minimum wage (see box **Economic Planning – Tinbergen**).

Government and Economic Life – Staley

The enormous increase in the economic role of the state over the last few years has the greatest possible importance for the future of international economic relationships. State economic activities have grown from such diverse roots as wartime needs, the fear of war and the race for rearmament and military self-sufficiency, the feelings of the man in the street on the subject of poverty in the midst of plenty, innumerable specific pressures from private interests, the idea of scientific management, the philosophy of collectivist socialism, the totalitariam philosophy of the state, the sheer pressure of economic emergency in the depression, and the acceptance of the idea that it is the state's business not only to see that nobody starves but also to ensure efficient running of the economic machine....

Governments have taken over industries of key importance, such as munition factories in France, have assumed the management of public utility services, as under the Central Electricity Board in Great Britian, and have set up public enterprises to prepare the way for development of whole regions and to provide "yardsticks" for private industries, as in the case of the Tennessee Valley Authority in the United States [46, pp. 128–129].

Economic Planning – Tinbergen

This study deals with the process of *central economic planning*, or economic planning by governments. It aims at a threefold treatment, which may be summarized as follows: (a) to describe the process of central planning, considered as one of the service industries of the modern economy; (b) to analyze its impact on the general economic process; (c) to indicate, as far as possible, the optimal extent and techniques of central planning [52, p. 3]

The orign of the planning techniques applied today clearly

> springs from two main sources: Russian communist planning and Western macroplanning....
>
> Western macroeconomic planning had a very different origin, namely the desire to understand the operation of the economy as a whole. It was highly influenced by the statistical concepts relevant to national or social accounts and by Keynesian concepts, combined with market analysis, which later developed into macroeconomic econometric models. There was still a basic belief that many detailed decisions could and should be left to the decentralized system of single enterprises and that guidance by the government might confine itself to indirect intervention with the help of a few instruments only [52, pp. 4–5].

While Tinbergen focuses on the role of econometrics in macroeconomic policy, agricultural policy has generated a variety of econometric applications. For example, the implementation of agricultural policies such as loan rates [42] results in an increase in the supply of crops such as corn and wheat. Econometric techniques are then used to estimate the effect of these programs on government expenditures (i.e., loan deficiency payments). The passage of the Energy Independence and Security Act of 2007 encouraged the production of biofuels by requiring that 15 billion gallons of ethanol be added to the gasoline consumed in the United States. This requirement resulted in corn prices significantly above the traditional loan rate for corn. The effect of ethanol on corn prices increased significantly with the drought in the U.S. Midwest in 2012. The combination of the drought and the ethanol requirement caused corn prices to soar, contributing to a significant increase in food prices. This interaction has spawned numerous debates, including pressure to reduce the ethanol requirements in 2014 by as much as 3 billion gallons. At each step of this policy debate, various econometric analyses have attempted to estimate the effect of policies on agricultural and consumer prices as well as government expenditures. In each case, these econometric applications were not intended to test economic theory, but to provide useful information to the policy process.

1.2 Mathematical Statistics and Modeling Economic Decisions

Apart from the use of statistical tools for inference, mathematical statistics also provides several concepts useful in the analysis of economic decisions under risk and uncertainty. Moss [32] demonstrates how probability theory

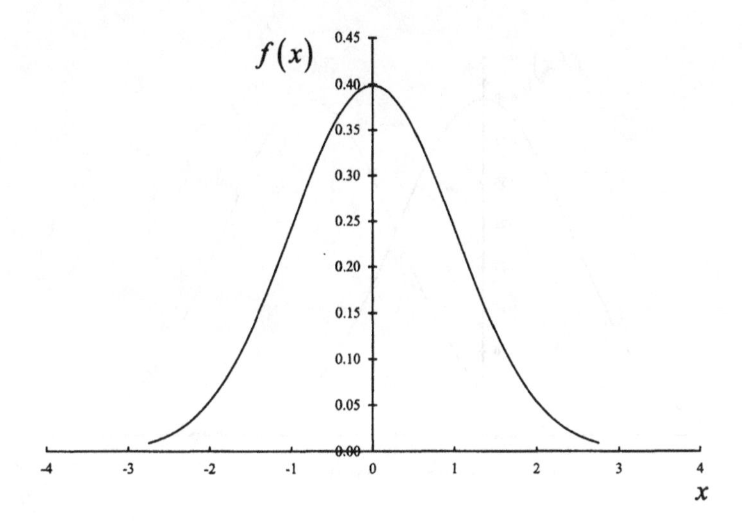

FIGURE 1.1
Standard Normal Density Function.

contributes to the derivation of the Expected Utility Hypothesis. Apart from the use of mathematical statistics in the development of theory, these tools are also important for the development of several important applied methodologies for dealing with risk and uncertainty, such as the Capital Asset Pricing Model, Stochastic Dominance, and Option Pricing Theory.

Skipping ahead a little bit, the normal distribution function depicts the probability density for a given outcome x as a function of the mean and variance of the distribution.

$$f\left(x;\mu,\sigma^2\right) = \frac{1}{\sigma\sqrt{2\pi}}\exp\left(-\frac{(x-\mu)^2}{2\sigma^2}\right). \tag{1.6}$$

Graphically, the shape of the function under the assumptions of the "standard normal" (i.e., $\mu = 0$ and $\sigma^2 = 1$) is depicted in Figure 1.1. This curve is sometimes referred to as the *Bell Curve*. Statistical inference typically involves designing a probabilistic measure for testing a sample of observations drawn from this data set against an alternative assumption, for example, $\mu = 0$ versus $\mu = 2$. The difference in these distributions is presented in Figure 1.2.

An alternative economic application involves the choice between the two distribution functions. For example, under what conditions does a risk averse producer prefer the alternative that produces each distribution?[1] Figure 1.3

[1] The optimizing behavior for risk averse producers typically involves a choice between combinations of expected return and risk. Under normality the most common measure of risk is the variance. In the scenario where the expected return (or mean) is the same, decision makers prefer the alternative that produces the lowest risk (or variance).

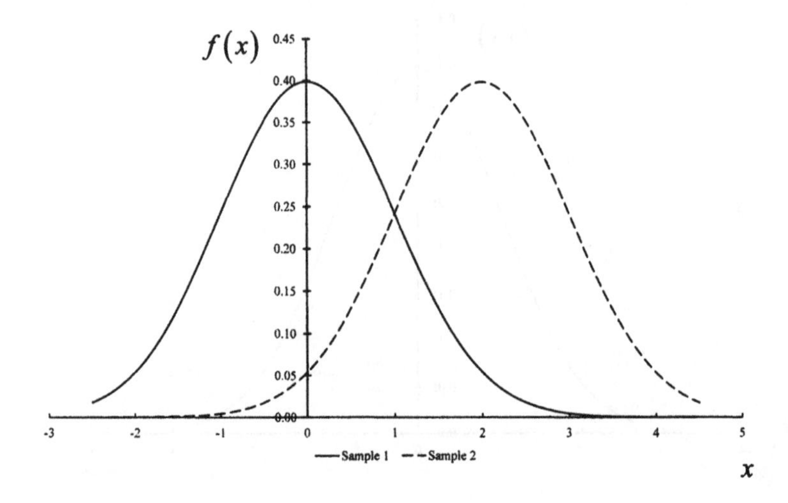

FIGURE 1.2
Normal Distributions with Different Means.

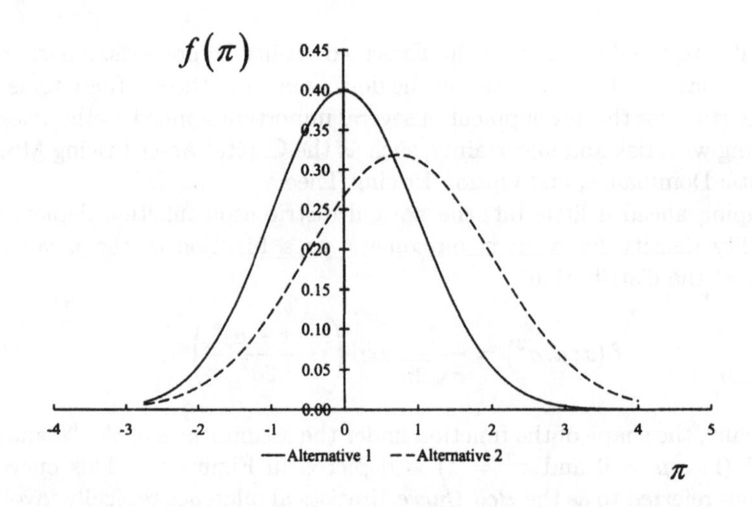

FIGURE 1.3
Alternative Normal Distributions.

presents the distribution functions of profit (π) for two alternative actions that a decision maker may choose. Alternative 1 has a mean of 0 and a variance of 1 (i.e., is standard normal) while the second distribution has a mean of 0.75 with a standard deviation of 1.25. There are a variety of ways to compare these

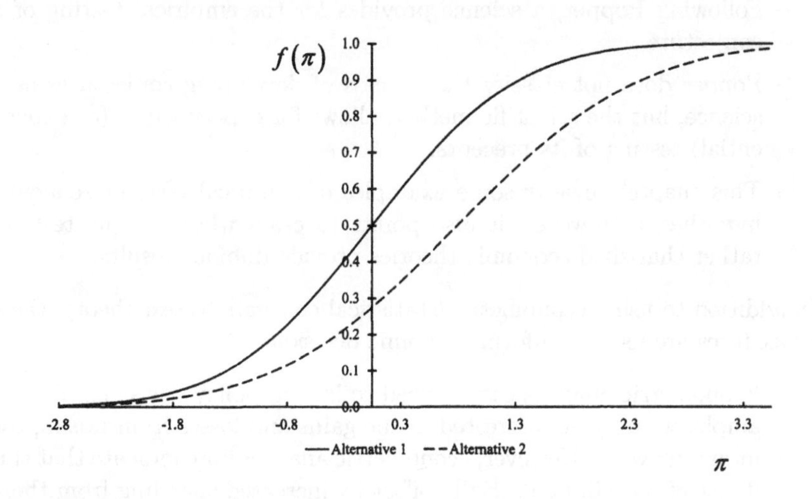

FIGURE 1.4
First Degree Stochastic Dominance – A Comparison of Cumulative Distribution Functions.

two alternatives; one is first degree stochastic dominance, which basically asks whether one alternative always has a higher probability of producing a higher return [32, pp. 150–152]. First degree stochastic dominance involves comparing the cumulative distribution function (i.e., the probability the random variable will be less than or equal to any value) for each alternative. As presented in Figure 1.4, Alternative 2 dominates (i.e., provides a higher return for the relative risk) Alternative 1.

1.3 Chapter Summary

- Mathematical statistics involves the rigorous development of statistical reasoning.
  - The goal of this textbook is to make the student think about statistics as more than a toolbox of techniques.
  - These mathematical statistic concepts form the basis of econometric formulations.
- Our analysis of mathematical statistics raises questions regarding our definition of economics as a science versus economics as a tool for decision makers.

  - Following Popper, a science provides for the empirical testing of a conjecture.
  - Popper does not classify the process of developing conjectures as a science, but the scientific method allows for experimental (or experiential) testing of its precepts.
  - This chapter reviews some examples of empirical tests of economic hypotheses. However, it also points to cases where simple tests of rather charished economic theories provide dubious results.

- In addition to using econometric/statistical concepts to test theory, these procedures are used to inform economic decisions.
  - Econometric analysis can be used to inform policy decisions. For example, we may be interested in the gains and losses from raising the minimum wage. However, econometric analysis may indicate that this direct effect will be partially offset by increased spending from those benefiting from the increase in the minimum wage.
  - In addition, mathematical statistics helps us model certain decisions such as producer behavior under risk and uncertainty.

1.4 Review Questions

1-1R. What are the two primary uses of econometrics?

1-2R. Review a recent issue of the *American Economic Review* and discuss whether the empirical applications test economic theory or provide estimates useful for policy analysis.

1-3R. Review a recent issue of the *American Economic Journal: Economic Policy* and discuss whether the empirical applications test economic theory or provide estimates useful for policy analysis.

1-4R. Discuss a scenario where mathematical statistics informs economic theory in addition to providing a means of scientific testing.

Part I

Defining Random Variables

2

Introduction to Statistics, Probability, and Econometrics

CONTENTS

A primary building block of statistical inference is the concept of probability. However, this concept has actually been the subject of significant debate over time. Hacking [16] provides a detailed discussion of the evolution of the concept of probability:

> Probability has two aspects. It is connected with the degree of belief warranted by evidence, and it is connected with the tendency, displayed by some chance devices, to produce stable relative frequencies [16, p. 1].

As developed by Hacking, the concept of probability has been around for thousands of years in the context of games of chance. This concept of probability follows the idea of the tendency of chance displayed by some device – dice or cards. The emergence of a formal concept of probability can be found in correspondence between the mathematical geniuses Pierre de Fermat (1601–1665) and Blaise Pascal (1623–1662).

The correspondence between Fermat and Pascal attempts to develop a rule for the division of an incomplete game. The incomplete game occurs when a multiple stage contest is interrupted (i.e., the game is completed when one individual rolls a certain value eight times but the play is interrupted after seven rolls).

> Let us say that I undertake to win a point with a single die in eight rolls; if we agree, after the money is already in the game, that I will not make my first roll, then according to my principle it is necessary

> that I take out 1/6 of the pot to be bought out, since I will be unable to roll to win in the first round (Letter of Fermat to Pascal [9]).

The first textbook on probability following this gaming approach was then produced by Christiaan Huygens [20] in 1657.

Typically, econometrics based its concept of probability loosely on this discrete gaming approach. As described by Tinbergen,

> A clear example which tells us more than abstract definitions is the number of times we can throw heads with one or more coins. If we throw with one coin, that number can be 0 or 1 on each throw; if we throw with three coins, it can be 0, 1, 2, or 3. The number of times each of these values appears is its frequency; the table of these frequencies is the frequency distribution. If the latter is expressed relatively, i.e., in figures the sum of the total of which is 1 and which are proportionate to the frequencies, we speak of probability distribution. The probability or relative frequency of the appearance of one certain result indicates which part of the observations leads to that outcome [51, pp. 60–61].

This pragmatic approach differs from more rigorous developments offered by Lawrence Klein [24, pp. 23–28].

Skipping some of the minutiae of the development of probability, there are three approaches.

1. The **frequency approach** – following Huygens, probability is simply defined by the relative count of outcomes.
2. **Personal probabilities** where individuals anticipate the likelihood of outcomes based on personal information. This approach is similar to the concept of utility developed in consumer theory. The formulation is primarily based on Leonard Savage [41] and Bruno de Finetti [10].
3. **Axiomatic probabilities** where the properties of a probability function are derived from basic conjectures. The axiomatic approach is typically attributed to Andrei Kolmogorov [39, pp. 137–143].

The frequency approach has the advantage of intuition. A variety of board games involve rolling dice, from Monopoly™to Risk™. The growth in popularity of state lotteries from the late 1980s through the early 2000s to fund various government programs has increased the popular knowledge of drawing without replacement. Finally, the growth of the casino industries in states such as Mississippi and on Indian reservations has extended participation in games of chance beyond the traditional bastions of Las Vegas and Reno, Nevada, and Atlantic City, New Jersey.

However, some central concepts of probability theory are difficult for a simple frequency motivation. For example, the frequency approach is intuitively appealing for discrete outcomes (i.e., the number of points depicted on the top of a die or whether a coin lands heads or tails up). The intuition of the frequentist vanishes when the variable of interest is continuous, such as the annual rainfall or temperature in a crop region. In these scenarios, the probability of any particular outcome is numerically zero. Further, there are some outcomes that are actually discrete but the possible number of outcomes is large enough that the frequency of outcomes approaches zero. For example, the observed Dow Jones Industrial Average reported in the news is a discrete number (value in hundredths of a dollar or cents). Similarly, for years stock prices were reported in sixteenths of a cent.

As one example, consider the rainfall in Sayre, Oklahoma, for months important for the production of hard red winter wheat (Table 2.1). Typically hard red winter wheat is planted in western Oklahoma in August or September, so the expected rainfall in this period is important for the crop to sprout. Notice that while we may think about rainfall as a continuous variable, the data presented in Table 2.1 is discrete (i.e., there is a countable number of hundreths of an inch of rainfall). In addition, we will develop two different ways of envisioning this data. From an observed sample framework we will look at the outcomes in Table 2.1 as equally likely like the outcomes of the roll of dice. From this standpoint, we can create the empirical distribution function (i.e., the table of outcomes and probabilities of those outcomes that we will develop more fully in Chapter 3) for rainfall in August and September presented in Table 2.2. An interesting outcome of the empirical distribution function is that the outcome of 7.26 inches of rainfall is twice as likely as any other outcome. Looking ahead at more empirical approaches to probability, suppose that we constructed a histogram by counting the number of times the rainfall occurred between inches (i.e., in three years the rainfall was between 1 and 2 inches). Note that the table excludes the rainfall of 17.05 inches in the 95/96 crop year. In the terms of probability, I am tentatively classifying this outcome as an outlier. This data is presented in Table 2.3. This table shows us that different rainfall amounts are really not equally likely, for example, the empirical probability that rainfall between 5 and 7 inches is more likely than between 3 and 4 inches (i.e., 0.170 is greater than 0.094). This general approach of constructing probability functions can be thought of as a frequentist application to continuous data.

In addition to the practical problem of continuous distributions, the frequentist approach did not provide a rigorous concept of the properties of probability:

> By the end of the seventeenth century the mathematics of many simple (and some not-so-simple) games of chance was well understood and widely known. Fermat, Pascal, Hugens, Leibniz, Jacob Bernoulli, and Arbutnot all examined the ways in which the mathematics of permutations and combinations could be employed in the enumeration of favorable cases in a variety of games of known properties. But this early work did not extend to the consideration of the problem: How, from the outcome of a game (or several outcomes of the same game),

TABLE 2.1
Rainfall in Sayre, Oklahoma, Important for Hard Red Winter Wheat

Crop Year	Aug-Sep	Oct-Dec	Jan-Mar	Apr-May	Crop Year	Aug-Sep	Oct-Dec	Jan-Mar	Apr-May
60/61	6.26	7.98	3.77	3.18	87/88	7.28	2.50	3.22	5.42
61/62	2.98	5.22	0.34	4.17	88/89	11.52	1.36	4.47	6.06
62/63	7.92	3.91	1.03	4.73	89/90	6.46	1.65	5.01	8.55
63/64	2.79	2.29	3.30	4.23	90/91	6.36	3.37	2.37	5.88
64/65	4.42	6.88	1.99	3.62	91/92	6.62	7.59	3.32	2.77
65/66	8.09	4.90	1.41	1.45	92/93	2.30	4.14	4.92	8.08
66/67	11.02	0.87	0.97	6.04	93/94	3.90	1.39	2.05	5.36
67/68	5.89	2.69	2.62	7.98	94/95	3.09	6.80	2.94	9.46
68/69	6.41	7.30	3.59	7.02	95/96	13.17	1.94	0.10	2.13
69/70	4.11	2.51	1.95	4.73	96/97	17.05	5.21	2.96	13.15
70/71	3.26	1.25	1.31	1.83	97/98	7.68	6.86	6.80	4.44
71/72	7.79	5.64	0.16	6.55	98/99	1.32	11.93	4.04	7.37
72/73	3.78	5.37	7.54	7.41	99/00	4.36	1.59	6.33	4.90
73/74	4.38	3.00	2.24	6.31	00/01	0.00	6.28	5.56	13.86
74/75	10.15	6.40	4.41	5.31	01/02	4.25	1.90	2.01	5.05
75/76	3.34	5.45	0.54	7.78	02/03	1.59	9.08	1.00	5.47
76/77	4.44	2.15	1.87	15.42	03/04	7.28	1.04	8.17	3.41
77/78	6.67	1.65	1.66	10.44	04/05	5.72	8.97	3.23	5.76
78/79	6.12	2.14	4.69	9.17	05/06	9.81	2.13	2.28	5.37
79/80	2.29	1.35	2.70	8.21	06/07	4.72	7.20	8.41	12.72
80/81	2.46	2.41	3.90	5.94	07/08	7.91	2.08	2.71	3.10
81/82	3.37	6.73	2.41	11.95	08/09	10.69	5.31	1.81	4.46
82/83	0.76	3.16	4.05	5.38	09/10	5.41	4.35	7.38	6.78
83/84	0.99	8.68	2.72	1.76	10/11	5.98	5.17	1.88	1.24
84/85	1.63	6.34	4.67	3.27	11/12	3.10	5.95	4.01	4.32
85/86	4.77	3.86	1.11	6.60	12/13	4.19	1.25	9.01	3.18
86/87	12.34	11.85	4.92	7.01					

TABLE 2.2
Empirical Probability Distribution of Rainfall

Rainfall	Count	Rainfall	Count	Rainfall	Count	Rainfall	Count
0.00	1	3.26	1	4.77	1	7.68	1
0.76	1	3.34	1	5.41	1	7.79	1
0.99	1	3.37	1	5.72	1	7.91	1
1.32	1	3.78	1	5.89	1	7.92	1
1.59	1	3.90	1	5.98	1	8.09	1
1.63	1	4.11	1	6.12	1	9.81	1
2.29	1	4.19	1	6.26	1	10.15	1
2.30	1	4.25	1	6.36	1	10.69	1
2.46	1	4.36	1	6.41	1	11.02	1
2.79	1	4.38	1	6.46	1	11.52	1
2.98	1	4.42	1	6.62	1	12.34	1
3.09	1	4.44	1	6.67	1	13.17	1
3.10	1	4.72	1	7.28	2	17.05	1

TABLE 2.3
Histogram of Rainfall in August and September in Sayre, Oklahoma

Rainfall	Count	Fraction of Sample
0–1	0	0.000
1–2	3	0.057
2–3	3	0.057
3–4	5	0.094
4–5	7	0.132
5–6	9	0.170
6–7	4	0.075
7–8	7	0.132
8–9	6	0.113
9–10	1	0.019
10–11	1	0.019
11–12	2	0.038
12–13	2	0.038
13–14	1	0.019
14–15	1	0.019

could one learn about the properties of the game and how could one quantify the uncertainty of our inferred knowledge of the properties? [47, p. 63]

Intuitively, suppose that you were playing a board game such as backgammon and your opponent rolls doubles 6 out of 16 rolls, advancing around the board and winning the game. Could you conclude that the die were indeed fair? What is needed is a systematic formulation of how probability works. For example, Zellner [54] defines the *direct probability* as a probability model where the form and nature of the probability structure are completely known (i.e., the probability of rolling a die). Given this formulation, the only thing that is unknown is the outcome of a particular roll of the dice. He contrasts this model with the *inverse probability*, where we observe the outcomes and attempt to say something about the probability generating process (see the **Zellner on Probability box**).

Zellner on Probability

On the other hand, problems usually encountered in science are not those of *direct probability* but those of *inverse probability.* That is, we usually observe data which are assumed to be the outcome or output of some probability process or model, the properties of which are not completely known. The scientist's problem is to infer or learn the properties of the probability model from observed data, a problem in the realm of inverse probability. For example, we may have data on individual's incomes and wish to determine whether they can be considered as drawn or generated from a normal probability distribution or by some other probability distribution. Questions like these involve considering alternative probability models and using observed data to try to determine from which hypothesized probability model the data probably came, a problem in the area of *statistical analysis of hypotheses.* Further, for any of the probability models considered, there is the problem of using data to determine or estimate the values of parameters appearing in it, a problem of *statistical estimation.* Finally, the problem of using probability models to make predictions about as yet unobserved data arises, a problem of *statistical prediction* [54, p. 69].

Three Views of the Interpretation of Probability

Objectivistic: views hold that some repetitive events, such as tosses of a penny, prove to be in reasonably close agreement with the mathematical concept of independently repeated random events, all with the same probability. According to such views, evidence for the quality of agreement between the behavior of the repetitive event and the mathematical concept, and for the magnitude of the probability that applies (in case any does), is to be obtained by observation of some repetitions of the event, and from no other source whatsoever.

Personalistic: views hold that probability measures the confidence that a particular individual has in the truth of a particular proposition, for example, the proposition that it will rain tomorrow. These views postulate that the individual concerned is in some ways "reasonable," but they do not deny the possibility that two reasonable individuals faced with the same evidence may have different degrees of confidence in the truth of the same proposition.

Necessary: views hold that probability measures the extent to which one set of propositions, out of logical necessity and apart from human opinion, confirms the truth of another. They are generally regarded by their holders as extensions of logic, which tells when one set of propositions necessitates the truth of another [41, p. 3].

2.1 Two Definitions of Probability for Econometrics

To begin our discussion, consider two fairly basic definitions of probability.

- *Bayesian* — probability expresses the degree of belief a person has about an event or statement by a number between zero and one.
- *Classical* — the relative number of times that an event will occur as the number of experiments becomes very large.

$$\lim_{N \to \infty} \mathrm{P}[O] = \frac{r_O}{N}. \tag{2.1}$$

The Bayesian concept of probability is consistent with the notion of a personalistic probability advanced by Savage and de Fenetti, while the classical probability follows the notion of an objective or frequency probability.

Intuitively, the basic concept of probability is linked to the notion of a random variable. Essentially, if a variable is deterministic, its probability is either one or zero – the result either happens or it does not (i.e., if $x = f(z) = z^2$ the probability that $x = f(2) = 4$ is one, while the probability that $x = f(2) = 5$ is zero). The outcome of a random variable is not certain. If x is a random variable it can take on different values. While we know the possible values that the variable takes on, we do not know the exact outcome before the event. For example, we know that flipping a coin could yield two outcomes – a head or a tail. However, we do not know what the value will be before we flip the coin. Hence, the outcome of the flip – head or tail – is a random variable. In order to more fully develop our notion of random variables, we have to refine our discussion to two general types of random variables: discrete random variables and continuous random variables.

A **discrete random variable** is some outcome that can only take on a fixed number of values. The number of dots on a die is a classic example of a discrete random variable. A more abstract random variable is the number of red rice grains in a given measure of rice. It is obvious that if the measure is small, this is little different from the number of dots on the die. However, if the measure of rice becomes large (a barge load of rice), the discrete outcome becomes a countable infinity, but the random variable is still discrete in a classical sense.

A **continuous random variable** represents an outcome that cannot be technically counted. Amemiya [1] uses the height of an individual as an example of a continuous random variable. This assumes an infinite precision of measurement. The normally distributed random variable presented in Figures 1.1 and 1.3 is an example of a continuous random variable. In our foregoing discussion of the rainfall in Sayre, Oklahoma, we conceptualized rainfall as a continuous variable while our measure was discrete (i.e., measured in a finite number of hundreths of an inch).

The exact difference between the two types of random variables has an effect on notions of probability. The standard notions of Bayesian or Classical probability fit the discrete case well. We would anticipate a probability of 1/6 for any face of the die. In the continuous scenario, the probability of any specific outcome is zero. However, the **probability density function** yields a measure of relative probability. The concepts of discrete and continuous random variables are then unified under the broader concept of a probability density function.

2.1.1 Counting Techniques

A simple method of assigning probability is to count how many ways an event can occur and assign an equal probability to each outcome. This methodology is characteristic of the early work on objective probability by Pascal, Fermat, and Huygens. Suppose we are interested in the probability that a die roll will be even. The set of all even events is $A = \{2, 4, 6\}$. The number of even events is $n(A) = 3$. The total number of die rolls is $S = \{1, 2, 3, 4, 5, 6\}$ or $n(S) = 6$. The probability of these countable events can then be expressed as

$$P[A] = \frac{n(A)}{n(S)} \tag{2.2}$$

where the probability of event A is simply the number of possible occurrences of A divided by the number of possible occurrences in the sample, or in this example $P[\text{even die rolls}] = 3/6 = 0.50$.

Definition 2.1. The number of permutations of taking r elements out of n elements is a number of distinct ordered sets consisting of r distinct elements which can be formed out of a set of n distinctive elements and is denoted P_r^n.

The first point to consider is that of factorials. For example, if you have two objects A and B, how many different ways are there to order the object? Two:

$$\{A, B\} \text{ or } \{B, A\}. \tag{2.3}$$

If you have three objects, how many ways are there to order the objects? Six:

$$\begin{gathered}\{A, B, C\}\ \{A, C, B\}\ \{B, A, C\}\ \{B, C, A\} \\ \{C, A, B\} \text{ or } \{C, B, A\}.\end{gathered} \tag{2.4}$$

The sequence then becomes – two objects can be drawn in two sequences, three objects can be drawn in six sequences (2×3). By inductive proof, four objects can be drawn in 24 sequences (6×4).

The total possible number of sequences is then for n objects $n!$ defined as:

$$n! = n\,(n-1)\,(n-2)\ldots 1. \tag{2.5}$$

Theorem 2.2. *The (partial) permutation value can be computed as*

$$P_r^n = \frac{n!}{(n-r)!}. \tag{2.6}$$

The term partial permutation is sometimes used to denote the fact that we are not completely drawing the sample (i.e., $r \leq n$). For example, consider the simple case of drawing two out of two possibilities:

$$P_1^2 = \frac{2!}{(2-1)!} = 2 \tag{2.7}$$

which yields the intuitive result that there are two possible values of drawing one from two (i.e., either A or B). If we increase the number of possible outcomes to three, we have

$$P_1^3 = \frac{3!}{(3-1)!} = \frac{6}{2} = 3 \tag{2.8}$$

which yields a similarly intuitive result that we can now draw three possible first values A, B, or C. Taking the case of three possible outcomes one step further, suppose that we draw two numbers from three possibilities:

$$P_2^3 = \frac{3!}{(3-2)!} = \frac{6}{1} = 6. \tag{2.9}$$

Table 2.4 presents these results. Note that in this formulation order matters. Hence $\{A, B\} \neq \{B, A\}$.

To develop the generality of these formulas, consider the number of permutations for completely drawing four possible numbers (i.e., $4! = 24$ possible

TABLE 2.4
Partial Permutation of Three Values

Low First	High First
$\{A,B\}$	$\{B,A\}$
$\{A,C\}$	$\{C,A\}$
$\{B,C\}$	$\{C,B\}$

TABLE 2.5
Permutations of Four Values

	A First	B First	C First	D First
1	$\{A,B,C,D\}$	$\{B,A,C,D\}$	$\{C,A,B,D\}$	$\{D,A,C,B\}$
2	$\{A,B,D,C\}$	$\{B,A,D,C\}$	$\{C,A,D,B\}$	$\{D,A,B,C\}$
3	$\{A,C,B,D\}$	$\{B,C,A,D\}$	$\{C,B,A,D\}$	$\{D,B,A,C\}$
4	$\{A,C,D,B\}$	$\{B,C,D,A\}$	$\{C,B,D,A\}$	$\{D,B,C,A\}$
5	$\{A,D,B,C\}$	$\{B,D,A,C\}$	$\{C,D,A,B\}$	$\{D,C,A,B\}$
6	$\{A,D,C,B\}$	$\{B,D,C,A\}$	$\{C,D,B,A\}$	$\{D,C,B,A\}$

sequences, as depicted in Table 2.5). How many ways are there to draw the first number?

$$P_1^4 = \frac{4!}{(4-1)!} = \frac{24}{6} = 4. \tag{2.10}$$

The results seem obvious – if there are four different numbers, then there are four different numbers you could draw on the first draw (i.e., see the four columns of Table 2.5). Next, how many ways are there to draw two numbers out of four?

$$P_2^4 = \frac{4!}{(4-2)!} = \frac{24}{2} = 12. \tag{2.11}$$

To confirm the conjecture in Equation 2.11, note that Table 2.5 is grouped by combinations of the first two numbers. Hence, we see that there are three unique combinations where A is first (i.e., $\{A,B\}$, $\{A,C\}$, and $\{A,D\}$). Given that the same is true for each column $4 \times 3 = 12$.

Next, consider the scenario where we don't care which number is drawn first – $\{A,B\} = \{B,A\}$. This reduces the total number of outcomes presented in Table 2.4 to three. Mathematically we could say that the number of outcomes K could be computed as

$$K = \frac{P_1^3}{2} = \frac{3!}{1! \times 2!} = 3. \tag{2.12}$$

Extending this result to the case of four different values, consider how many different outcomes there are for drawing two numbers out of four if we don't care about the order. From Table 2.5 we can have six (i.e., $\{A,B\} = \{B,A\}$,

$\{A,C\} = \{C,A\}$, $\{A,D\} = \{D,A\}$, $\{B,C\} = \{C,B\}$, $\{B,D\} = \{D,B\}$, and $\{C,D\} = \{D,C\}$). Again, we can define this figure mathematically as

$$K = \frac{P_1^4}{2} = \frac{4!}{(4-2)!2!} = 6. \tag{2.13}$$

This formulation is known as a **combinatorial**. A more general form of the formulation is given in Definition 2.3.

Definition 2.3. The number of combinations of taking r elements from n elements is the number of distinct sets consisting of r distinct elements which can be formed out of a set of n distinct elements and is denoted C_r^n.

$$C_r^n = \begin{pmatrix} n \\ r \end{pmatrix} = \frac{n!}{(n-r)!r!}. \tag{2.14}$$

Apart from their application in probability, combinatorials are useful for binomial arithmetic.

$$(a+b)^n = \sum_{k=0}^{n} \begin{pmatrix} n \\ k \end{pmatrix} a^k b^{n-k}. \tag{2.15}$$

Taking a simple example, consider $(a+b)^3$.

$$(a+b)^3 = \begin{pmatrix} 3 \\ 0 \end{pmatrix} a^{(3-0)} b^0 + \begin{pmatrix} 3 \\ 1 \end{pmatrix} a^{(3-1)} b^1 + \begin{pmatrix} 3 \\ 2 \end{pmatrix} a^{(3-2)} b^2 + \begin{pmatrix} 3 \\ 3 \end{pmatrix} a^{(3-3)} b^3. \tag{2.16}$$

Working through the combinatorials, Equation 2.16 yields

$$(a+b)^3 = a^3 + 3a^2b + 3ab^2 + b^3 \tag{2.17}$$

which can also be drived using Pascal's triangle, which will be discussed in Chapter 5. As a direct consequence of this formulation, combinatorials allow for the extension of the Bernoulli probability form to the more general binomial distribution.

To develop this more general formulation, consider the example from Bierens [5, Chap. 1]; assume we are interested in the game Texas lotto. In this game, players choose a set of 6 numbers out of the first 50. Note that the ordering does not count so that 35, 20, 15, 1,5, 45 is the same as 35, 5, 15, 20, 1, 45. How many different sets of numbers can be drawn? First, we note that we could draw any one of 50 numbers in the first draw. However, for the second draw we can only draw 49 possible numbers (one of the numbers has been eliminated). Thus, there are 50×49 different ways to draw two numbers. Again, for the third draw, we only have 48 possible numbers left. Therefore, the total number of possible ways to choose 6 numbers out of 50 is

$$\prod_{j=1}^{5}(50-j) = \prod_{k=45}^{50} k = \frac{\prod_{k=1}^{50} k}{\prod_{k=1}^{50-6} k} = \frac{50!}{(50-6)!}. \tag{2.18}$$

Finally, note that there are 6! ways to draw a set of 6 numbers (you could draw 35 first, or 20 first, ...). Thus, the total number of ways to draw an unordered set of 6 numbers out of 50 is

$$\binom{50}{6} = \frac{50!}{6!(50-6)!} = 15{,}890{,}700. \tag{2.19}$$

This description of lotteries allows for the introduction of several definitions important to probability theory.

Definition 2.4. *Sample space* The set of all possible outcomes. In the Texas lotto scenario, the sample space is all possible 15,890,700 sets of 6 numbers which could be drawn.

Definition 2.5. *Event* A subset of the sample space. In the Texas lotto scenario, possible events include single draws such as $\{35, 20, 15, 1, 5, 45\}$ or complex draws such as all possible lotto tickets including $\{35, 20, 15\}$. Note that this could be $\{35, 20, 15, 1, 2, 3\}, \{35, 20, 15, 1, 2, 4\}, \ldots$.

Definition 2.6. *Simple event* An event which cannot be expressed as a union of other events. In the Texas lotto scenario, this is a single draw such as $\{35, 20, 15, 1, 5, 45\}$.

Definition 2.7. *Composite event* An event which is not a simple event.

Formal development of probability requires these definitions. The sample space specifies the possible outcomes for any random variable. In the roll of a die the sample space is $\{1, 2, 3, 4, 5, 6\}$. In the case of a normal random variable, the sample space is the set of all real numbers $x \in (-\infty, \infty)$. An event in the roll of two dice could be the number of times that the values add up to 4 – $\{1, 3\}, \{2, 2\}, \{3, 1\}$. The simple event could be a single dice roll for the two dice – $\{1, 3\}$.

2.1.2 Axiomatic Foundations

In our gaming example, the most basic concept is that each outcome $s_i = 1, 2, 3, 4, 5, 6$ is equally likely in the case of the six-sided die. Hence, the probability of each of the events is $P[s_i] = 1/6$. That is, if the die is equally weighted, we expect that each side is equally likely. Similarly, we assume that a coin landing heads or tails is equally likely. The question then arises as to whether our framework is restricted to this equally likely mechanism. Suppose we are interested in whether it is going to rain tomorrow. At one level, we could say that there are two events – it could rain tomorrow or not. Are we bound to the concept that these events are equally likely and simply assume that each event has a probability of 1/2? Such a probability structure would not make a very good forecast model.

The question is whether there is a better way to model the probability of raining tomorrow. The answer is yes. Suppose that in a given month over

TABLE 2.6
Outcomes of a Simple Random Variable

Sample	Samples		
Draw	1	2	3
1	1	0	0
2	0	0	1
3	1	0	1
4	0	0	1
5	1	0	0
6	0	1	0
7	1	1	1
8	1	0	1
9	1	1	0
10	1	0	1
11	0	0	1
12	0	0	1
13	0	1	0
14	1	1	1
15	0	1	0
16	1	1	1
17	1	0	0
18	1	1	1
19	1	1	1
20	1	1	1
21	1	0	1
22	0	0	0
23	1	1	1
24	1	1	0
25	1	1	0
26	1	1	1
27	1	1	0
28	0	1	1
29	1	1	1
30	1	0	1
Total	21	17	19
Percent	0.700	0.567	0.633

the past thirty years that it rained five days. We could conceptualize a game of chance, putting five black marbles and twenty five white marbles into a bag. Drawing from the bag with replacment (putting the marble back each time) could be used to represent the probability of raining tomorrow. Notice that the chance of drawing each individual marble remains the same – like the counting exercise at 1/30. However, the relative difference is the number of marbles in the sack. It is this difference in the relative number of marbles in the sack that yields the different probability measure.

TABLE 2.7
Probability of the Simple Random Sample

Observation	Draw	Probability
1	1	p
2	0	$1-p$
3	1	p
4	0	$1-p$
5	1	p

It is the transition between these two concepts that gives rise to more sophisticated specifications of probability than simple counting mechanics. For example, consider the blend of the two preceding examples. Suppose that I have a random outcome that yields either a zero or a one (heads or tails). Suppose that I want to define the probability of a one. As a starting place, I could assign a probability of $1/2$ – equal probability. Consider the first column of draws in Table 2.6. The empirical evidence from these draws yields 21 ones (heads) or a one occurs 0.70 of the time. Based on this draw, would you agree with your initial assessment of equal probability? Suppose that we draw another thirty observations as depicted in column 2 of Table 2.6. These results yield 17 heads. In this sample 57 percent of the outcomes are ones. This sample is closer to equally likely, but if we consider both samples we have 38 ones out of 60 or 63.3 percent heads.

The question is how to define a set of common mechanics to compare the two alternative views (i.e., equally versus unequally likely). The mathematical basis is closer to the marbles in the bag than to equally probable. For example, suppose that we define the probability of heads as p. Thus, the probability of drawing a white ball is p while the probability of drawing a black ball is $1-p$. The probability of drawing the first five draws in Table 2.6 are then given in Table 2.7.

As a starting point, consider the first event. To rigorously develop a notion of the probability, we have to define the sample space. To define the sample space, we define the possible events. In this case there are two possible events – 0 or 1 (or $E = 1 \, \text{or} \, 0$). The sample space defined on these events can then be represented as $S = \{0, 1\}$. Intuitively, if we define the probability of $E = 1$ as p, then by definition of the sample space the probability of $E = 0$ is $1 - p$ because one of the two events must occur. Several aspects of the last step cannot be dismissed. For example, we assume that one and only one event must occur – the events are exclusive (a 0 and a 1 cannot both occur) and exhaustive (either a 0 or a 1 must occur). Thus, we denote the probability of a 1 occurring to be p and the probability of 0 occurring to be q. If the events are exclusive and exhaustive,

$$p + q = 1 \Rightarrow q = 1 - p \tag{2.20}$$

because one of the events must occur. In addition, to be a valid probability we need $p \geq 0$ and $1 - p \geq 0$. This is guaranteed by $p \in [0, 1]$.

Next, consider the first two draws from Table 2.7. In this case the sample space includes four possible events – $\{0,0\}$, $\{0,1\}$, $\{1,0\}$, and $\{1,1\}$. Typically, we aren't concerned with the order of the draw so $\{0,1\} = \{1,0\}$. However, we note that there are two ways to draw this event. Thus, following the general framework from Equation (2.20),

$$\sum_{r=0}^{2} \binom{2}{r} p^{(2-r)} q^r = p^2 + 2pq + q^2 \Rightarrow p^2 + 2p(1-p) + (1-p)^2 . \quad (2.21)$$

To address the exclusive and exhaustive nature of the event space, we need to guarantee that the probabilities sum to one – at least one event must occur.

$$p^2 + 2p(1-p) + (1-p)^2 = p^2 + 2p - 2p^2 + 1 - 2p + p^2 = 1. \quad (2.22)$$

In addition, the restriction that $p \in [0, 1]$ guarantees that each probability is positive.

By induction, the probability of the sample presented in Table 2.7 is

$$\mathrm{P}[S|p] = p^3 (1-p)^2 \quad (2.23)$$

for a given value of p. Note that for any value of $p \in [0, 1]$

$$\sum_{r=1}^{5} \binom{5}{r} p^{(5-r)} (1-p)^r = 1, \quad (2.24)$$

or a valid probability structure can be defined for any value p on the sample space.

These concepts offer a transition to a more rigorous way of thinking about probability. In fact, the distribution functions developed in Equations 2.20 through 2.24 are typically referred to as **Bernoulli distributions** for Jacques Bernoulli, who offered some of the very first rigorous proofs of probability [16, pp. 143–164]. This rigorous development is typically refered to as an **axiomatic development** of probability. The starting point for this axiomatic development is set theory.

Subset Relationships

As described in Definitions 2.4, 2.5, 2.6 and 2.7, events or outcomes of random variables are defined as elements or subsets of the set of all possible outcomes. Hence, we take a moment to review set notation.

(a) $A \subset B \Leftrightarrow x \in A \Rightarrow x \in B$.

(b) $A = B \Leftrightarrow A \subset B \,\text{and}\, B \subset A$.

(c) *Union:* The union of A and B, written $A \cup B$, the set of elements that belong either to A or B.

$$A \cup B = \{x : x \in A \,\text{or}\, x \in B\}. \tag{2.25}$$

(d) *Intersection:* The intersection of A and B, written $A \cap B$, is the set of elements that belong to both A and B.

$$A \cap B = \{x : x \in B \,\text{and}\, x \in B\}. \tag{2.26}$$

(e) *Complementation:* The complement of A, written A^C, is the set of all elements that are not in A.

$$A^C \in \{x : x \notin A\}. \tag{2.27}$$

Combining the subset notations yields Theorem 2.8.

Theorem 2.8. *For any three events A, B, and C defined on a sample space S,*

(a) Commutativity: $A \cup B = B \cup A$, $A \cap B = B \cap A$.

(b) Associativity: $A \cup (B \cup C) = (A \cup B) \cup C$, $A \cap (B \cup C) = (A \cup B) \cup C$.

(c) Distributive Laws: $A \cap (B \cup C) = (A \cup B) \cap (A \cup C)$, $A \cup (B \cap C) = (A \cap B) \cup (A \cap C)$.

(d) DeMorgan's Laws: $(A \cup B)^C = A^C \cup B^C, (A \cap B)^C = A^C \cup B^C$.

Axioms of Probability

A set $\{\omega_{j_1}, \ldots \omega_{j_k}\}$ of different combinations of outcomes is called an event. These events could be simple events or compound events. In the Texas lotto case, the important aspect is that the event is something you could bet on (for example, you could bet on three numbers in the draw 35, 20, 15). A collection of events F is called a family of subsets of sample space Ω. This family consists of all possible subsets of Ω including Ω itself and the null set $\emptyset$. Following the betting line, you could bet on all possible numbers (covering the board) so that Ω is a valid bet. Alternatively, you could bet on nothing, or $\emptyset$ is a valid bet.

Next, we will examine a variety of closure conditions. These are conditions that guarantee that if one set is contained in a family, another related set must also be contained in that family. First, we note that the family is closed under complementarity: If $A \in F$ then $A^c \in \Omega | A \in F$. In this case $A^c \in \Omega | A \in F$ denotes all elements of Ω that are not contained in A (i.e., $A^c = \{x : x \in \Omega | x \notin A\}$). Second, we note that the family is closed under union: If $A, B \in F$ then $A \cup B \in F$.

Definition 2.9. A collection F of subsets of a nonempty set Ω satisfying closure under complementarity and closure under union is called an algebra [5].

Adding closure under infinite union is defined as: If $A_j \in F$ for $j = 1, 2, 3, ...$ then $\cup_{j=1}^{\infty} A_j \in F$.

Definition 2.10. A collection F of subsets of a nonempty set Ω satisfying closure under complementarity and infinite union is called a σ-algebra (sigma-algebra) or a Borel Field [5].

Building on this foundation, a probability measure is the measure which maps from the event space into real number space on the [0,1] interval. We typically think of this as an odds function (i.e., what are the odds of a winning lotto ticket? 1/15,890,700). To be mathematically precise, suppose we define a set of events $A = \{\omega_1, ...\omega_j\} \in \Omega$, for example, we choose n different numbers. The probability of winning the lotto is $\mathrm{P}[A] = n/N$. Our intuition would indicate that $\mathrm{P}[\Omega] = 1$, or the probability of winning given that you have covered the board is equal to one (a certainty). Further, if you don't bet, the probability of winning is zeros or $\mathrm{P}[\emptyset] = 0$.

Definition 2.11. Given a sample space Ω and an associated σ-algebra F, a *probability function* is a function $\mathrm{P}[A]$ with domain F that satisfies

- $\mathrm{P}(A) \geq 0$ for all $A \in F$.
- $\mathrm{P}(\Omega) = 1$.

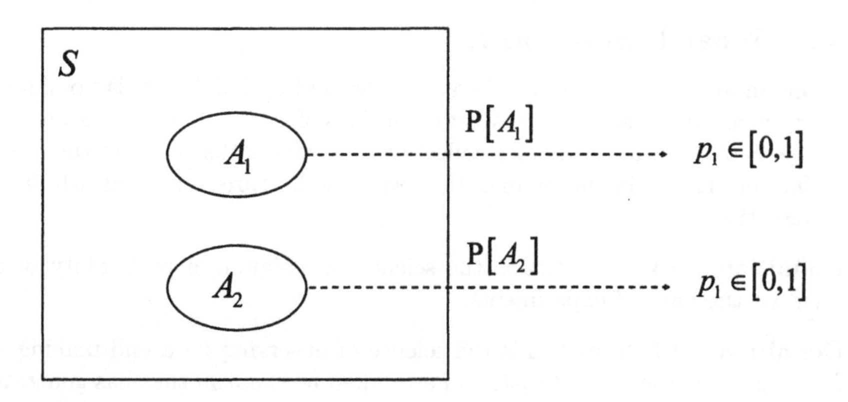

FIGURE 2.1
Mapping from Event Space to Probability Space.

- If $A_1, A_2, ... \in F$ are pairwise disjoint, then $\mathrm{P}\left(\cup_{i=1}^{\infty}\right) = \sum_{i=1}^{\infty} \mathrm{P}\left(A_i\right)$.

Breaking this down a little at a time, $\mathrm{P}[A]$ is a probability measure that is defined on an event space. The concept of a measure will be developed more fully in Chapter 3, but for our current uses, the measure assigns a value to an outcome in event space (see Figure 2.1). This value is greater than or equal to zero for any outcome in the algebra. Further, the value of the measure for the entire sample space is 1. This implies that some possible outcome will occur. Finally, the measure is additive over individual events. This definition is related to the required axioms of probability

$$\mathrm{P}\left[\bigcup_{i=1}^{\infty} A_i\right] = \sum_{i=1}^{\infty} \mathrm{P}\left[A_i\right]. \tag{2.28}$$

Stated slightly differently, the basic axioms of probability are:

Definition 2.12. Axioms of Probability:

1. $\mathrm{P}[A] \geq 0$ for any event A.
2. $\mathrm{P}[S] = 1$ where S is the sample space.
3. If $\{A\}\, i = 1, 2, \ldots$ are mutually exclusive (that it $A_i \cap A_j = \emptyset$ for all $i \neq j$, then $\mathrm{P}[A_1 \cap A_2 \cap \ldots] = \mathrm{P}[A_1] + \mathrm{P}[A_2] + \cdots$.

Thus, any function obeying these properties is a probability function.

2.2 What Is Statistics?

Given an understanding of random variables and probability, it is possible to offer a definition of **statistics**. Most definitions of statistics revolve around the synthesis of data into a smaller collection of numbers that contain the meaningful information in the sample. Here we consider three standard definitions of statistics.

Definition 2.13. Statistics is the science of assigning a probability of an event on the basis of experiments.

Definition 2.14. Statistics is the science of observing data and making inferences about the characteristics of a random mechanism that has generated the data.

Definition 2.15. Statistics is the science of estimating the probability distribution of a random variable on the basis of repeated observations drawn from the same random variable.

These definitions highlight different facets of statistics. Each definition contains a notion of probability. Two of the definitions make reference to estimation. As developed in Kmenta's [26] definition in Chapter 1, estimation can imply description or inference. In addition, one of the definitions explicitly states that statistics deals with experiments.

2.3 Chapter Summary

- Probability is a primary building block of statistical inference.
- Most of the early development of probability theory involved games of chance (aleatoric, from alea, a dice game).
- In general, probability theory can be justified using three approaches:
  - **Frequency** – based on the relative number of times an event occurs.
  - **Personal Probability** – based on personal belief. This approach is similar to the construction of utility theory.
  - **Axiomatic Probability** – based on the mathematics of measure theory.
- Most econometric applications involve the estimation of unknown parameters. As developed by Zellner, a **direct probability** is a probability model where we know the probability structure completely. For example, we know the probability structure for the unweighted die. An alternative approach is the **inverse probability** formulation, where we observe the outcomes of the probability model and wish to infer something about the true nature of the probability model. Most econometric applications involve an inverse probability formulation.
- While there are three different approaches to probability theory, most econometric applications are interested in two broad categories of probability:
  - **Bayesian** – the probability structure based on the degree of belief.
  - **Classical** – where we are interested in the empirical or observed frequency of events.
- The concept of a probability is related to the notion of a random variable. This concept is best described by contrasting the notion of a deterministic outcome with a random outcome.
  - We always know the outcome of a deterministic process (or function).

  - The outcome of a random variable may take on at least two different values. The exact outcome is unknowable before the event occurs.
  - Counting techniques provide a mechanism for developing classical probabilities. These models are related to the frequency approach.
  - The **Sample Space** is the set of all possible outcomes.
  - An **Event** can either be simple (i.e., containing a single outcome) or composite (i.e., including several simple events). The event of an even numbered die roll is complex. It contains the outcomes $s = \{2, 4, 6\}$.

- The axiomatic development of probability theory allows us to generalize models of random events which allow for tests of consistency for random variables.

2.4 Review Questions

2-1R. What are the three different approaches to developing probability?

2-2R. Consider two definitions of the same event – whether it rains tomorrow. We could talk about the probability that it will rain tomorrow or the rainfall observed tomorrow. Which event is more amenable to the frequency approach to probability?

2-3R. Consider two physical mechanisms – rolling an even number given a six-sided die and a flipped coin landing heads up. Are these events similar (i.e., do they have the same probability function)?

2.5 Numerical Exercises

2-1E. What is the probability that you will roll an even number given the standard six-sided die?

a. Roll the die 20 times. Did you observe the anticipated number of even numbered outcomes?

b. Continue rolling until you have rolled the die 40 times. Is the number of outcomes closer to the theoretical number of even-numbered rolls?

2-2E. Continuing from Exercise 2-1E, what is the probability of rolling a 2 or a 4 given a standard six-sided die?

a. Roll the die 20 times. Did you observe the anticipated number of 2s and 4s?

b. Continue rolling until you have rolled the die 40 times. Is the number of outcomes closer to the theoretical number of rolls?

c. Are the rolls of 2 or 4 closer to the theoretical results than in Exercise 2-1E?

2-3E. Construct an empirical model for rainfall in the October–December time period using a frequency approach using intervals of 1.0 inch of rainfall.

2-4E. How many ways are there to draw 2 events from 5 possibilities? Hint: $S = \{A, B, C, D, E\}$.

2-5E. What is the probability of $s = \{C, D\}$ when the order is not important?

2-6E. Consider a random variable constructed by rolling a six-sided die and flipping a coin. Taking $x = \{1, 2, 3, 4, 5, 6\}$ to be the outcome of the die roll and $y = \{1 \text{ if heads}, -1 \text{ if tails}\}$ to be the outcome of the coin toss, construct the random variable $z = x \times y$.

- What is the probability that the value of y will be between -2 and 2?
- What is the probability that the value of y will be greater than 2?

3

Random Variables and Probability Distributions

CONTENTS

Much of the development of probability in Chapter 2 involved probabilities in the abstract. We briefly considered the distribution of rainfall, but we were largely interested in flipping coins or rolling dice. These examples are typically referred to as aleatoric – involving games of chance. In this chapter we develop somewhat more complex versions of probability which form the basis for most econometric applications. We will start by developing the uniform distribution that defines a frequently used random variable. Given this basic concept, we then develop several probability relationships. We then discuss more general specifications of random variables and their distributions.

3.1 Uniform Probability Measure

I think that Bierens's [5] discussion of the uniform probability measure provides a firm basis for the concept of a probability measure. First, we follow the conceptual discussion of placing ten balls numbered 0 through 9 into a container. Next, we draw an infinite sequence of balls out of the container, replacing the ball each time. In Excel^{TM}, we can mimic this sequence using the function floor(rand()*10,1). This process will give a sequence of random numbers such as presented in Table 3.1. Taking each column, we can generate three random numbers: $\{0.741483, 0.029645, 0.302204\}$. Note that each of these sequences is contained in the unit interval $\Omega = [0,1]$. The primary point of the demonstration is that the number drawn $\{x \in \Omega = [0,1]\}$ is a probability measure. Taking $x = 0.741483$ as the example, we want to prove that $\mathrm{P}\left([0, x = 0.741483]\right) = 0.741483$. To do this we want to work out the probability of drawing a number less than 0.741483. As a starting point, what is the probability of drawing the first number in Table 3.1 less than 7? It is $7 \sim \{0,1,2,3,4,5,6\}$. Thus, without considering the second number, the probability of drawing a number less than 0.741483 is somewhat greater than 7/10. Next, we consider drawing a second number given that the first number drawn is greater than or equal to 7. As a starting point, consider the scenario where the number drawn is equal to seven. This occurs 1/10 of the time. Note that the two scenarios are disjoint. If the first number drawn is less than seven, it is not equal to seven. Thus, we can rely on the summation rule of probabilities:

$$\text{If } A_i \cap A_j = \emptyset \text{ then } \mathrm{P}\left[\bigcup_{k=1}^{n} A_k\right] = \sum_{k=1}^{n} \mathrm{P}\left[A_k\right]. \tag{3.1}$$

The probability of drawing a number less than 0.74 is the sum of drawing the first number less than 7 and the second number less than 4 given that the first number drawn is 7. The probability of drawing the second number less than 4 is $4/10 \sim \{0,1,2,3\}$. Given that the first number equal to 7 only

TABLE 3.1
Random Draws of Single Digits

Ball Drawn	Draw 1	Draw 2	Draw 3
1	7	0	3
2	4	2	0
3	1	9	2
4	4	6	2
5	8	4	0
6	3	5	4

occurs 1/10 of the time, the probability of the two events is

$$\mathrm{P}\left([0, x = 0.74]\right) = \frac{7}{10} + \frac{4}{10}\left(\frac{1}{10}\right) = \frac{7}{10} + \frac{4}{100} = 0.74. \tag{3.2}$$

Continuing to iterate this process backward, we find that $\mathrm{P}\left([0, x = 0.741483]\right) = 0.741483$. Thus, for $x \in [0,1]$ we have $\mathrm{P}\left([0, x]\right) = x$.

Before we complete Bieren's discussion, let us return to Definition 2.12. We know that a function meets our axioms for a probability function if (1) the value of the function is non-negative for any event, (2) the probability of the sample space is equal to one, and (3) if a set of events are mutually exclusive their probabilities are additive. For our purposes in econometrics, it is typically sufficient to conceptualize probability as a smooth function $F(x_1, x_2)$ where x_1 and x_2 are two points defined on the real number line. Given that we define an event ω such that $\omega \Rightarrow x \in [x_1, x_2]$, the probability of ω is defined as

$$F[x_1, x_2] = \int_{x_1}^{x_2} f(x)\, dx \tag{3.3}$$

where $f(x) \geq 0$ for all $x \in X$ implied by Ω (where $\omega \in \Omega$) and

$$\int_X f(x)\, dx = 1. \tag{3.4}$$

Hence, given that $f(x) \geq 0 \forall x \in X$ we meet the first axiom of Definition 2.12. Second, the definition of $f(x)$ in Equation 3.4 implies the second axiom. And, third, given that we can form mutually exclusive events by partitioning the real number line, this specificaiton meets the third axiom. Thus, the uniform distribution defined as

$$\mathrm{U}[0,1] \Rightarrow f(x) = 1 \,\text{for}\, x \in [0,1] \,\text{and}\, 0 \,\text{otherwise} \tag{3.5}$$

meets the basic axioms for a valid probability function.

While the definition of a probabilty function in Equation 3.3 is sufficient for most econometric applications, more rigorous proofs and formulations are frequently developed in the literature. In order to understand these formulations, consider the Reimann sum, which is used in most undergraduate calculus texts to justify the integral

$$\int_{x_1}^{x_N} f(x)\, dx = \lim_{K \to \infty} \sum_k |x_k - x_{k-1}|\, f(x_k)\, , k = 1, \cdots (N-1)/K. \tag{3.6}$$

Taking some liberty with the mathematical proof, the concept is that as the number of intervals K becomes large, the interval of approximation $x_k - x_{k-1}$ becomes small and the Riemann sum approaches the antiderivative of the function. In order to motivate our development of the Lebesgue integral, consider a slightly more rigorous specification of the Riemann sum. Specifically, we define two formulations of the Riemann sum:

$$\begin{aligned} S_1(K) &= \sum_k |x_k - x_{k-1}| \sup_{[x_{k-1},x_k]} f(x) \\ S_2(K) &= \sum_k |x_k - x_{k-1}| \inf_{[x_{k-1},x_k]} f(x) \end{aligned} \quad (3.7)$$

where $S_1(K)$ is the upper bound of the integral and $S_2(K)$ is the lower bound of the integral. Notice that sup (supremum) and inf (infimum) are similar to max (maximum) and min (minimum). The difference is that the maximum and minimum values are inside the range of the function while the supremum and infimum may be limits (i.e., greatest lower bounds and least upper bounds). Given the specification in Equation 3.7 we know that $S_1(K) \geq S_2(K)$. Further, we can define a residual

$$\epsilon(K) = S_1(K) - S_2(K). \quad (3.8)$$

The real value of the proof is that we can make $\epsilon(K)$ arbitrarily small by increasing K.

As complex as the development of the Riemann sum appears, it is simplified by the fact that we only consider simple intervals on the real number line. The axioms for a probability (measure) defined in Definitions 2.9 and 2.10 refer to a sigma-algebra. As a starting point, we need to develop the concept of a measure a little more rigorously. Over the course of a student's education, most become so comfortable with the notion of measuring physical phenomena that they take it for granted. However, consider the problem of learning to count, basically the development of a number system. Most of us were introduced to the number system by counting balls or marbles. For example, in Figure 3.1 we conceptualize a measure ($\mu(A)$), defined as the number of objects in set S).[1] In this scenario, the defined set is the set of all objects A so the measure can be defined as

$$\mu(A) \to R^1_+ \Rightarrow \mu(A) = 18. \quad (3.9)$$

Notice that if we change the definition of the set slightly, the mapping changes. Instead of A, suppose that I redefined the set as that set of circles — B. The measure then becomes $\mu(B) = 15$.

Implicitly, this counting measure has several imbedded assumptions. For example, the count is always positive. (actually the count is always a natural number). In addition, the measure is additive. I can divide the sets in S into circles (set B) and triangles (C, which has a count of $\mu(C) = 3$). Note that

[1] To be terribly precise, the count function could be defined as

$$\mu(A) = \sum_{i \in A} 1$$

where $n(A) = \mu(A)$ from Chapter 2.

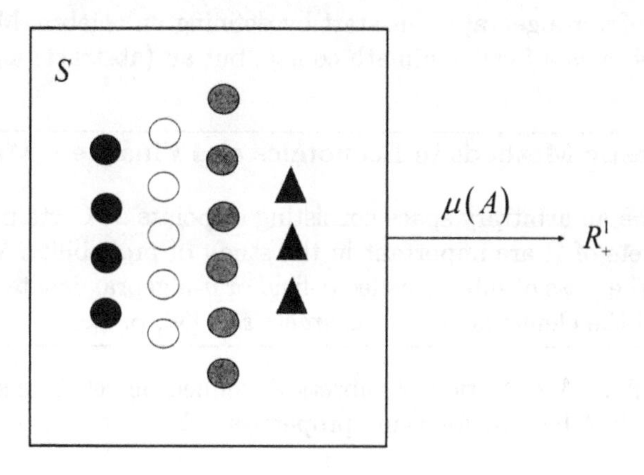

FIGURE 3.1
Defining a Simple Measure.

the set of circles and triangles is mutually exclusive – no element in the set is both a circle and a triangle. A critical aspect of a measure is that it is additive.

$$\mu(B \cup C) = \mu(B) + \mu(C) \Rightarrow 15 + 3 = 18. \tag{3.10}$$

Stochastic Process – Doob

The theory of probability is concerned with the measure properties of various spaces, and with the mutual relations of measurable functions defined on those spaces. Because of the applications, it is frequently (although not always) appropriate to call these spaces *sample spaces* and their measurable sets *events*, and these terms should be borne in mind in applying the results.

The following is the precise mathematical setting.... It is supposed that there is some basic space Ω, and a certain basic collection of sets of points of Ω. These sets will be called *measurable sets*; it is supposed that the class of measurable sets is a Borel field. It is supposed that there is a function $\mathrm{P}\{.\}$, defined for all measurable sets, which is a probability measure, that is, $\mathrm{P}\{.\}$ is a completely additive non-negative set function, with value 1 on the whole space. The number $\mathrm{P}\{\Lambda\}$ will be called the *probability* or *measure* of Λ [11, p. 2].

The concept of a measure is then a combination of the set on which the measure is defined and the characteristics of the measure itself. In general, the properties of a set that make a measure possible are defined by the characteristics of a Borel set, which is a specific class of σ-algebras. To build up the concept of a σ-algebra, let us start by defining an algebra. Most students think of algebra as a freshman math course, but an (abstract) algebra can be defined as

> **Stochastic Methods in Economics and Finance – Malliaris**
>
> Let Ω be an arbitrary space consisting of points ω. Certain classes of subsets of Ω are important in the study of probability. We now define the class of subsets called σ-field or σ-algebra. denoted F.... We call the elements of F *measurable sets* [30, p. 2].

Definition 3.1. A collection of subsets A defined on set X is said to be an algebra in X if A has the following properties.

i. $X \in A$.

ii. $X^C \in A \Rightarrow X^c$ is the complement of X relative to A.

iii. If A, $B \in A$ then $A \cup B \in X$ [6, p. 7].

Notice that the set defined in Figure 3.1 is an algebra.

Definition 3.2. A collection of sets M of subsets X is said to be a σ-algebra in X if $A_n \in M$ for all $n \in N_+$; then $\cup_{n=1}^{\infty} A_n \in M$ [6, p. 7].

In Defintion 3.2 A_n is a sequence of subsets and N_+ denotes the set of all natural numbers. Basically, a σ-algebra is an abstract algebra that is closed under infinite union. In the case of the set depicted in Figure 3.1, the set of all possible unions is finite. However, the algebra is closed under all possible unions so it is a σ-algebra. The Borel set is then the σ-algebra that contains the smallest number of sets (in addition, it typically contains the whole set X and the null set).

Given this somewhat lengthy introduction, we can define the σ-algebra and Borel set for Bierens's [5] discussion as

Definition 3.3. The σ-algebra generated by the collection

$$C = \{(a, b) : \forall a < b,\ a, b \in \Re\} \tag{3.11}$$

of all open intervals in Re is called the Euclidean Borel field, denoted B, and its members are called Borel sets.

In this case, we have defined $a = 0$ and $b = 1$.

Also, note that for any $x \in [0, 1]$, $\mathrm{P}[\{x\}\ ] = \mathrm{P}([x, x]) = 0$. This has the advantage of eliminating the lower end of the range. Specifically,

$$\mathrm{P}([0, x]) = \mathrm{P}([0]) + \mathrm{P}((0, x])\,. \tag{3.12}$$

Further, for $a < b$, $a, b \in [0,1]$

$$P([a,b]) = P((a,b]) = P([a,b)) = P((a,b)) = b - a. \tag{3.13}$$

In the Bierens formulation

$$F_0 = \{(a,b) : [a,b], (a,b], [a,b), \forall a, b \in [0,1], a < b, \text{ and their countable union}\} \tag{3.14}$$

This probability measure is a special case of the Lebesgue measure.

Building on the uniform distribution, we next define the Lebesgue measure as a function λ that measures the length of the interval (a, b) on any Borel set B in R.

$$\lambda(B) = \inf_{B \subset \bigcup_{j=1}^{\infty}(a_j,b_j)} \sum_{j=1}^{\infty} \lambda((a_j, b_j)) = \inf_{B \subset \bigcup_{j=1}^{\infty}(a_j,b_j)} \sum_{j=1}^{\infty} (b_j - a_j). \tag{3.15}$$

It is the total length of the Borel set taken from the outside. Based on the Lebesgue measure, we can then define the Lebesgue integral based on the basic definition of the Reimann integral.

$$\int_a^b f(x)\,dx = \sup \sum_{m=1}^{n} \left(\inf_{x \in I_m} f(x) \right) \lambda(I_m). \tag{3.16}$$

Note that the result in Equation 3.16 is similar in concept to the simple forms of the Riemann sum presented in Equation 3.6. Replacing the interval of the summation, the Lebegue integral becomes

$$\int_A f(x)\,dx = \sup \sum_{m=1}^{n} \left(\inf_{x \in B_n} f(x) \right) \lambda(B_m). \tag{3.17}$$

Hence, the probability measure in Equation 3.17 is a more general version of the integral.

Specifically, the real number line which forms the basis of the Riemann sum is but one of the possible Borel sets that we can use to construct a probability measure. In order to flush out this concept, consider measuring the pH (acidity) of a pool using Phenol red. Phenol red changes color with the level of pH, ranging from yellow when the water is relatively acidic to a red/purple when the water is alkaline. This concept of acidity is mapped onto a pH scale running from 6.8 when the result is yellow to 8.2 when the result is red/purple. This mapping is constructed using a kit which tells me the pH level associated with each color. In this case we have a measure outside the typical concept of counting – it does not necessarily fit on the real number line. The mechanics given above insure that we can develop a probability measure associated with the pH level (i.e., without first mapping through the numeric pH measure).

3.2 Random Variables and Distributions

Now we have established the existence of a probability measure based on a specific form of σ-algebras called Borel fields. The question is, can we extend this rather specialized formulation to broader groups of random variables? Of course, or this would be a short textbook. As a first step, let's take the simple coin-toss example. In the case of a coin there are two possible outcomes (heads or tails). These outcomes completely specify the sample space. To add a little structure, we construct a random variable X that can take on two values $X = 0 \text{ or } 1$ (as depicted in Table 2.6). If $X = 1$ the coin toss resulted in a head, while if $X = 0$ the coin toss resulted in a tail. Next, we define each outcome based on an event space ω:

$$\begin{aligned} \mathrm{P}(X=1) &= \mathrm{P}(\{\omega \in \Omega : X(\omega) = 1\}) = \mathrm{P}([H]) \\ \mathrm{P}(X=0) &= \mathrm{P}(\{\omega \in \Omega : X(\omega) = 0\}) = \mathrm{P}([T]). \end{aligned} \tag{3.18}$$

In this case the physical outcome of the experiment is either a head ($\omega =$ heads) or a tail ($\omega =$ tails). These events are "mapped into" number space – the measure of the event is either a zero or a one.

The probability function is then defined by the random event ω. Defining ω as a uniform random variable from our original example, one alternative would be to define the function as

$$X(\omega) = 1 \text{ if } \omega \leq 0.50. \tag{3.19}$$

This definition results in the standard 50-50 result for a coin toss. However, it admits more general formulations. For example, if we let

$$X(\omega) = 1 \text{ if } \omega \leq 0.40 \tag{3.20}$$

the probability of heads becomes 40 percent.

Given this intuition, the next step is to formally define a random variable. Three alternative definitions should be considered

Definition 3.4. A *random variable* is a function from a sample space S into the real numbers.

Definition 3.5. A *random variable* is a variable that takes values according to a certain probability.

Definition 3.6. A *random variable* is a real-valued function defined over a sample space.

In this way a random variable is an abstraction. We assumed that there was a random variable defined on some sample space like flipping a coin. The flipping of the coin is an outcome in an abstract space (i.e., a Borel set).

$$S = \{s_1, s_2, \cdots s_n\}. \tag{3.21}$$

We then define a numeric value to this set of random variables.

$$\begin{array}{ccc} X : S \rightarrow R_{+}^{1} & \text{or} & X(\omega) : \Omega \rightarrow R_{+}^{1} \\ x_i = X(s_i) & & x_i = X(\omega_i) . \end{array} \tag{3.22}$$

There are two ways of looking at this tranformation. First, the Borel set is simply defined as the real number line (remember that the real number line is a valid Borel set). Alternatively, we can view the transformation as a two step mapping. For example, a measure can be used to define the quantity of wheat produced per acre. Thus, we are left with two measures of the same phenomena — the quantity of wheat produced per acre and the probability of producing that quantity of wheat. The probability function (or measure) is then defined based on that random variable for either case defined as

$$\begin{array}{c} \mathrm{P}_X(X = x_i) = \mathrm{P}(\{s_i \in S : X(s_i) = x_i\}) \\ \mathrm{P}(X(\omega) = x_i) = \mathrm{P}(\{\omega \in \Omega : X(\omega) = x_i\}) . \end{array} \tag{3.23}$$

Using either justification, for the rest of this text we are simply going to define a random variable as either a discrete ($x_i = 1, 2, \cdots N$) or real number ($x = (-\infty, \infty)$).

3.2.1 Discrete Random Variables

Several of the examples used thus far in the text have been discrete random variables. For example, the coin toss is a simple discrete random variable where the outcome can take on a finite number of values – $X = \{Tails, Heads\}$ or in numeric form $X = \{0, 1\}$. Using this intuition, we can then define a discrete random variable as

Definition 3.7. A *discrete random variable* is a variable that takes a countable number of real numbers with certain probability.

In addition to defining random variables as either discrete or continuous we can also define random variables as either univariate or multivariate. Consider the dice rolls presented in Table 3.2. Anna rolled two six-sided dice (one blue and one red) while Alex rolled one eight-sided die and one-six sided die. Conceptually, the die rolled by each individual is a bivariate discrete set of random variables as defined in Definition 3.8.

Definition 3.8. A *bivariate discrete random variable* is a variable that takes a countable number of bivariate points on the plane with certain probability.

For example, the pair $\{2, 1\}$ is the tenth outcome of Anna's rolls. In most board games the sum of the outcomes of the two dice is the important number – the number of spaces moved in Monopoly™. However, in other games the outcome may be more complex. For example, the outcome may be whether a player suffers damage defined by whether the eight-sided die is greater than three while the amount of damage suffered is determined by the six-sided

TABLE 3.2
Anna and Alex's Dice Rolls

	Anna		Alex	
Roll	Blue	Red	Eight Sided	Six Sided
1	6	4	5	6
2	6	3	8	9
3	5	4	1	1
4	5	3	7	6
5	3	5	5	1
6	5	1	4	5
7	3	6	6	1
8	4	4	5	1
9	5	2	5	5
10	2	1	2	5
11	4	2	6	4
12	2	5	3	1
13	5	3	1	4
14	3	2	1	3
15	1	4	6	6
16	2	3	6	6
17	3	3	3	2
18	5	4	2	3
19	3	3	1	3
20	6	6	7	2

die. Thus, we may be interested in defining a secondary random variable (the number of spaces moved as the sum of the result of the blue and red die or the amount of damage suffered by a character of a board game based on a more complex protocal) based on the outcomes of the bivariate random variables. However, at the most basic level we are interested in a bivariate random variable.

3.2.2 Continuous Random Variables

While discrete random variables are important in some econometric applications, most econometric applications are based on continuous random variables such as the price of consumption goods or the quantity demanded and supplied in the market place. As discussed in Chapter 2, defining a continuous random variable as some subset on the real number line complicates the definition of probability. Because the number of real numbers for any subset of the real number line is infinite, the standard counting definition of probability used by the frequency approach presented in Equation 2.2 implies a zero probability. Hence, it is necessary to develop probability using the concept of a **probability density function** (or simply the **density function**) as presented in Definition 3.9.

Definition 3.9. If there is a non-negative function $f(x)$ defined over the whole line such that

$$\mathrm{P}\left(x_1 \leq X \leq x_2\right) = \int_{x_1}^{x_2} f(x)dx \tag{3.24}$$

for any x_1 and x_2 satisfying $x_1 \leq x_2$, then X is a continuous random variable and $f(x)$ is called its density function.

By the second axiom of probability (see definition 2.12)

$$\int_{-\infty}^{\infty} f(x)dx = 1. \tag{3.25}$$

The simplest example of a continuous random variable is the uniform distribution

$$f\left(x\right) = \begin{cases} 1 \text{ if } 0 \leq x \leq 1 \\ 0 \text{ otherwise.} \end{cases} \tag{3.26}$$

Using the definition of the uniform distribution function in Equation 3.26, we can demonstrate that the probability of the continuous random variable defined in Equation 3.24 follows the required axioms for probability. First, $f\left(x\right) \geq 0$ for all x. Second, the total probability equals one. To see this, consider the integral

$$\begin{aligned} \int_{-\infty}^{\infty} f(x)dx &= \int_{-\infty}^{0} f(x)dx + \int_{0}^{1} f(x)dx + \int_{1}^{\infty} f(x)dx \\ &= \int_{0}^{1} f(x)dx = \int_{0}^{1} dx = \left(x|_0^1\right. + C = (1-0) + C. \end{aligned} \tag{3.27}$$

Thus the total value of the integral is equal to one if $C = 0$.

The definition of a continuous random variable, like the case of the univariate random variable, can be extended to include the possibility of a bivariate continuous random variable. Specifically, we can extend the univariate uniform distribution in Equation 3.26 to represent the density function for the bivariate outcome $\{x, y\}$

$$f\left(x, y\right) = \begin{cases} 1 \text{ if } 0 \leq x \leq 1, 0 \leq y \leq 1 \\ 0 \text{ otherwise.} \end{cases} \tag{3.28}$$

The fact that the density function presented in Equation 3.28 conforms to the axioms of probability are left as an exercise.

Definition 3.10. If there is a non-negative function $f(x, y)$ defined over the whole plane such that

$$\mathrm{P}\left(x_1 \leq X \leq x_2, y_1 \leq Y \leq y_2\right) = \int_{y_1}^{y_2} \int_{x_1}^{x_2} f\left(x, y\right) dxdy \tag{3.29}$$

for x_1, x_2, y_1, and y_2 satisfying $x_1 \leq x_2$, $y_1 \leq y_2$, then (X, Y) is a bivariate continuous random variable and $f\left(X, Y\right)$ is called the joint density function.

Much of the work with distribution functions involves integration. In order to demonstrate a couple of solution techniques, I will work through some examples.

Example 3.11. If $f(x,y) = xy\exp(-x-y)$, $x > 0$, $y > 0$ and 0 otherwise, what is $P(X > 1, Y < 1)$?

$$\mathrm{P}(X > 1, Y < 1) = \int_0^1 \int_1^\infty xye^{-(x+y)}dxdy. \tag{3.30}$$

First, note that the integral can be separated into two terms:

$$\mathrm{P}(X > 1, Y < 1) = \int_1^\infty xe^{-1}dx \int_0^1 ye^{-y}dy. \tag{3.31}$$

Each of these integrals can be solved using integration by parts:

$$\begin{aligned} d(uv) &= v\,du + u\,dv \\ v\,du &= d(uv) - u\,dv \\ \textstyle\int v\,du &= uv - \textstyle\int u\,dv. \end{aligned} \tag{3.32}$$

In terms of a proper integral we have

$$\int_a^b v\,du = (uv|_a^b - \int_a^b u\,dv. \tag{3.33}$$

In this case, we have

$$\begin{aligned} &\int_1^\infty xe^{-x}dx \Rightarrow \begin{cases} v = x,\ dv = 1 \\ du = e^{-x},\ u = -e^{-x} \end{cases} \\ &\int_1^\infty xe^{-x}dx = (-xe^{-x}|_1^\infty + \int_1^\infty e^{-x}dx = 2e^{-1} = 0.74. \end{aligned} \tag{3.34}$$

Working on the second part of the integral,

$$\begin{aligned} \int_0^1 ye^{-y}dy &= (-ye^{-1}|_0^1 + \int_0^1 e^{-y}dy \\ &= (-ye^{-1}|_0^1 + (-e^{-y}|_0^1 \\ &= (-e^{-1} + 0) + (-e^{-1} + 1). \end{aligned} \tag{3.35}$$

Putting the two parts together,

$$\begin{aligned} \mathrm{P}(X > 1, Y < 1) &= \int_1^\infty xe^{-x}dx \int_0^1 ye^{-y}dy \\ &= (0.735)(0.264) = 0.194. \end{aligned} \tag{3.36}$$

Definition 3.12. A *T-variate random variable* is a variable that takes a countable number of points on the T-dimensional Euclidean space with certain probabilities.

Following our development of integration by parts, we have attempted to keep the calculus at an intermediate level throughout this textbook. However, the development of certain symbolic computer programs may be useful to students. Appendix A presents a brief discussion of two such symbolic programs – Maxima (an open source program) and *Mathematica* (a proprietary program).

3.3 Conditional Probability and Independence

In order to define the concept of a conditional probability, it is necessary to discuss joint probabilities and marginal probabilities. A joint probability is the probability of two random events. For example, consider drawing two cards from the deck of cards. There are $52 \times 51 = 2{,}652$ different combinations of the first two cards from the deck. The marginal probability is the overall probability of a single event or the probability of drawing a given card. The conditional probability of an event is the probability of that event given that some other event has occurred. Taking the roll of a single die, for example – what is the probability of the die being a one if you know that the face number is odd? (1/3). However, note that if you know that the roll of the die is a one, the probability of the roll being odd is 1.

As a starting point, consider the requirements (axioms) for a conditional probability to be valid.

Definition 3.13. Axioms of Conditional Probability:

1. $\mathrm{P}(A|B) \geq 0$ for any event A.

2. $\mathrm{P}(A|B) = 1$ for any event $A \supset B$.

3. If $\{A_i \cap B\}$ $i = 1, 2, \ldots$ are mutually exclusive, then

$$\mathrm{P}(A_1 \cup A_2 \cup \ldots) = \mathrm{P}(A_1|B) + \mathrm{P}(A_2|B) + \cdots \tag{3.37}$$

4. If $B \supset H$, $B \supset G$, and $\mathrm{P}(G) \neq 0$ then

$$\frac{\mathrm{P}(H|B)}{\mathrm{P}(G|B)} = \frac{\mathrm{P}(H)}{\mathrm{P}(G)}. \tag{3.38}$$

Note that Axioms 1 through 3 follow the general probability axioms with the addition of a conditional term. The new axiom (Axiom 4) states that two events conditioned on the same probability set have the same relationship as the overall (as we will develop shortly – marginal) probabilities. Intuitively, the conditioning set brings in no additional information about the relative likelihood of the two events.

Theorem 3.14 provides a formal definition of conditional probability.

Theorem 3.14. $\mathrm{P}(A|B) = \mathrm{P}(A \cap B)/\mathrm{P}(B)$ *for any pair of events* A *and* B *such that* $\mathrm{P}(B) > 0$.

Taking this piece by piece – $\mathrm{P}(A \cap B)$ is the probability that both A and B will occur (i.e., the joint probability of A and B). Next, $\mathrm{P}(B)$ is the probability that B will occur. Hence, the conditional probability $\mathrm{P}(A|B)$ is defined as the joint probability of A and B given that we know that B has occurred. Some texts refer to Theorem 3.14 as Bayes' theorem; however, in this text we will define Bayes' theorem as depicted in Theorem 3.15.

Theorem 3.15 (Bayes' Theorem). *Let Events* $A_1, A_2, \ldots A_n$ *be mutually exclusive events such that* $\mathrm{P}(A_1 \cup A_2 \cup \cdots A_n) = 1$ *and* $\mathrm{P}(A_i) > 0$ *for each* i. *Let* E *be an arbitrary event such that* $\mathrm{P}(E) > 0$. *Then*

$$\mathrm{P}(A_i|E) = \frac{\mathrm{P}(E|A)\,\mathrm{P}(A_i)}{\sum_{j=1}^{n} \mathrm{P}(E|A_j)\,\mathrm{P}(A_j)}. \tag{3.39}$$

While Equation 3.39 appears different from the specification in Theorem 3.14, we can demonstrate that they are the same concept. First, let us use the relationship in Theorem 3.14 to define the probability of the joint event $E \cap A_i$.

$$\mathrm{P}(E \cap A_i) = \mathrm{P}(E|A_i)\,\mathrm{P}(A_i). \tag{3.40}$$

Next, if we assume that events A_1, A_2, $\cdots$ are mutually exclusive and exhaustive, we can rewrite the probability of event E as

$$\mathrm{P}(E) = \sum_{i=1}^{n} \mathrm{P}(E|A_i)\,\mathrm{P}(A_i). \tag{3.41}$$

Combining the results of Equations 3.40 and 3.41 yields the friendlier version of Bayes' theorem found in Thereom 3.14:

$$\mathrm{P}(A_i|E) = \frac{\mathrm{P}(E \cap A_i)}{\mathrm{P}(E)}. \tag{3.42}$$

Notice the direction of the conditional statement – if we know that event E has occurred, what is the probability that event A_i will occur?

Given this understanding of conditional probability, it is possible to define **statistical independence**. One random variable is independent of the probability of another random variable if

Definition 3.16. Events A and B are said to be independent if $\mathrm{P}(A) = \mathrm{P}(A|B)$.

Hence, the random variable A is independent of the random variable B if knowing the value of B does not change the probability of A. Extending the scenario to the case of three random variables:

Definition 3.17. Events A, B, and C are said to be mutually independent if the following equalities hold:

a) $\mathrm{P}(A \cap B) = \mathrm{P}(A)\,\mathrm{P}(B)$

b) $\mathrm{P}(A \cap C) = \mathrm{P}(A)\,\mathrm{P}(C)$

c) $\mathrm{P}(B \cap C) = \mathrm{P}(B)\,\mathrm{P}(C)$

d) $\mathrm{P}(A \cap B \cap C) = \mathrm{P}(A)\,\mathrm{P}(B)\,\mathrm{P}(C)$

3.3.1 Conditional Probability and Independence for Discrete Random Variables

In order to develop the concepts of conditional probability and independence, we start by analyzing the discrete bivariate case. As a starting point, we define the marginal probability of a random variable as the probability that a given value of one random variable will occur (i.e., $X = x_i$) regardless of the value of the other random variable. For this discussion, we simplify our notation slightly so that $\mathrm{P}[X = x_i \cap Y = y_j] = \mathrm{P}[X = x_i, Y = y_j] = \mathrm{P}[x_i, y_j]$. The **marginal distribution** for x_i can then be defined as

$$\mathrm{P}[x_i] = \sum_{j=1}^{m} \mathrm{P}[x_i, y_j]. \tag{3.43}$$

Turning to the binomial probability presented in Table 3.3, the marginal probability that $X = x_1$ (i.e., $X = 0$) can be computed as

$$\mathrm{P}[x_1] = \mathrm{P}[x_1|\, y_1] + \mathrm{P}[x_1|\, y_2] + \cdots \mathrm{P}[x_1|\, y_6]$$

$$= 0.01315 + 0.04342 + 0.05790 + 0.03893 + 0.01300 + 0.00158 = 0.16798. \tag{3.44}$$

By repetition the marginal value for each X_i and Y_j is presented in Table 3.3.

Applying a discrete form of Bayes' theorem,

$$\mathrm{P}[x_i|\, y_j] = \frac{\mathrm{P}(x_i, y_j)}{\mathrm{P}(y_j)} \tag{3.45}$$

we can compute the conditional probability of $X = 0$ given $Y = 2$ as

$$\mathrm{P}[x_1|\, y_3] = \frac{0.581}{0.3456} = 0.16881. \tag{3.46}$$

TABLE 3.3
Binomial Probability

	y						
x	0	1	2	3	4	5	Marginal Probability
0	0.01315	0.04342	0.05790	0.03893	0.01300	0.00158	0.16798
1	0.02818	0.09304	0.12408	0.08343	0.02786	0.00339	0.35998
2	0.02417	0.07979	0.10640	0.07155	0.02389	0.00290	0.30870
3	0.01039	0.03430	0.04574	0.03075	0.01027	0.00125	0.13270
4	0.00222	0.00733	0.00978	0.00658	0.00220	0.00027	0.02838
5	0.00018	0.00059	0.00079	0.00053	0.00018	0.00000	0.00227
Marginal Probability	0.07829	0.25847	0.34469	0.23177	0.07740	0.00939	

TABLE 3.4
Binomial Conditional Probabilities

X	P $[X, Y = 2]$	P $[Y = 2]$	P $[X \mid Y = 2]$	P $[X]$
0	0.05790	0.34469	0.16798	0.16798
1	0.12408	0.34469	0.35998	0.35998
2	0.10640	0.34469	0.30868	0.30870
3	0.04574	0.34469	0.13270	0.13270
4	0.00978	0.34469	0.02837	0.02838
5	0.00079	0.34469	0.00229	0.00227

Table 3.4 presents the conditional probability for each value of X given $Y = 2$.

Next, we offer a slightly different definition of independence for the discrete bivariate random variable.

Definition 3.18. Discrete random variables are said to be independent if the events $X = x_i$ and $Y = y_j$ are independent for all i, j. That is to say, $\mathrm{P}(x_i, y_j) = \mathrm{P}(x_i)\,\mathrm{P}(y_j)$.

To demonstrate the consistency of Definition 3.18 with Definition 3.16, note that

$$\mathrm{P}[x_i] = \mathrm{P}[x_i \mid y_j] \Rightarrow \mathrm{P}[x_i] = \frac{\mathrm{P}[x_i, y_j]}{\mathrm{P}[y_j]}. \tag{3.47}$$

Therefore, multiplying each side of the last equality in Equation 3.47 yields $\mathrm{P}[x_i] \times \mathrm{P}[y_j] = \mathrm{P}[x_i, y_j]$.

Thus, we determine independence by whether the $\mathrm{P}[x_i, y_j]$ values equal $\mathrm{P}[x_i] \times \mathrm{P}[y_j]$. Taking the first case, we check to see that

$$\mathrm{P}[x_1] \times \mathrm{P}[y_1] = 0.1681 \times 0.0778 = 0.0131 = \mathrm{P}[x_1, y_1]. \tag{3.48}$$

Carrying out this process for each cell in Table 3.3 confirms the fact that X and Y are independent. This result can be demonstrated in a second way (more consistent with Definition 3.18). Note that the $\mathrm{P}[X \mid Y = 2]$ column in Table 3.4 equals the $\mathrm{P}[X]$ column – the conditional is equal to the marginal in all cases.

Next, we consider the discrete form of the uncorrelated normal distribution as presented in Table 3.5. Again, computing the conditional distribution of X such that $Y = 2$ yields the results in Table 3.6.

Theorem 3.19. *Discrete random variables X and Y with the probability distribution given in Table 3.1 are independent if and only if every row is proportional to any other row, or, equivalently, every column is proportional to any other column.*

Finally, we consider a discrete form of the correlated normal distribution in Table 3.7. To examine whether the events are independent, we compute the conditional probability for X when $Y = 2$ and compare this conditional

TABLE 3.5
Uncorrelated Discrete Normal

x	y = 0	1	2	3	4	5	Marginal Probability
0	0.00610	0.01708	0.02520	0.01707	0.00529	0.00080	0.07154
1	0.02058	0.05763	0.08503	0.05761	0.01786	0.00271	0.24142
2	0.03193	0.08940	0.13191	0.08936	0.02770	0.00421	0.37451
3	0.02054	0.05752	0.08488	0.05750	0.01783	0.00271	0.24098
4	0.00547	0.01531	0.02259	0.01530	0.00474	0.00072	0.06413
5	0.00063	0.00177	0.00261	0.00177	0.00055	0.00008	0.00741
Marginal Probability	0.08525	0.23871	0.35222	0.23861	0.07397	0.01123	

TABLE 3.6
Uncorrelated Normal Conditional Probabilities

X	$\mathrm{P}[X, Y=2]$	$\mathrm{P}[Y=2]$	$\mathrm{P}[X \mid Y=2]$	$\mathrm{P}[X]$
0	0.02520	0.35222	0.07155	0.07154
1	0.08503	0.35222	0.24141	0.24142
2	0.13191	0.35222	0.37451	0.37451
3	0.08488	0.35222	0.24099	0.24098
4	0.02259	0.35222	0.06414	0.06413
5	0.00261	0.35222	0.00741	0.00741

distribution with the marginal distribution of X. The results presented in Table 3.8 indicate that the random variables are not independent.

3.3.2 Conditional Probability and Independence for Continuous Random Variables

The development of conditional probability and independence for continuous random variables follows the same general concepts as discrete random variables. However, constructing the conditional formulation for continuous variables requires some additional mechanics. Let us start by developing the **conditional density** function.

Definition 3.20. Let X have density $f(x)$. The *conditional density* of X given $a \leq X \leq b$, denoted by $f(x \mid a \leq X \leq b)$, is defined by

$$f(x \mid a \leq X \leq b) = \frac{f(x)}{\int_a^b f(x)dx} \text{ for } a \leq x \leq b, \qquad (3.49)$$

$$= 0 \quad \text{otherwise.}$$

Notice that Definition 3.20 defines the conditional probability for a single continuous random variable conditioned on the fact that the random variable is in a specific range ($a \leq X \leq b$). This definition can be expanded slightly by considering any general range of the random variable X ($X \in S$).

Definition 3.21. Let X have the density $f(x)$ and let S be a subset of the real line such that $\mathrm{P}(X \in S) > 0$. Then the conditional density of X given $X \in S$, denoted by $f(x \mid S)$, is defined by

$$f(x \mid S) = \frac{f(x)}{\mathrm{P}(X \in S)} \text{ for } x \in S \qquad (3.50)$$

$$= 0 \quad \text{otherwise.}$$

TABLE 3.7
Correlated Discrete Normal

	y						
x	0	1	2	3	4	5	Marginal Probability
0	0.01326	0.02645	0.02632	0.01082	0.00191	0.00014	0.07890
1	0.02647	0.06965	0.08774	0.04587	0.00991	0.00093	0.24057
2	0.02603	0.08749	0.13529	0.08733	0.02328	0.00274	0.36216
3	0.01086	0.04563	0.08711	0.06964	0.02304	0.00334	0.23962
4	0.00187	0.00984	0.02343	0.02320	0.00950	0.00178	0.06962
5	0.00013	0.00088	0.00271	0.00332	0.00172	0.00040	0.00916
Marginal Probability	0.07862	0.23994	0.36260	0.24018	0.06936	0.00933	

TABLE 3.8
Correlated Normal Conditional Probabilities

X	P[$X,Y=2$]	P[$Y=2$]	P[$X|Y=2$]	P[X]
0	0.02632	0.36260	0.07259	0.07890
1	0.08774	0.36260	0.24197	0.24057
2	0.13529	0.36260	0.37311	0.36216
3	0.08711	0.36260	0.24024	0.23962
4	0.02343	0.36260	0.06462	0.06962
5	0.00271	0.36260	0.00747	0.00916

To develop the conditional relationship between two continuous random variables (i.e., $f(x|\,y)$) using the general approach to conditional density functions presented in Definitions 3.20 and 3.21, we have to define the marginal density (or marginal distribution) of continuous random variables.

Theorem 3.22. *Let $f(x,y)$ be the joint density of X and Y and let $f(x)$ be the marginal density of X. Then*

$$f(x) = \int_{-\infty}^{\infty} f(x,y)\,dy. \tag{3.51}$$

Going back to the distribution function from Example 3.11, we have

$$f(x,y) = xye^{-(x+y)}. \tag{3.52}$$

To prove that this is a proper distribution function, we limit our consideration to non-negative values of x and y (i.e., $f(x,y) \geq 0$ if $x,\, y \geq 0$). From our previous discussion it is also obvious that

$$\begin{aligned}\int_0^{\infty}\int_0^{\infty} f(x,y)\,dxdy &= \left(\int_0^{\infty} xe^{-x}dx\right)\left(\int_0^{\infty} ye^{-y}dy\right)\\ &= \left(-\left(xe^{-x}\right|_0^{\infty} + \int_0^{\infty} e^{-x}dx\right)\left(-\left(ye^{-y}\right|_0^{\infty} + \int_0^{\infty} e^{-y}dy\right)\\ &= \left(-\left(\infty\cdot 0 - 0\cdot 1\right) - (0-1)\right)\left(-\left(\infty\cdot 0 - 0\cdot 1\right) - (0-1)\right) = 1.\end{aligned} \tag{3.53}$$

Thus, this is a proper density function. The marginal density function for x follows this formulation:

$$\begin{aligned}f(x) &= \int_0^{\infty} f(x,y)\,dy = \left(xe^{-x}\right)\int_0^{\infty} ye^{-y}dy\\ &= \left(xe^{-x}\right)\left(-\left(ye^{-y}\right|_0^{\infty} + \int_0^{\infty} e^{-y}dy\right)\\ &= xe^{-x}.\end{aligned} \tag{3.54}$$

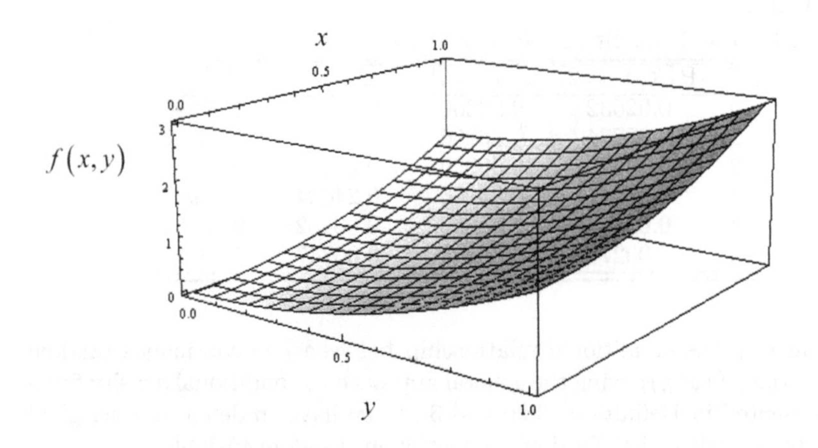

FIGURE 3.2
Quadratic Probability Density Function.

Example 3.23. Consider the continuous bivariate distribution function

$$f(x,y) = \frac{3}{2}\left(x^2 + y^2\right) \; for \, x, y \in [0,1] \tag{3.55}$$

which is depicted graphically in Figure 3.2. First, to confirm that Equation 3.55 is a valid distribution function,

$$\begin{aligned} \frac{3}{2}\int_0^1 \left(x^2+y^2\right)dx &= \frac{3}{2}\left(\frac{1}{3}x^3 + \frac{1}{3}y^3\right|_0^1 \\ &= \frac{3}{2}\left(\frac{1}{3}+\frac{1}{3}\right) = 1. \end{aligned} \tag{3.56}$$

Further, $f(x,y) \geq 0$ for all $x, y \in [0,1]$. To prove this rigorously we would show that $f(x,y)$ is at a minimum at $\{x,y\} = \{0,0\}$ and that the derivative of $f(x,y)$ is positive for all $x, y \in [0,1]$.

This example has a characteristic that deserves discussion. Notice that $f(1,1) = 3 > 1$; thus, while the axioms of probability require that $f(x,y) \geq 0$, the function can assume almost any positive value as long as it integrates to one. Departing from the distribution function in Equation 3.55 briefly, consider the distribution function $g(z) = 2$ for $z \in [0, 1/2]$. This is a uniform distribution function with a more narrow range than the U$[0,1]$. It is valid because $g(z) \geq 0$ for all z and

$$\int_0^{1/2} 2dz = 2\left(z\right|_0^{1/2} = 2\left(\frac{1}{2} - 0\right) = 1. \tag{3.57}$$

Hence, even though a distribution function has values greater than one, it may still be a valid density function.

Returning to the density function defined in Equation 3.55, we derive the marginal density for x:

$$\begin{aligned} f(x) &= \int_0^1 \frac{3}{2}\left(x^2+y^2\right) dy \\ &= \frac{3}{2}x^2 \int_0^1 dy + \frac{3}{2}\int_0^1 y^2 dy \\ &= \frac{3}{2}x^2 \left(y|_0^1 + \frac{3}{2}\left(\frac{1}{3}y^3\right|_0^1 \\ &= \frac{3}{2}x^2 + \frac{1}{2}. \end{aligned} \tag{3.58}$$

While the result of Equation 3.58 should be a valid probability density function by definition, it is useful to make sure that the result conforms to the axioms of probability (e.g., it provides a check on your mathematics). First, we note that $f(x) \geq 0$ for all $x \in [0,1]$. Technically, $f(x) = 0$ if $x \notin [0,1]$. Next, to verify that the probability is one for the entire sample set,

$$\begin{aligned} \int_0^1 \left(\frac{3}{2}x^2 + \frac{1}{2}\right) &= \frac{3}{2}\int_0^1 x^2 dx + \frac{1}{2}\int_0^1 dx \\ &= \frac{3}{2}\left(\frac{1}{3}x^3\right|_0^1 + \frac{1}{2}\left(x|_0^1\right. \\ &= \frac{1}{2} + \frac{1}{2} = 1. \end{aligned} \tag{3.59}$$

Thus, the marginal distribution function from Equation 3.58 meets the criteria for a probability measure.

Next, we consider the bivariate extension of Definition 3.21.

Definition 3.24. Let (X, Y) have the joint density $f(x, y)$ and let S be a subset of the plane which has a shape as in Figure 3.3. We assume that $\mathrm{P}[(X,Y) \in S] > 0$. Then the conditional density of X given $(X, Y) \in S$, denoted $f(x|S)$, is defined by

$$f(x|S) = \begin{cases} \dfrac{\int_{h(x)}^{g(x)} f(x,y)\, dy}{\mathrm{P}[(X,Y) \in S]} & \text{for } a \leq x \leq b, \\ 0 \quad \text{otherwise.} \end{cases} \tag{3.60}$$

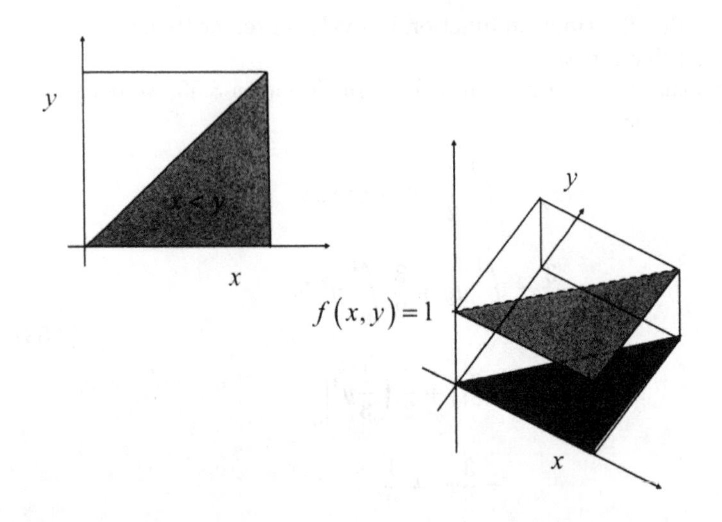

FIGURE 3.3
Conditional Distribution for a Region of a Bivariate Uniform Distribution.

Building on Definition 3.24, consider the conditional probability of $x < y$ for the bivariate uniform distribution as depicted in Figure 3.3.

Example 3.25. Suppose $f(x,y) = 1$ for $0 \leq x \leq 1$, $0 \leq y \leq 1$, and 0 otherwise. Obtain $f(x|X < Y)$.

$$\begin{aligned} f(x|X<Y) &= \frac{\int_x^1 dy}{\int_0^1 \int_x^1 dy dx} \\ &= \frac{\left(y\right|_x^1}{\int_0^1 \left(y\right|_x^1 dx} \\ &= \frac{1-x}{\int_0^1 (1-x)\,dx} \\ &= \frac{1-x}{\left(x-\frac{1}{2}x^2\right|_0^1} \\ &= 2(1-x). \end{aligned} \tag{3.61}$$

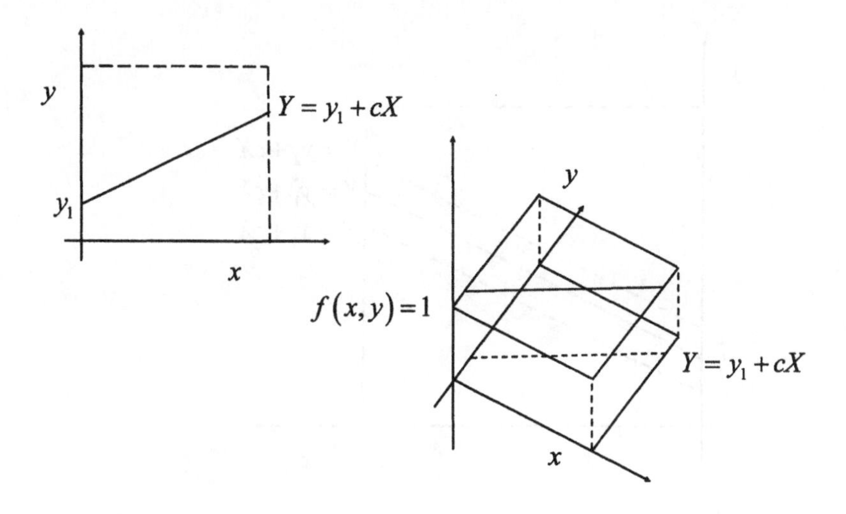

FIGURE 3.4
Conditional Distribution of a Line for a Bivariate Uniform Distribution.

Notice the downward sloping nature of Equation 3.61 is consistent with the area of the projection in the upper right diagram of Figure 3.3. Initially each increment of y implies a fairly large area of probability (i.e., the difference $1 - x$). However, as y increases, this area declines.

Suppose that we are interested in the probability of X along a linear relationship $Y = y_1 + cX$. As a starting point, consider the simple bivariate uniform distribution that we have been working with where $f(x, y) = 1$. We are interested in the probability of the line in that space presented in Figure 3.4. The conditional probability that X falls into $[x_1, x_2]$ given $Y = y_1 + cX$ is defined by

$$\begin{gathered}\mathrm{P}\left(x_1 \leq X \leq x_2 | Y = y_1 + cX\right) = \\ \lim_{y_2 \to y_1} \mathrm{P}\left(x_1 \leq X \leq x_2 | y_1 + cX \leq Y \leq y_2 + cX\right)\end{gathered} \tag{3.62}$$

for all x_1, x_2 satisfying $x_1 \leq x_2$. Intuitively, as depicted in Figure 3.5, we are going to start by bounding the line on which we want to define the conditional probability (i.e., $Y = y_1 + cX \leq Y = y_1^* + cX \leq Y = y_2 + CX$). Then we are going to reduce the bound $y_1 \to y_2$, leaving the relationship for y_1^*. The conditional density of X given $Y = y_1 + cX$, denoted by $f(x|Y = y_1 + cX)$, if it exists, is defined to be a function that satisfies

$$\mathrm{P}\left(x_1 \leq X \leq x_2 | Y = y_1 + cX\right) = \int_{x_1}^{x_2} f\left(x | Y = y_1 + cX\right) dx. \tag{3.63}$$

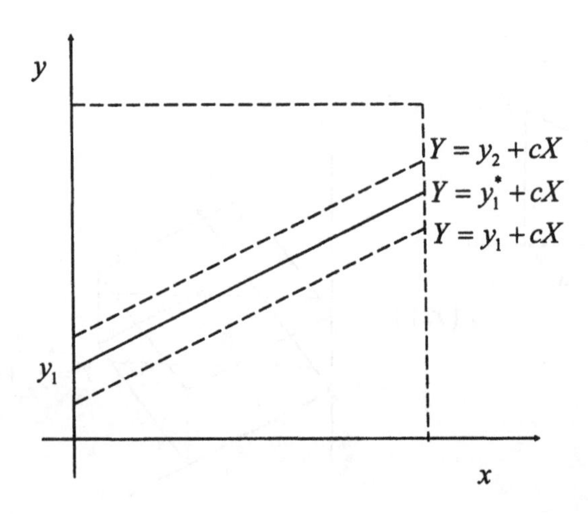

FIGURE 3.5
Bounding the Conditional Relationship.

In order to complete this proof we will need to use the **mean value theorem of integrals.**

Theorem 3.26. *Let $f(x)$ be a continuous function defined on the closed interval $[a, b]$. Then there is some number X in that interval $(a \leq X \leq b)$ such that*

$$\int_a^b f(x)\,dx = (b-a)\,f(X). \tag{3.64}$$

[48, p. 45]

The intuition for this proof is demonstrated in Figure 3.6. We don't know what the value of X is, but at least one X satisfies the equality in Equation 3.64.

Theorem 3.27. *The conditional density $f(x|Y = y_1 + cX)$ exists and is given by*

$$f(x|Y = y_1 + cX) = \frac{f(x, y_1 + cx)}{\int_{-\infty}^{\infty} f(x, y + cx)\,dx} \tag{3.65}$$

provided the denominator is positive.

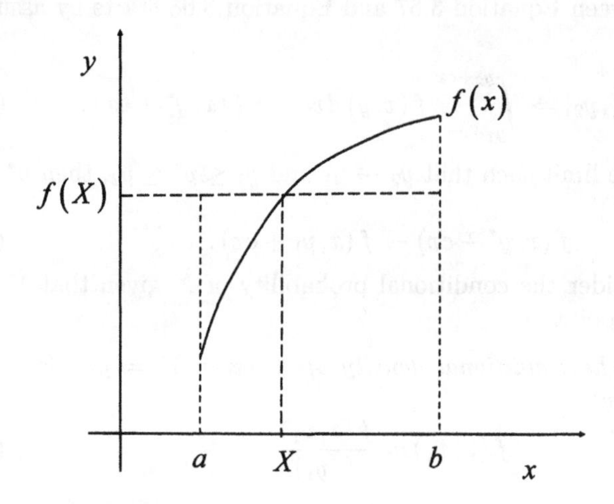

FIGURE 3.6
Mean Value of Integral.

Proof. We have

$$\lim_{y_2 \to y_1} P(x_1 \le X \le x_2 | y_1 + cX \le Y \le y_2 + cX)$$

$$= \lim_{y_2 \to y_1} \frac{\displaystyle\int_{x_1}^{x_2} \int_{y_1+cx}^{y_2+cx} f(x,y)\, dydx}{\displaystyle\int_{-\infty}^{\infty} \int_{y_1+cx}^{y_2+cx} f(x,y)\, dxdy}. \tag{3.66}$$

Thus, by the mean value of integration,

$$\lim_{y_2 \to y_1} \frac{\displaystyle\int_{x_1}^{x_2} \int_{y_1+cx}^{y_2+cx} f(x,y)\, dydx}{\displaystyle\int_{-\infty}^{\infty} \int_{y_1+cx}^{y_2+cx} f(x,y)\, dxdy} = \lim_{y_1 \to y_2} \frac{\displaystyle\int_{x_1}^{x_2} f(x, y^* + cx)\, dx}{\displaystyle\int_{-\infty}^{\infty} f(x, y^* + cx)\, dx} \tag{3.67}$$

where $y_1 \le y^* \le y_2$. As $y_2 \to y_1$, $y^* \to y_1$, hence

$$\lim_{y_1 \to y_2} \frac{\displaystyle\int_{x_1}^{x_2} f(x, y^* + cx)\, dx}{\displaystyle\int_{-\infty}^{\infty} f(x, y^* + cx)\, dx} = \frac{\displaystyle\int_{x_1}^{x_2} f(x, y_1 + cx)\, dx}{\displaystyle\int_{-\infty}^{\infty} f(x, y_1 + cx)\, dx}. \tag{3.68}$$

□

The transition between Equation 3.67 and Equation 3.68 starts by assuming that for some

$$y^* \in [y_1, y_2] \Rightarrow \int_{y_1+cx}^{y_2+cx} f(x,y)\, dxdy = f(x, y^* + cx). \tag{3.69}$$

Thus, if we take the limit such that $y_2 \to y_1$ and $y_1 \leq y^* \leq y_2$, then $y^* \to y_1$ and

$$f(x, y^* + cx) \to f(x, y_1 + cx). \tag{3.70}$$

Finally, we consider the conditional probability of X given that Y is restricted to a single point.

Theorem 3.28. *The conditional density of X given $Y = y_1$, denoted by $f(x|\, y_1)$, is given by*

$$f(x|\, y_1) = \frac{f(x, y_1)}{f(y_1)}. \tag{3.71}$$

Note that a formal statement of Theorem 3.28 could follow Theorem 3.27, applying the mean value of the integral to a range of X.

One would anticipate that continuous formulations of independence could follow the discrete formulation such that we attempt to show that $f(x) = f(x|\, y)$. However, independence for continuous random variables simply relates to the separability of the joint distribution function.

Definition 3.29. Continuous random variables X and Y are said to be independent if $f(x,y) = f(x)\, f(y)$ for all x and y.

Again returning to Example 3.11,

$$f(x,y) = xy \exp[-(x+y)] = (x \exp[-x])(y \exp[-y]). \tag{3.72}$$

Hence, X and Y are independent. In addition, the joint uniform distribution function is independent because $f(x,y) = 1 = g(x)\, h(y)$ where $g(x) = h(y) = 1$. This simplistic definition of independence can be easily extended to T random variables.

Definition 3.30. A finite set of continuous random variables $X, Y, Z, \cdots$ are said to be mutually independent if

$$f(x, y, z, \cdots) = g(x)\, h(y)\, i(z) \cdots. \tag{3.73}$$

A slightly more rigorous statement of independence for bivariate continuous random variables is presented in Theorem 3.31.

Theorem 3.31. *Let S be a subset of the plane such that $f(x,y) > 0$ over S and $f(x,y) = 0$ outside of S. Then X and Y are independent if and only if S is a rectangle (allowing $-\infty$ or ∞ to be an end point) with sides parallel to the axes and $f(x,y) = g(x)/h(y)$ over S, where $g(x)$ and $h(y)$ are some functions of x and y, respectively. Note that $g(x) = cf(x)$ for some c, $h(y) = c^{-1} f(y)$.*

3.4 Cumulative Distribution Function

Another transformation of the density function is the cumulative density function, which gives the probability that a random variable is less than some specified value.

Definition 3.32. The cumulative distribution function of a random variable X, denoted $F(x)$, is defined by

$$F(x) = \mathrm{P}(X < x) \tag{3.74}$$

for every real x. In the case of a discrete random variable

$$F(x_i) = \sum_{x_j \leq x_i} \mathrm{P}(x_j). \tag{3.75}$$

In the case of a continuous random variable

$$F(x) = \int_{-\infty}^{x} f(t)\, dt. \tag{3.76}$$

In the case of the uniform distribution

$$F(x) = \int_{-\infty}^{x} dt \Rightarrow \left\{ \begin{array}{c} 0 \text{ if } x \leq 0 \\ (t|_0^x \text{ if } 0 < x < 1 \\ 1 \text{ if } x > 1. \end{array} \right. \tag{3.77}$$

The cumulative distribution function for the uniform distribution is presented in Figure 3.7.

We will develop the normal distribution more fully over the next three sections, but certain aspects of the normal distribution add to our current discussion. The normal distribution is sometimes referred to as the **bell curve** because its density function, presented in Figure 3.8, has a distinctive bell shape. One of the vexing characteristics of the normal curve is the fact that its anti-derivative does not exist. What we know about the integral of the normal distribution we know because we can integrate it over the range $(-\infty, \infty)$. Given that the anti-derivative of the normal does not exist, we typically rely on published tables for finite integrals. The point is that Figure 3.9 presents an empirical cumulative distribution for the normal density function.

3.5 Some Useful Distributions

While there are an infinite number of continuous distributions, a small number of distributions account for most applications in econometrics.

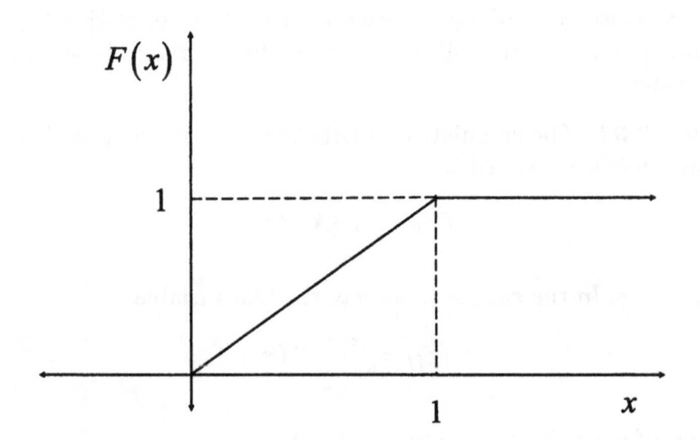

FIGURE 3.7
Cumulative Distribution of the Uniform Distribution.

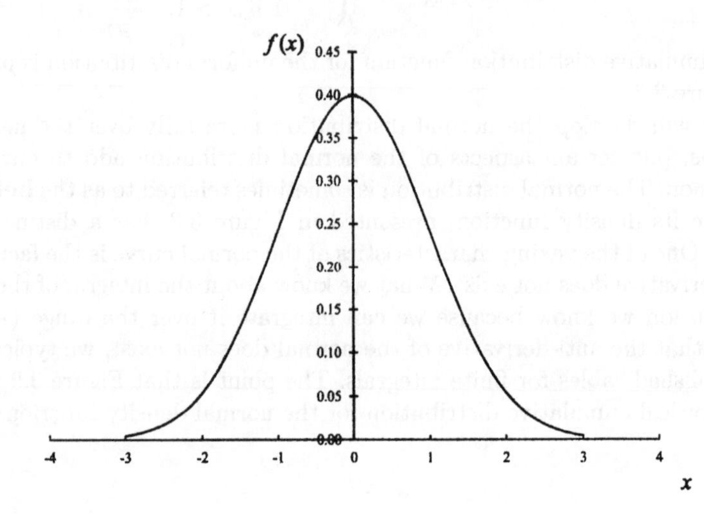

FIGURE 3.8
Normal Distribution Probability Density Function.

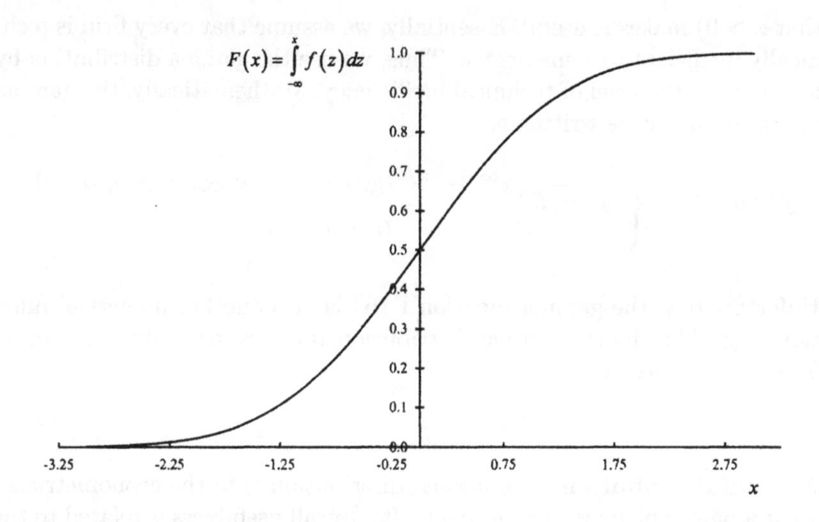

FIGURE 3.9
Normal Cumulative Distribution Function.

A. **Uniform distribution**: In our discussion we have made frequent use of the U [0, 1] distribution. A slightly more general form of this distribution can be written as

$$f(x|a,b) = \begin{cases} \dfrac{1}{b-a} & \text{if } a \leq x \leq b \\ 0 & \text{otherwise.} \end{cases} \tag{3.78}$$

Apart from the fact that it is relatively easy to work with, the uniform distribution is important for a wide variety of applications in econometrics and applied statistics. Interestingly, one such application is sampling theory. Given that the cumulative density function for any distribution "maps" into the unit interval (i.e., $F(x) \to [0,1]$), one way to develop sample information involves drawing a uniform random variable ($z \sim \mathrm{U}[0,1]$ read as z is distributed U [0, 1]) and determining the value of x associated with that probability (i.e., $x = F^{-1}(z)$ where $F^{-1}(z)$ is called the inverse mapping function).

B. **Gamma distribution**: The gamma distribution has both pure statistical uses and real applications to questions such as technical inefficiency or crop insurance problems. Taking the statistical applications first, the χ^2 distribution is a form of a gamma distribution. The χ^2 distribution is important because if x is a standard normal distribution (i.e., a normal distribution with a mean of zero and a standard deviation of one) $x^2 \sim \chi^2$. Thus, variances tend to have a χ^2 distribution. From a technical inefficiency or crop yield perspective, the one-sided nature (i.e., the fact that $x > 0$) makes it useful. Essentially, we assume that every firm is technically inefficient to some degree. Thus, we use the gamma distribution by letting x be the level of technical inefficiency. Mathematically, the gamma distribution can be written as

$$f(x|\alpha,\beta) = \begin{cases} \dfrac{1}{\Gamma(\alpha)\beta^{\alpha}} x^{\alpha-1} e^{-x/\beta} & \text{given } 0 < x < \infty,\ \alpha > 0,\ \beta > 0 \\ 0 \text{ otherwise.} \end{cases} \tag{3.79}$$

Unfortunately, the gamma function $\Gamma(\alpha)$ is also another numerical function (e.g., like the cumulative distribution function presented in Section 3.4). It is defined as

$$\Gamma(\alpha) = \int_0^{\infty} t^{\alpha-1} e^t dt. \tag{3.80}$$

C. **Normal distribution**: The normal distribution is to the econometrician what a pair of pliers is to a mechanic. Its overall usefulness is related to the **central limit theorem**, which essentially states that averages tend to be normally distributed. Given that most estimators including ordinary least squares are essentially weighted averages, most of our parameter estimates tend to be normally distributed. In many cases throughout this textbook we will rely on this distribution. Mathematically, this distribution can be written as

$$f(x|\mu,\sigma^2) = \frac{1}{\sqrt{2\pi}\sigma} \exp\left[-\frac{(x-\mu)^2}{2\sigma^2}\right] \text{ for all } -\infty < x < \infty \tag{3.81}$$

where μ is the mean and σ^2 is the variance.

D. **Beta distribution**: The bounds on the beta distribution make it useful for estimation of the **Bernoulli distribution**. The Bernoulli distribution is the standard formulation for two-outcome random variables like coin tosses.

$$P[x_i] = p^{x_i}\left(1 - p^{1-x_i}\right),\ x_i = \{0,1\},\ 0 \le p \le 1. \tag{3.82}$$

Given the bounds of the probability of a "heads," several Bayesian estimators of p often use the beta distribution, which is mathematically written

$$f(p|\alpha,\beta) = \begin{cases} \dfrac{1}{B(\alpha,\beta)} p^{\alpha-1}(1-p)^{\beta-1}, & \text{for } 0 < p < 1,\ \alpha > 0,\ \beta > 0 \\ 0 \ \text{ otherwise.} \end{cases} \tag{3.83}$$

The beta function $(B(\alpha,\beta))$ is defined using the gamma function:

$$B(\alpha,\beta) = \frac{\Gamma(\alpha)\Gamma(\beta)}{\Gamma(\alpha+\beta)}. \tag{3.84}$$

3.6 Change of Variables

Change of variables is a technique used to derive the distribution function for one random variable by transforming the distribution function of another random variable.

Theorem 3.33. *Let $f(x)$ be the density of X and let $Y = \phi(X)$, where ϕ is a monotonic differentiable function. Then the density $g(y)$ of Y is given by*

$$g(y) = f\left[\phi^{-1}(y)\right] \times \left|\frac{d\phi^{-1}(y)}{dy}\right|. \tag{3.85}$$

The term **monotonic** can be simplified to a "one-to-one" function. In other words, each x is associated with one y over the range of a distribution function. Hence, given an x, a single y is implied (e.g., the typical definition of a function) and for any one y, a single x is implied.

Example 3.34. Suppose $f(x) = 1$ for $0 < x < 1$ and 0 otherwise. Assuming $Y = X^2$, what is the distribution function $g(y)$ for Y? First, it is possible to show that $Y = X^2$ is a monotonic or one-to-one mapping over the relevant range. Given this one-to-one mapping, it is possible to derive the inverse function:

$$\phi(x) = x^2 \Rightarrow \phi^{-1}(y) = \sqrt{y}. \tag{3.86}$$

Following the definition:

$$g(y) = 1\left|\frac{1}{2}y^{-\frac{1}{2}}\right| = \frac{1}{2\sqrt{y}}. \tag{3.87}$$

Extending the formulation in Equation 3.33, we can envision a more complex mapping that is the sum of individual one-to-one mappings.

Theorem 3.35. *Suppose the inverse of $\phi(x)$ is multivalued and can be written as*

$$x_i = \psi_i(y) \quad i = 1, 2, \ldots n_y. \tag{3.88}$$

Note that n_y indicates the possibility that the number of values of x varies with y. Then the density $g(y)$ of Y is given by

$$g(y) = \sum_{i=1}^{n_y} \frac{f\left[\psi_i(y)\right]}{\left|\phi'\left[\psi_i(y)\right]\right|} \tag{3.89}$$

where $f(.)$ is the density of X and ϕ' is the derivative of ϕ.

One implication of Theorem 3.35 is for systems of simultaneous equations. Consider a very simplified demand system:

$$\begin{bmatrix} 1 & 1/2 \\ 1 & -1/3 \end{bmatrix} \begin{bmatrix} y_1 \\ y_2 \end{bmatrix} = \begin{bmatrix} 10 & 4 \\ -3 & 0 \end{bmatrix} \begin{bmatrix} x_1 \\ x_2 \end{bmatrix} \tag{3.90}$$

where y_1 is the quantity supplied and demanded, y_2 is the price of the good, x_1 is the price of an alternative good, and x_2 is consumer income. We develop the matrix formulations in Chapter 10, but Equation 3.90 can be rewritten

$$\begin{aligned} y_1 &= \frac{11}{5}x_1 + \frac{8}{5}x_2 \\ y_2 &= \frac{78}{5}x_1 + \frac{24}{5}x_2. \end{aligned} \tag{3.91}$$

The question is then – if we know something about the distribution of x_1 and x_2, can we derive the distribution of y_1 and y_2? The answer of course is yes.

Theorem 3.36. *Let $f(x_1, x_2)$ be the joint density of a bivariate random variable (X_1, X_2) and let (Y_1, Y_2) be defined by a linear transformation*

$$\begin{aligned} Y_1 &= a_{11}X_1 + a_{12}X_2 \\ Y_2 &= a_{21}X_1 + a_{22}X_2. \end{aligned} \tag{3.92}$$

Suppose $a_{11}a_{22} - a_{12}a_{21} \neq 0$ so that the equations can be solved for X_1 and X_2 as

$$\begin{aligned} X_1 &= b_{11}Y_1 + b_{12}Y_2 \\ X_2 &= b_{21}Y_1 + b_{22}Y_2. \end{aligned} \tag{3.93}$$

Then the joint density $g(y_1, y_2)$ of (Y_1, Y_2) is given by

$$g(y_1, y_2) = \frac{f(b_{11}y_1 + b_{12}y_2, b_{21}y_1 + b_{22}y_2)}{|a_{11}a_{22} - a_{12}a_{21}|} \tag{3.94}$$

where the support of g, that is, the range of (y_1, y_2) over which g is positive, must be appropriately determined.

Theorem 3.36 is used to derive the **Full Information Maximum Likelihood** for systems with endogenous variables (i.e., variables with values that are determined inside the system like the quantity supplied and demanded, and the price that clears the market). Appendix B presents the maximum likelihood formulation for a system of equations.

3.7 Derivation of the Normal Distribution Function

As stated in Section 3.5, the normal distribution forms the basis for many problems in applied econometrics. However, the formula for the normal distribution is abstract to say the least. As a starting point for understanding the normal distribution, consider the change in variables application presented in Example 3.37.

Example 3.37. Assume that we want to compute the probability of an event that occurs within the unit circle given the standard bivariate uniform distribution function ($f(x,y) = 1$ with $Y^2 \leq 1 - X^2$ given $0 \leq x, y \leq 1$). The problem can be rewritten slightly – $Y \leq \sqrt{1-X^2} \rightarrow Y^2 + X^2 \leq 1$. Hence, the problem can be written as

$$P\left(X^2 + Y^2 < 1\right) = \int_0^1 \left(\int_0^{\sqrt{1-x^2}} dy\right) dx = \int_0^1 \sqrt{1-x^2} dx. \tag{3.95}$$

As previously stated, we will solve this problem using integration by change in variables. By trigonometric identity $1 = \sin^2(x) + \cos^2(x)$. Therefore, $\sin^2(x) = 1 - \cos^2(x)$. The change in variables is then to let $x = \cos(t)$. The integration by change in variables is then

$$\int_{x_1}^{x_2} f(x)\, dx = \int_{t_1}^{t_2} f[\phi(t)]\, \phi'(t)\, dt \tag{3.96}$$

such that $t_1 = \phi^{-1}(x_1)$ and $t_2 = \phi^{-1}(x_2)$. Thus, in explicit form our transformation becomes

$$\int_{x_1}^{x_2} \sqrt{1-x^2} dx = \int_{t_1}^{t_2} \sqrt{1-\cos^2(t)} \frac{\partial \cos(t)}{\partial t} dt. \tag{3.97}$$

Given that

$$\begin{gathered} 1 = \sin^2(t) + \cos^2(t) \text{ implies } \sin(t) = \sqrt{1-\cos^2(t)} \text{ and} \\ \frac{\partial \cos(t)}{\partial t} = -\sin(t) \end{gathered} \tag{3.98}$$

the transformed integral becomes

$$\int_{t_1}^{t_2} \sin(t) \times -\sin(t)\, dt = \int_{t_1}^{t_2} -\sin^2(t)\, dt. \tag{3.99}$$

To complete the transformation we derive the bounds of integration. However, notice that $t_1 = \cos^{-1}(0) = \pi/2$ (or 90 degrees) while $t_2 = \cos^{-1}(1) = 0$, implying

$$\int_{\pi/2}^{0} -\sin^2(t)\, dt \tag{3.100}$$

or that the order of the bounds of the integral are opposite from the standard case. The solution is to reverse the order of integration:

$$\int_{\pi/2}^{0} -\sin^2(t)\, dt = \int_0^{\pi/2} \sin^2(t)\, dt. \tag{3.101}$$

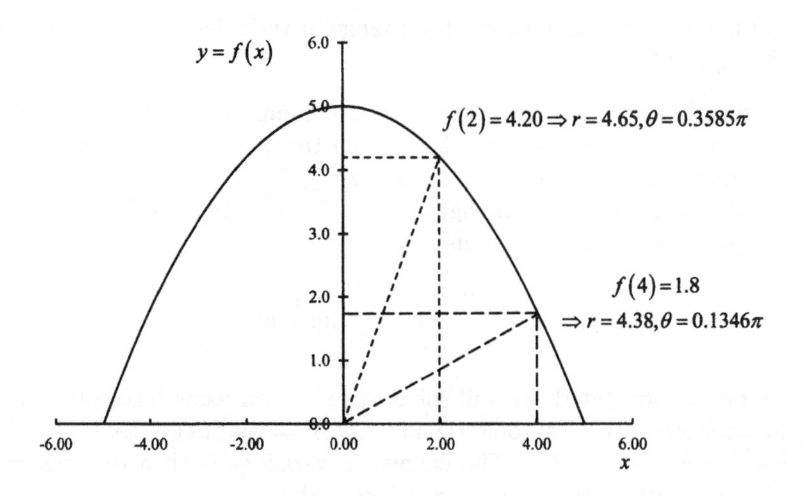

FIGURE 3.10
Simple Quadratic Function.

The value of the integral is then

$$\int_0^{\pi/2} \sin^2(t)\,dt = \left[-\frac{1}{2}\sin(t)\cos(t) + \frac{1}{2}t\right|_0^{\pi/2} = \frac{\pi}{4}. \tag{3.102}$$

While Example 3.37 appears simple enough, it opens the door to some very powerful tools of functional analysis. Specifically, while most transformations appear minor (i.e., taking the square root of a variable in Example 3.34 or a linear transformation of two variables in Theorem 3.36) more radical transformations of the variable space are possible. One such transformation is the polar functional form.

Refer to the quadratic function

$$y = f(x) = 5 - \frac{x^2}{5} \quad x \in [-5, 5] \tag{3.103}$$

depicted in Figure 3.10. Consider the point $f(4.0) = 1.8$; the length of the ray from the origin to that point on the function can be computed as

$$r(4, 1.8) = \sqrt{4^2 + 1.8^2} = 4.38. \tag{3.104}$$

We can also compute the value of the inscribed angle. To do this we start by noting that

$$\tan(\theta(4, 1.8)) = \frac{1.8}{4} = 0.45 \Rightarrow \theta(4, 1.8) = \tan^{-1}(0.45) = 0.1346\pi. \tag{3.105}$$

Repeating the process for $f(2) = 4.2$ yields $r(2, 4.2) = 4.20$ and $\theta(2, 4.2) =$

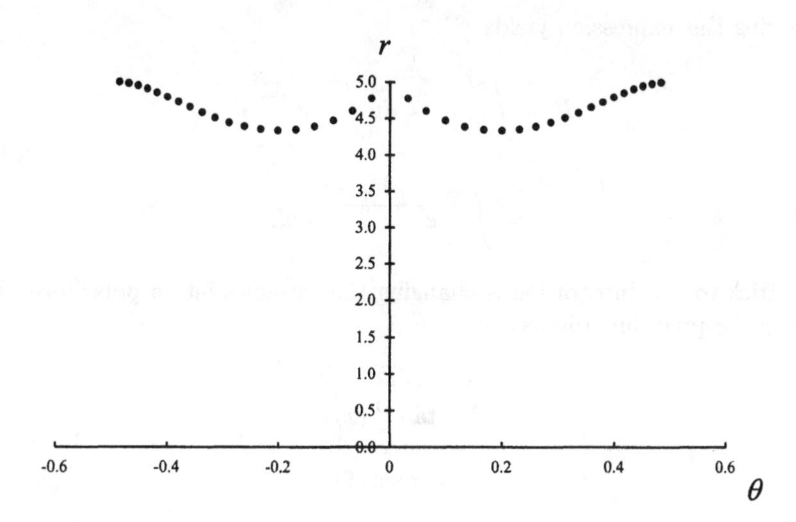

FIGURE 3.11
Polar Transformation of Simple Quadratic.

0.3585π. Applying this transformation at a sequence of points between $x = -5$ and $x = 5$ yields the transformation presented in Figure 3.11. Given that we can define a function $g(\theta, r(\theta))$ that defines these points, we have an alternative respresentation of the simple quadratic function in Equation 3.103. For example, we could approximate the function as

$$r(\theta) = 4.6584. \tag{3.106}$$

The approximation error could be reduced by adding additional terms (i.e., $r(\theta) = a + b\theta + c\theta^2$ – see the discussion in Appendix C).

Intuitively, we could integrate the simple quadratic following Example 3.37, but it is obvious that such an integration would be more trouble than it is worth. However, the polar transformation simplifies some complex integrals such as the normal density function.

To develop the normal distribution, we start with the standard normal (i.e., $x \sim \mathrm{N}(0,1)$), which can be written as

$$f(x) = \frac{1}{\sqrt{2\pi}} e^{-\frac{x^2}{2}}. \tag{3.107}$$

First, we need to demonstrate that the distribution function does integrate to one over the entire sample space, which is $-\infty$ to ∞. This is typically accomplished by proving the constant. Let us start by assuming that

$$I = \int_{-\infty}^{\infty} e^{-\frac{y^2}{2}} dy. \tag{3.108}$$

Squaring this expression yields

$$\begin{aligned} I^2 &= \int_{-\infty}^{\infty} e^{-\frac{y^2}{2}} dy \int_{-\infty}^{\infty} e^{-\frac{x^2}{2}} dx \\ &= \int_{-\infty}^{\infty} e^{-\frac{y^2+x^2}{2}} dydx. \end{aligned} \tag{3.109}$$

The trick to this integration is changing the variables into a polar form. Following the preceding discussion,

$$\begin{aligned} r &= \sqrt{x^2+y^2} \\ \theta &= \tan^{-1}(x/y) \\ y &= r\cos(\theta) \\ x &= r\sin(\theta). \end{aligned} \tag{3.110}$$

We apply the change in variable technique to change the integral into the polar space. First, we transform the variables of integration

$$dydx = rdrd\theta. \tag{3.111}$$

Folding these two results together we get

$$I^2 = \int_0^{2\pi} \int_0^{\infty} re^{-\frac{r^2}{2}} drd\theta = \int_0^{2\pi} d\theta = 2\pi. \tag{3.112}$$

A couple of points about the result in Equation 3.112; first note that

$$\frac{\partial e^{-\frac{r^2}{2}}}{\partial r} = -\frac{2r}{2} e^{-\frac{r^2}{2}} \Rightarrow \int re^{-\frac{r^2}{2}} dr = e^{-\frac{r^2}{2}}. \tag{3.113}$$

Second, the distance function is non-negative by definition (i.e., $\sqrt{x^2+y^2} \geq 0$). Hence, the range of the inner integral in Equation 3.112 is $r \in [0, \infty)$. Taking the square root of each side yields

$$I = \sqrt{2\pi}. \tag{3.114}$$

Thus, we know that

$$\int_{-\infty}^{\infty} \frac{1}{\sqrt{2\pi}} e^{-\frac{y^2}{2}} dy = 1. \tag{3.115}$$

The expression in Equation 3.115 is referred to as the standard normal. A more general form of the normal distribution function can be derived by defining a transformation function. Defining

$$\begin{aligned} y &= a + bx \\ x &= \frac{y-a}{b}. \end{aligned} \tag{3.116}$$

by the change in variable technique, we have

$$f(x) = \frac{1}{\sqrt{2\pi}} e^{-\frac{(y-a)^2}{2b^2}} \left| \frac{1}{b} \right|. \tag{3.117}$$

As presented in Section 3.5 we typically denote a as μ (i.e., the mean) and b as σ^2 (i.e., the variance).

3.8 An Applied Sabbatical

In the past, farmers received assistance during disasters (i.e., drought or floods) through access to concessionary credit. Increasingly, during the last 10 years of the 20th century, agricultural policy in the United States shifted toward market-based crop insurance. This insurance was supposed to be actuarially sound so that producers would make decisions that were consistent with maximizing economic surplus. Following the discussion of [36], the loss of a crop insurance event could be parameterized as

$$L = AC - AR \tag{3.118}$$

where C is the level of coverage (i.e., the number of bushels guaranteed under the insurance policy, typically 10, 20, or 40 percent of some expected level of yield), A is the probability of that level of yield, R is the expected value of the yield given that an insured event has occurred, L is the insurance indemnity or actuarially fair value of the insurance. Given these definitions, the insurance indemnity becomes

$$L = \int_{\infty}^{C} (C - y)\, dF(y). \tag{3.119}$$

This loss is in yield space; it ignores the price of the output. Apart from the question of prices, a critical part of the puzzle is the distribution function

$$dF(y) = f(y)\, dy. \tag{3.120}$$

Differences in the functional form of the distribution function imply different insurance premiums for producers. The goal of the selection of a distribution function is for the distribution function to match the actual distribution function of crop yields. Differences between the actual distribution function and the empirical form used to estimate the premium leads to an economic loss. If a distribution systematically understates the probability of lower return, farmers could make an arbitrage gain by buying crop insurance. If a distribution systematically overstates the probability of a lower return, farmers would not buy the insurance (it is not a viable instrument).

The divergence between the relative probabilities is a function of the flexibility of the distribution's moments (developed more fully in Chapter 4).

- **Expected value**: First moment

$$\mu_1 = \int_{-\infty}^{\infty} x f(x)\, dx. \tag{3.121}$$

- **Variance**: Second central moment

$$\mu_2^C = \int_{-\infty}^{\infty} (x - \mu_1)^2 f(x)\, dx. \tag{3.122}$$

- **Skewness**: Third central moment

$$\mu_c^C = \int_{-\infty}^{\infty} (x - \mu_1)^3 f(x)\, dx. \tag{3.123}$$

- **Kurtosis**: Fourth central moment

$$\mu_4^C = \int_{-\infty}^{\infty} (x - \mu_1)^4 f(x)\, dx. \tag{3.124}$$

Each distribution implies a certain level of flexibility between the moments. For the normal distribution, all odd central moments are equal to zero, which implies that the distribution function is symmetric. In addition, all even moments are a function of the second central moment (i.e., the variance). Moss and Shonkwiler [33] propose a distribution function that has greater flexibility based on the normal (specifically in the third and fourth moments). This new distribution (presented in Figure 3.12) is accomplished by parametric transformation to normality. The distribution is called an inverse hyperbolic sine transformation:

$$e_t = \frac{\ln\left(\theta e_t + \left[(\theta e_t)^2 + 1\right]^{\frac{1}{2}}\right)}{\theta} \tag{3.125}$$
$$\epsilon_t = z_t - \delta.$$

Norwood, Roberts, and Lusk [37] evaluate the goodness of yield distributions using a variant of Kullback's [28] information criteria:

$$I = \int_{-\infty}^{\infty} f(X|\theta_f) \ln\left(\frac{f(X|\theta_f)}{g(X|\theta_g)}\right) \geq 0. \tag{3.126}$$

Like most informational indices, this index reaches a minimum of zero if the two distribution functions are identical everywhere. Otherwise, a positive number reflects the magnitude of the divergence. The Norwood, Roberts, and Lusk [37] model then suggests that a variety of models can be tested against each other by comparing their out-of-sample measure. This measure is actually constructed byletting the probability of an out-of-sample forecast equal $1/N$

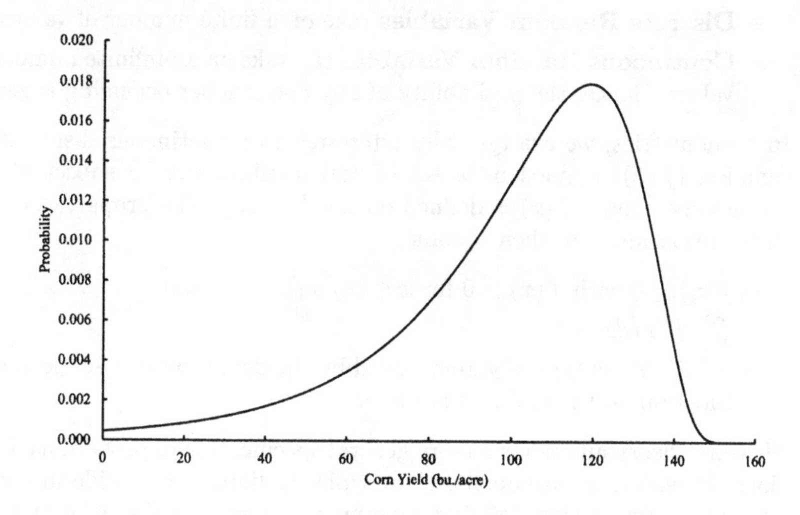

FIGURE 3.12
Inverse Hyperbolic Sine Transformation of the Normal Distribution.

where N is the number of out-of-sample draws.

$$\begin{aligned}\hat{I} &= \sum_{i=1}^{N} \frac{1}{N} \ln \left(\frac{1/N}{g\left(X|\theta_g\right)} \right) \\ &= \sum_{i=1}^{N} \left[\frac{1}{N} \ln \left(\frac{1}{N} \right) - \frac{1}{N} \ln \left(g\left(X|\theta_g\right) \right) \right] \\ &= \frac{N}{N} \ln \left(\frac{1}{N} \right) - \frac{1}{N} \sum_{i=1}^{N} \ln \left(g\left(X|\theta_g\right) \right) \\ &\Rightarrow \hat{I} \propto \tilde{I} = -\frac{1}{N} \sum_{i=1}^{N} \ln \left(g\left(X|\theta_g\right) \right).\end{aligned} \tag{3.127}$$

Ignoring the constants, the measure of goodness becomes negative. The more negative the number, the less good is the distributional fit. Norwood, Roberts, and Lusk then construct a number of out-of-sample measures of I.

3.9 Chapter Summary

- **Random Variables** are real numbers whose outcomes are elements of a sample space following a probability function.

- **Discrete Random Variables** take on a finite number of values.
- **Continuous Random Variables** can take on an infinite number of values. Hence, the probability of any one number occurring is zero.

- In econometrics, we are typically interested in a continuous distribution function ($f(x)$) defined on a set of real numbers (i.e., a subset of the real number line – $f(x)$ is defined on $x \in [x_0, x_1]$). The properties of the distribution function then become:
  - $f(x) \geq 0$ with $f(x) > 0$ for $x \in [x_0, x_1]$.
  - $\int_{x_0}^{x_1} f(x)\, dx = 1$.
  - Additivity is typically guaranteed by the definition of a single valued function on the real number line.

- Measure theory allows for a more general specification of probability functions. However, in econometrics we typically limit our considerations to simplier specifications defining random variables as subsets of the real number space.

- If two (or more) random variables are independent, their joint probability density function can be factored $f(x, y) = f_1(x) f_2(y)$.

- The conditional relationship between two continuous random variables can be written as

$$f(x|\, y) = \frac{f(x, y_0)}{\left. \int_{-\infty}^{\infty} f(x, y)\, dx \right|_{y=y_0}}. \tag{3.128}$$

3.10 Review Questions

3-1R. Demonstrate that the outcomes of dice rolls meet the criteria for a Borel set.

3-2R. Construct the probability of damage given the outcome of two die, one with eight sides and one with six sides. Assume that damage occurs when the six-sided die is 5 or greater, while the amount of damage is given by the outcome of the eight-sided die. How many times does damage occur given the outcomes from Table 3.2? What is the level of that damage? Explain the outcome of damage using a set-theoretic mapping.

3-3R. Explain why the condition that $f(x, y) = f_1(x) f_2(y)$ is the same relationship for independence as the condition that the marginal distribution for x is equal to its conditional probability given y.

TABLE 3.9
Discrete Distribution Functions for Bivariate Random Variables

Outcome for Y	Outcome for X 1	2	3
	Density 1		
1	0.075	0.150	0.075
2	0.100	0.200	0.100
3	0.075	0.150	0.075
	Density 2		
1	0.109	0.130	0.065
2	0.087	0.217	0.087
3	0.065	0.130	0.109

3.11 Numerical Exercises

3-1E. What is the probability of rolling a number less than 5 given that two six-sided dice are rolled?

3-2E. Is the function

$$f(x) = \frac{3}{100}\left(5 - \frac{x^2}{5}\right) \tag{3.129}$$

a valid probability density function for $x \in [-5, 5]$?

3-3E. Derive the cumulative density function for the probability density function in Equation 3.129.

3-4E. Is the function

$$f(x) = \begin{cases} \frac{x}{2}, & x \in [0,1] \\ \frac{3}{8}, & x \in [2,4] \\ 0 & \text{otherwise} \end{cases} \tag{3.130}$$

a valid probabilty density function?

3-5E. Given the probability density function in Equation 3.129, what is the probability that $x \leq -0.75$?

3-6E. Given the probability density function in Equation 3.130, what is the probability that $0.5 \leq x \leq 3$?

3-7E. Are the density functions presented in Table 3.9 independent?

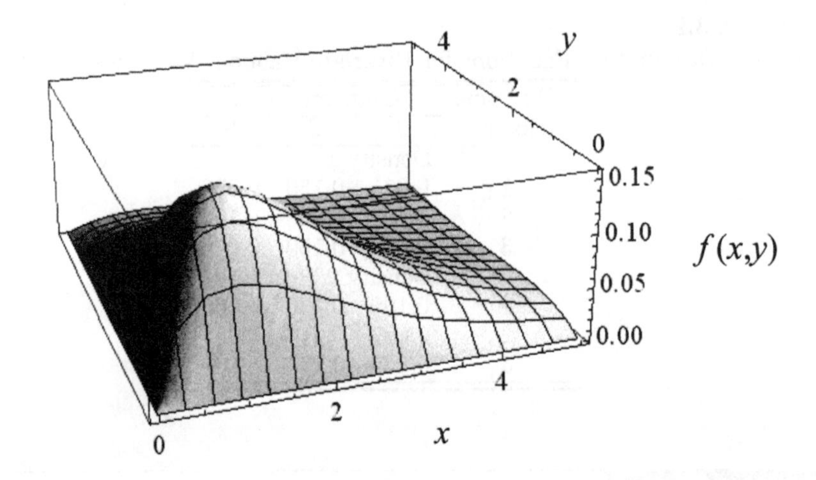

FIGURE 3.13
Continuous Joint Distribution.

3-8E. Consider the joint probability density function

$$f(x,y) = 0.9144xy\exp\left(-x - y + 0.05\sqrt{xy}\right) \tag{3.131}$$

as depicted in Figure 3.13. Are x and y independent?

3-9E. Consider the joint distribution function

$$f(x,y) = \frac{1}{4} \text{ for } x \in [0,2]\ y \in [0,2]. \tag{3.132}$$

Suppose that we transform the distribution by letting $z = \sqrt{x}$. Derive the new distribution function and plot the new distribution function over the new range.

3-10E. For Equation 3.132 compute the conditional probability density function for x given that $y \geq x$.

4

Moments and Moment-Generating Functions

CONTENTS

Given our discussion of random variables and their distributions in Chapter 3, we can now start to define statistics or functions that summarize the information for a particular random variable. As a starting point, Chapter 4 develops the definition of the moments of the distribution. Moment is a general term for the expected kth power of the distribution

$$E\left[x^k\right] = \int_{\infty}^{\infty} x^k f(x)\,dx. \tag{4.1}$$

The first moment of a distribution (i.e., $k = 1$) is typically referred to as the mean of the distribution. Further, the variance of a distribution can be derived using the mean and the second moment of the distribution. As a starting point, we need to develop the concept of an expectation rigorously.

4.1 Expected Values

In order to avoid circularity, let us start by defining the expectation of a function ($g(X)$) of a random variable X in Definition 4.1.

Definition 4.1. The expected value or mean of a random variable $g(X)$ denoted $E[g(x)]$ is

$$E[g(X)] = \begin{cases} \int_{\infty}^{\infty} g(x) f(x) \, dx \text{ if } x \text{ is continuous} \\ \sum_{x \in X} g(x) f_x \text{ if } x \text{ is discrete.} \end{cases} \tag{4.2}$$

Hence, the expectation is a weighted sum – the result of the function is weighted by the distribution or density function. To demonstrate this concept, consider the mean of the distribution (i.e., $g(X) = X$):

$$E[x] = \int_{-\infty}^{\infty} x f(x) \, dx. \tag{4.3}$$

For example, derive the mean of the exponential distribution

$$\begin{aligned} E[x] &= \int_{0}^{\infty} \frac{1}{\lambda} x e^{-\lambda x} dx \\ &= -\left(x e^{-\frac{x}{\lambda}} \right|_0^{\infty} + \int_0^{\infty} e^{-\lambda x} dx \\ &= -\left(\lambda e^{-\frac{x}{\lambda}} \right|_0^{\infty} = \lambda. \end{aligned} \tag{4.4}$$

The exponential distribution function can be used to model the probability of first arrival. Thus, $f(x = 1)$ would be the probability that the first arrival will occur in one hour.

Given the definition of the expectation from Definition 4.1, the properties of the expectation presented in Theorem 4.2 follow.

Theorem 4.2. *Let X be a random variable and let a, b, and c be constants. Then for any functions $g_1(X)$ and $g_2(X)$ whose expectations exist:*

1. $E[a g_1(X) + b g_2(X) + c] = a E[g_1(X)] + b E[g_2(X)] + c$.
2. *If $g_1(X) \geq 0$ for all X, then $E[g_1(X)] \geq 0$.*
3. *If $g_1(X) \geq g_2(X)$ for all X, then $E[g_1(X)] \geq E[g_2(X)]$.*
4. *If $a \leq g_1(X) \leq b$ for all X, then $a \leq E[g_1(X)] \leq b$.*

Result 1 follows from the linearity of the expectation. Result 2 can be further strengthened in Jensen's inequality (i.e., $E[g(x)] \geq g[E[x]]$ if $\partial g(x)/\partial x \geq 0$ for all x). Result 3 actually follows from result 1 (i.e., if $a = 1$ and $b = -1$ then $g_1(x) - g_2(x) = 0 \Rightarrow E[g_1(x)] - E[g_2(x)] = 0$).

A critical concept in computing expectations is whether either the sum (in the case of discrete random variables) or the integral (in the case of continuous random variables) is bounded. To introduce the concept of boundedness, consider the expectation of a discrete random variable presented in Defintion 4.3. In this discussion we denote the sum of the positive values of x_i as $\sum_+$ and the sum over negative values as $\sum_-$.

TABLE 4.1
Expected Value of a Single-Die Roll

Number	Probability	$x_i\mathrm{P}(x_i)$
1	0.167	0.167
2	0.167	0.333
3	0.167	0.500
4	0.167	0.667
5	0.167	0.833
6	0.167	1.000
Total		3.500

Definition 4.3. Let X be a discrete random variable taking the value x with probability $\mathrm{P}(x_i)$, $i = 1, 2, \cdots$. Then the expected value (expectation or mean) of X, denoted $\mathrm{E}[X]$, is defined to be $\mathrm{E}[X] = \sum_{i=1}^{\infty} x_i\mathrm{P}(x_i)$ if the series converges absolutely.

1. We can write $\mathrm{E}[X] = \sum_+ x_i\mathrm{P}(x_i) + \sum_- x_i\mathrm{P}(x_i)$ where in the first summation we sum for i such that $x_i > 0$ ($\sum_+ x_i > 0$) and in the second summation we sum for i such that $x_i < 0$ ($\sum_- x_i < 0$).
2. If $\sum_+ x_i\mathrm{P}(x_i) = \infty$ and $\sum_- x_i\mathrm{P}(x_i) = -\infty$ then $\mathrm{E}[X]$ does not exist.
3. If $\sum_+ x_i\mathrm{P}(x_i) = \infty$ and $\sum_- x_i\mathrm{P}(x_i)$ is finite then we say $\mathrm{E}[X] = \infty$.
4. If $\sum_- x_i\mathrm{P}(x_i) = -\infty$ and $\sum_+ x_i\mathrm{P}(x_i)$ is finite then we say that $\mathrm{E}[X] = -\infty$.

Thus the second result states that $-\infty + \infty$ does not exist. Results 3 and 4 imply that if either the negative or positive sum is finite, the expected value is determined by $-\infty$ or ∞, respectively.

Consider a couple of aleatory or gaming examples.

Example 4.4. Given that each face of the die is equally likely, what is the expected value of the roll of the die? As presented in Table 4.1, the expected value of a single die roll with values $x_i = \{1, 2, 3, 4, 5, 6\}$ and each value being equally likely is 3.50.

An interesting aspect of Example 4.4 is that the expected value is not part of the possible outcomes – it is impossible to roll a 3.5.

Example 4.5. What is the expected value of a two-die roll? This time the values of x_i are the integers between 2 and 12 and the probabilities are no longer equal (as depicted in Table 4.2). This time the expected value is 7.00.

The result in Example 4.5 explains many of the dice games from casinos.

Turning from the simple game expectations, consider an application from risk theory.

TABLE 4.2
Expected Value of a Two-Die Roll

Die 1	Die 2	Number	$x_i\mathrm{P}(x_i)$	Die 1	Die 2	Number	$x_i\mathrm{P}(x_i)$
1	1	2	0.056	1	4	5	0.139
2	1	3	0.083	2	4	6	0.167
3	1	4	0.111	3	4	7	0.194
4	1	5	0.139	4	4	8	0.222
5	1	6	0.167	5	4	9	0.250
6	1	7	0.194	6	4	10	0.278
1	2	3	0.083	1	5	6	0.167
2	2	4	0.111	2	5	7	0.194
3	2	5	0.139	3	5	8	0.222
4	2	6	0.167	4	5	9	0.250
5	2	7	0.194	5	5	10	0.278
6	2	8	0.222	6	5	11	0.306
1	3	4	0.111	1	6	7	0.194
2	3	5	0.139	2	6	8	0.222
3	3	6	0.167	3	6	9	0.250
4	3	7	0.194	4	6	10	0.278
5	3	8	0.222	5	6	11	0.306
6	3	9	0.250	6	6	12	0.333
Total							7.000

Example 4.6. Expectation has several applications in risk theory. In general, the expected value is the value we expect to occur. For example, if we assume that the crop yield follows a binomial distribution, as depicted in Figure 4.1, the expected return on the crop given that the price is \$3 and the cost per acre is \$40 becomes \$95 per acre, as demonstrated in Table 4.3.

Notice that the expectation in Example 4.6 involves taking the expectation of a more general function (i.e., not simply taking the expected value of $k = 1$).

$$\mathrm{E}\left[p_X X - C\right] = \sum_i \left[p_x x_i - C\right] \mathrm{P}\left(x_i\right) \tag{4.5}$$

or the expected value of profit.

In the parlance of risk theory, the expected value of profit for the wheat crop is termed the actuarial value or fair value of the game. It is the value that a risk neutral individual would be willing to pay for the bet [32].

Another point about this value is that it is sometimes called the population mean as opposed to the sample mean. Specifically, the sample mean is an observed quantity based on a sample drawn from the random generating function. The sample mean is defined as

$$\bar{x} = \frac{1}{N}\sum_{i=1}^{N} x_i. \tag{4.6}$$

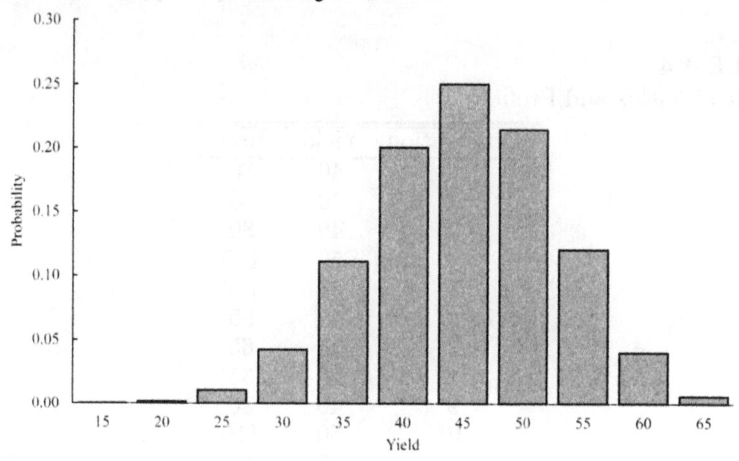

FIGURE 4.1
Wheat Yield Density Function.

Table 4.4 presents a sample of 20 observations drawn from the theoretical distribution above. Note that the sample mean for yield is smaller than the population mean (33.75 for the sample mean versus 45.00 for the population mean). It follows that the sample mean for profit is smaller than the population mean for profit.

Another insight from the expected value and gambling is the Saint Petersburg paradox. The Saint Petersburg paradox involves the valuation of gambles with an infinite value. The simplest form of the paradox involves the value of a series of coin flips. Specifically, what is the expected value of a bet that pays off \$2 if the first toss is a head and 2 times that amount for each subsequent

TABLE 4.3
Expected Return on an Acre of Wheat

X	$\text{P}[X]$	$x_i \text{P}[x_i]$	$(p_x x_i - C)\,\text{P}(x_i)$
15	0.0001	0.0016	0.0005
20	0.0016	0.0315	0.0315
25	0.0106	0.2654	0.3716
30	0.0425	1.2740	2.1234
35	0.1115	3.9017	7.2460
40	0.2007	8.0263	16.0526
45	0.2508	11.2870	23.8282
50	0.2150	10.7495	23.6490
55	0.1209	6.6513	15.1165
60	0.0403	2.4186	5.6435
65	0.0060	0.3930	0.9372
	Total	45.0000	95.0000

TABLE 4.4
Sample of Yields and Profits

Observation	Yield	Profit
1	40	80
2	40	80
3	40	80
4	50	110
5	50	110
6	45	95
7	35	65
8	25	35
9	40	80
10	50	110
11	30	50
12	35	65
13	40	80
14	25	35
15	45	95
16	35	65
17	35	65
18	40	80
19	30	50
20	45	95
Mean	38.75	76.25

head? If the series of coin flips is HHHT, the payoff is \$8 (i.e., $2 \times 2 \times 2$ or 2^3). In theory, the expected value of this bet is infinity, but no one is willing to pay an infinite price.

$$\mathrm{E}[G] = \sum_{i=1}^{\infty} 2^i 2^{-i} = \sum_{i=1}^{\infty} 1 = \infty. \tag{4.7}$$

This unwillingness to pay an infinite price for the gamble led to expected utility theory.

Turning from discrete to continuous random variables, Definition 4.7 presents similar results for continuous random variables as developed for discrete random variables in Definition 4.3.

Definition 4.7. Let X be a continuous random variable with density $f(x)$. Then, the expected value of X, denoted $\mathrm{E}[X]$, is defined to be $\mathrm{E}[X] = \int_{-\infty}^{\infty} x f(x)\, dx$ if the integral is absolutely convergent.

1. If $\int_0^{\infty} x f(x)\, dx = \infty$ and $\int_{-\infty}^{0} x f(x)\, dx = -\infty$, we say that the expectation does not exist.

2. If $\int_0^{\infty} x f(x)\, dx = \infty$ and $\int_{-\infty}^{0} x f(x)\, dx$ is finite, then $\mathrm{E}[X] = \infty$.

3. If $\int_{-\infty}^{0} x f(x)\, dx = -\infty$ and $\int_0^{\infty} x f(x)\, dx$ is finite, then we write $\mathrm{E}[X] = -\infty$.

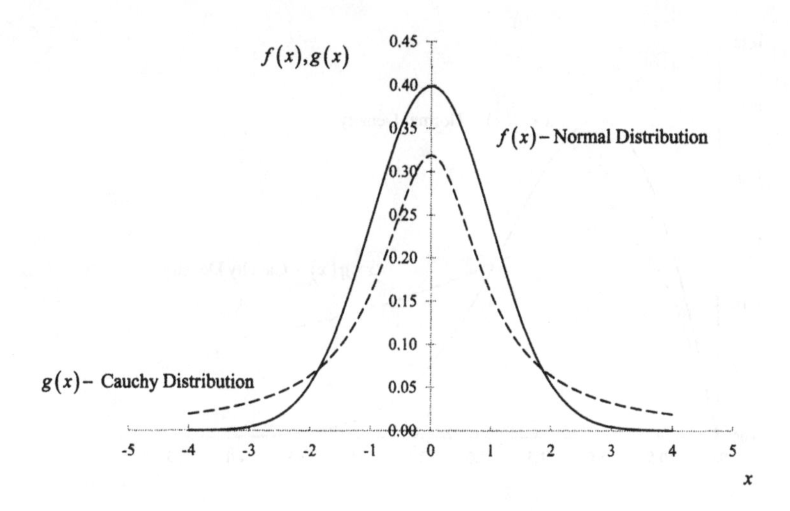

FIGURE 4.2
Standard Normal and Cauchy Distributions.

As in our discussion of the boundedness of the mean in the discrete probability distributions, Definition 4.7 provides some basic conditions to determine whether the continuous expectations are bounded. To develop the point, consider the integral over the positive tails of two distribution functions: the standard normal and Cauchy distribution functions. The mathematical form of the standard normal distribution is given by Equation 3.107, while the Cauchy distribution can be written as

$$g(x) = \frac{1}{\pi\left(1+x^2\right)}. \tag{4.8}$$

Figure 4.2 presents the two distributions; the solid line depicts the standard normal distribution while the broken line depicts the Cauchy distribution. Graphically, these distributions appear to be similar. However, to develop the boundedness we need to evaluate

$$\begin{aligned} x \times f(x) &\Rightarrow \int_0^x z \times \frac{1}{\sqrt{2\pi}} e^{-z^2/2} dz \\ x \times g(x) &\Rightarrow \int_0^x z \times \frac{1}{\pi(1+z)} dz. \end{aligned} \tag{4.9}$$

Figure 4.3 presents the value of the variable times each density function. This comparison shows the mathematics behind the problem with boundedness — the value for the normal distribution (depicted as the solid line) converges much more rapidly than the value for the Cauchy distribution (depicted with

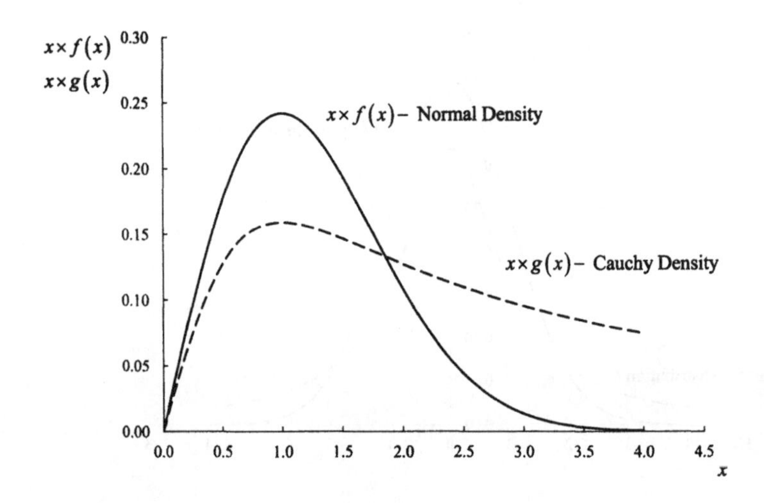

FIGURE 4.3
Function for Integration.

the broken line). The integral for each of the distributions (i.e., the left-hand sides of Equation 4.9) are presented in Figure 4.4. Confirming the intuition from Figure 4.3, the integral for the normal distribution is bounded (reaching a maximum around 0.40) while the integral for the Cauchy distribution is unbounded (increasing almost linearly throughout its range). Hence, the

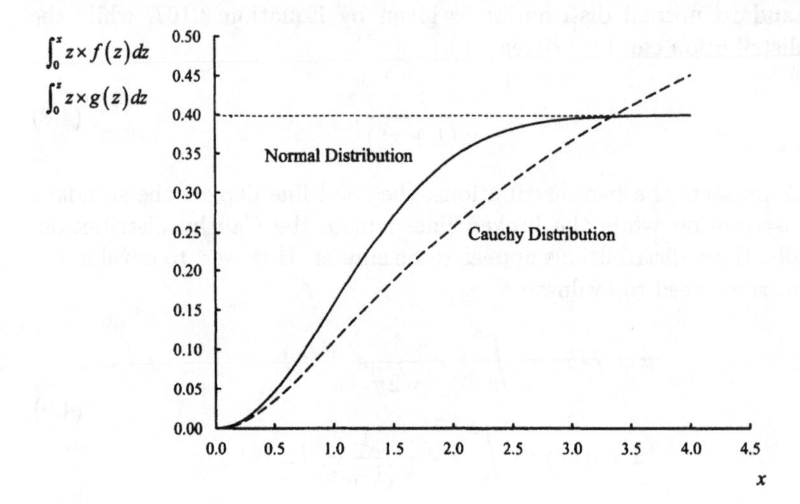

FIGURE 4.4
Integrals of the Normal and Cauchy Expectations.

expectation for the normal exists, but the expectation of the Cauchy distribution does not exist (i.e., $-\infty + \infty$ does not exist).

Moving to expectations of bivariate random variables, we start by defining the expectation of a bivariate function defined on a discrete random variable in Theorem 4.8.

Theorem 4.8. *Let (X, Y) be a bivariate discrete random variable taking value (x_i, y_j) with probability $\mathrm{P}(x_i, y_j)$, $i, j = 1, 2, \cdots$ and let $\phi(x_i, y_i)$ be an arbitrary function. Then*

$$\mathrm{E}[\phi(X,Y)] = \sum_{i=1}^{\infty}\sum_{j=1}^{\infty} \phi(x_i, y_j)\,\mathrm{P}(x_i, y_j). \tag{4.10}$$

Implicit in Theorem 4.8 are similar boundedness conditions discussed in Definition 4.3. Similarly, we can define the expectation of a bivariate continuous random variable as

Theorem 4.9. *Let (X, Y) be a bivariate continuous random variable with joint density function $f(x, y)$, and let $\phi(x, y)$ be an arbitrary function. Then*

$$\mathrm{E}[\phi(X,Y)] = \int_{-\infty}^{\infty}\int_{-\infty}^{\infty} \phi(x, y)\, f(x, y)\, dx dy. \tag{4.11}$$

Next, consider a couple of special cases. First, consider the scenario where the bivariate function is a constant ($\phi(x, y) = \alpha$).

Theorem 4.10. *If $\phi(x, y) = \alpha$ is a constant, then $\mathrm{E}[\alpha] = \alpha$.*

Next, we consider the expectation of a linear function of the two random variables ($\phi(x, y) = \alpha x + \beta y$).

Theorem 4.11. *If X and Y are random variables and α and β are constants,*

$$\mathrm{E}[\alpha X + \beta Y] = \alpha \mathrm{E}[X] + \beta \mathrm{E}[Y]. \tag{4.12}$$

Finally, we consider the expectation of the multiple of two independent random variables ($\phi(x, y) = xy$ where x and y are independent).

Theorem 4.12. *If X and Y are independent random variables, then $\mathrm{E}[XY] = \mathrm{E}[X]\,\mathrm{E}[Y]$.*

The last series of theorems is important to simplify decision making under risk. In the crop example we have

$$\pi = pX - C \tag{4.13}$$

where π is profit, p is the price of the output, X is the yield level and C is the cost per acre. The distribution of profit along with its expected value is dependent on the distribution of p, X, and C. In the example above, we assume that p and C are constant at $\tilde{p}$ and $\tilde{C}$. The expected value of profit is then

$$E[\pi = \phi(p, X, C)] = E[pX - C] = \tilde{p}E[X] - \tilde{C}. \tag{4.14}$$

As a first step, assume that cost is a random variable; then

$$E[\pi = \phi(p, X, C)] = E[pX - C] = \tilde{p}E[x] - E[C]. \tag{4.15}$$

Next, assume that price and yield are random, but cost is constant:

$$E[\pi = \phi(p, X, C)] = E[pX] - \tilde{C} = \int_{-\infty}^{\infty}\int_{-\infty}^{\infty} px f(x, y)\, dx\, dp - \tilde{C}. \tag{4.16}$$

By assuming that p and X are independent (e.g., the firm level assumptions),

$$E[\pi = \phi(p, X, C)] = E[p]\, E[X] - \tilde{C}. \tag{4.17}$$

4.2 Moments

Another frequently used function of random variables is the moments of the distribution function

$$\mu_r(X) = E[X^r] = \int_{-\infty}^{\infty} x^r f(x)\, dx \tag{4.18}$$

where r is a non-negative integer. From this definition, it is obvious that the mean is the first moment of the distribution function. The second moment is defined as

$$\mu_2(X) = E\left[X^2\right] = \int_{-\infty}^{\infty} x^2 f(x)\, dx. \tag{4.19}$$

The higher moments can similarly be represented as moments around the mean or central moment:

$$\tilde{\mu}_r(X) = E[X - E[X]]^r. \tag{4.20}$$

The first, second, third, and fourth moments of the uniform distribution can be derived as

$$\begin{aligned}
\mu_1(X) &= \int_0^1 x(1)\, dx = \frac{1}{2}\left(x^2\right|_0^1 = \frac{1}{2} \\
\mu_2(X) &= \int_0^1 x^2 dx = \frac{1}{3}\left(x^3\right|_0^1 = \frac{1}{3} \\
\mu_3(X) &= \int_0^1 x^3 dx = \frac{1}{4}\left(x^4\right|_0^1 = \frac{1}{4} \\
\mu_4(X) &= \int_0^1 x^4 dx = \frac{1}{5}\left(x^5\right|_0^1 = \frac{1}{5}.
\end{aligned} \tag{4.21}$$

The variance is then the second central moment.

Definition 4.13. The second central moment of the distribution defines the variance of the distribution

$$V(X) = E[X - E[x]]^2 = E[X^2] - E[X]^2. \tag{4.22}$$

The last equality is derived by

$$\begin{aligned} E[X - E[X]]^2 &= E[(X - E[X])(X - E[X])] \\ &= E\left[X^2 - 2XE[X] + E[X]^2\right] \\ &= E[X^2] - 2E[X]E[X] + E[X]^2 \\ &= E[X^2] - E[X]^2. \end{aligned} \tag{4.23}$$

Put another way, the variance can be derived as

$$V(X) = \sigma^2 = \mu_2 - (\mu_1)^2 = \tilde{\mu}_2. \tag{4.24}$$

From these definitions, we see that for the uniform distribution

$$V(X) = \mu_2 - (\mu_1)^2 = \frac{1}{3} - \left(\frac{1}{2}\right)^2 = \frac{1}{3} - \frac{1}{4} = \frac{1}{12}. \tag{4.25}$$

This can be verified directly by

$$\begin{aligned} V(X) &= \int_0^1 \left(x - \frac{1}{2}\right)^2 dx \\ &= \int_0^1 \left(x^2 - x + \frac{1}{4}\right) dx \\ &= \int_0^1 x^2 dx - \int_0^1 x dx + \frac{1}{4}\int_0^1 dx \\ &= \frac{1}{3} - \frac{1}{3} + \frac{1}{4} = \frac{4}{12} - \frac{6}{12} + \frac{3}{12} = \frac{1}{12}. \end{aligned} \tag{4.26}$$

4.3 Covariance and Correlation

The most frequently used moments for bivariate and multivariate random variables are their means (again the first moment of each random variable) and covariances (which are the second own moments and cross moments with the other random variables). Consider the covariance between two random variables X and Y presented in Definition 4.14.

Definition 4.14. The covariance between two random variables X and Y can be defined as

$$\begin{aligned} \mathrm{Cov}(X,Y) &= \mathrm{E}\left[(X-\mathrm{E}[X])(Y-\mathrm{E}[Y])\right] \\ &= \mathrm{E}\left[XY - X\mathrm{E}[Y] - \mathrm{E}[x]Y + \mathrm{E}[X][Y]\right] \\ &= \mathrm{E}[XY] - \mathrm{E}[X]\mathrm{E}[Y] - \mathrm{E}[X]\mathrm{E}[Y] + \mathrm{E}[X]\mathrm{E}[Y] \\ &= \mathrm{E}[XY] - \mathrm{E}[X]\mathrm{E}[Y]. \end{aligned} \tag{4.27}$$

Note that this is simply a generalization of the standard variance formulation. Specifically, letting $Y \to X$ yields

$$\begin{aligned} \mathrm{Cov}(XX) &= \mathrm{E}[XX] - \mathrm{E}[X]\mathrm{E}[X] \\ &= \mathrm{E}\left[X^2\right] - \left(\mathrm{E}[X]\right)^2. \end{aligned} \tag{4.28}$$

Over the next couple of chapters we will start developing both theoretical and empirical statistics. Typically, theoretical statistics assume that we know the parameters of a known distribution. For example, assume that we know that a pair of random variables are distributed bivariate normal and that we know the parameters of that distribution. However, when we compute empirical statistics we assume that we do not know the underlying distribution or parameters. The difference is the weighting function. When we assume that we know the distribution and the parameters of the distribution, we will use the known probabilty density function to compute statistics such as the variance and covariance coefficients. In the case of empirical statistics, we typically assume that observations are equally likely (i.e., typically weighting by $1/N$ where there are N observations).

These assumptions have implications for our specification of the variance/covariance matrix. For example, assume that we are interested in the theoretical variance/covariance matrix. Assume that the joint distribution function can be written as $f(x,y|\,\theta)$ where θ is a known set of parameters. Using a slight extension of the moment specification, we can compute the first moment of both x and y as

$$\begin{aligned} \mu_x(\theta) &= \int_{x_a}^{x_b}\int_{y_a}^{y_b} x f(x,y|\,\theta)\,dxdy \\ \mu_y(\theta) &= \int_{x_a}^{x_b}\int_{y_a}^{y_b} y f(x,y|\,\theta)\,dxdy. \end{aligned} \tag{4.29}$$

where $x \in [x_a, x_b]$ and $y \in [y_a, y_b]$. Note that we are dropping the subscript 1 to denote the first moment and use the subscript to denote the variable

$(\mu_x(\theta) \leftarrow \mu_1(\theta))$. Given these means (or first moments), we can the define the variance for each variable as

$$\begin{aligned}\sigma_x^2(\theta) = \sigma_{xx}(\theta) &= \int_{x_a}^{x_b}\int_{y_a}^{y_b} (x - \mu_x(\theta))^2 f(x, y|\,\theta)\, dxdy \\ \sigma_y^2(\theta) = \sigma_{yy}(\theta) &= \int_{x_a}^{x_b}\int_{y_a}^{y_b} (y - \mu_y(\theta))^2 f(x, y|\,\theta)\, dxdy.\end{aligned} \tag{4.30}$$

The covariance coefficient can then be expressed as

$$\sigma_{xy}(\theta) = \int_{x_a}^{x_b}\int_{y_a}^{y_b} (x - \mu_x(\theta))(y - \mu_y(\theta)) f(x, y|\,\theta)\, dxdy. \tag{4.31}$$

Notice that the covariance function is symmetric (i.e., $\sigma_{xy}(\theta) = \sigma_{yx}(\theta)$). The coefficients from Equations 4.30 and 4.31 can be used to populate the variance/covariance matrix conditional on θ as

$$\Sigma(\theta) = \begin{bmatrix} \sigma_{xx}(\theta) & \sigma_{xy}(\theta) \\ \sigma_{xy}(\theta) & \sigma_{yy}(\theta) \end{bmatrix}. \tag{4.32}$$

Example 4.15 presents a numerical example of the covariance matrix using a discrete (theoretical) distribution function.

Example 4.15. Assume that the probability for a bivariate discrete random variable is presented in Table 4.5. We start by computing the means

$$\begin{aligned}\mu_x &= (0.167 + 0.083 + 0.167) \times 1 + (0.083 + 0.000 + 0.083) \\ &\quad \times 0 + (0.167 + 0.083 + 0.167) \times -1 \\ &= 0.417 \times 1 + 0.167 \times 0 + 0.417 \times -1 = 0. \\ \mu_y &= (0.167 + 0.083 + 0.167) \times 1 + (0.083 + 0.000 + 0.083) \\ &\quad \times 0 + (0.167 + 0.083 + 0.167) \times -1 \\ &= 0.417 \times 1 + 0.167 \times 0 + 0.417 \times -1 = 0.\end{aligned} \tag{4.33}$$

TABLE 4.5
Discrete Sample

	Y			Marginal
X	−1	0	1	Probability
−1	0.167	0.083	0.167	0.417
0	0.083	0.000	0.083	0.167
1	0.167	0.083	0.167	0.417
Marginal Probability	0.417	0.167	0.417	

Given that the means for both variables are zero, we can compute the variance and covariance as

$$V[X] = \sum_{i\in\{1,0,1\}} \sum_{j\in\{1,0,1\}} x_i^2 P(x_i, y_i) = (-1)^2 \times 0.167 + (0)^2 \times 0.083 + \cdots$$
$$(1)^2 \times 0.167 = 0.834$$

$$V[Y] = \sum_{i\in\{1,0,1\}} \sum_{j\in\{1,0,1\}} y_i^2 P(x_i, y_i) = (-1)^2 \times 0.167 + (0)^2 \times 0.083 + \cdots$$
$$(1)^2 \times 0.167 = 0.834$$

$$\begin{aligned} \mathrm{Cov}[X,Y] = \sum_{i\in\{1,0,1\}} \sum_{j\in\{1,0,1\}} x_i y_i P(x_i, y_i) = -1 \times -1 \times 0.167 + \\ -1 \times 0 \times 0.083 + \cdots 1 \times 1 \times 0.167 = 0. \end{aligned} \quad (4.34)$$

Thus, the variance matrix becomes

$$\Sigma = \begin{bmatrix} 0.834 & 0.000 \\ 0.000 & 0.834 \end{bmatrix}. \quad (4.35)$$

The result in Equation 4.35 allows for an additional definition of independence – two distributions are independent if their covariance is equal to zero.

From a sample perspective, we can compute the variance and covariance as

$$\begin{aligned} s_{xx} &= \frac{1}{N}\sum_{i=1}^{N} x_i^2 - \bar{x}^2 \\ s_{yy} &= \frac{1}{N}\sum_{i=1}^{N} y_i^2 - \bar{y}^2 \\ s_{xy} &= \frac{1}{N}\sum_{i=1}^{N} x_i y_i - \bar{x}\bar{y} \end{aligned} \quad (4.36)$$

where $\bar{x} = 1/N \sum_{i=1}^{N} x_i$ and $\bar{y} = 1/N \sum_{i=1}^{N} y_i$. Typically, a lower case Roman character is used to denote individual sample statistics. Substituting the sample measures into the variance matrix yields

$$\begin{aligned} S = \begin{bmatrix} s_{xx} & s_{xy} \\ s_{yx} & s_{yy} \end{bmatrix} &= \begin{bmatrix} \frac{1}{N}\sum_{i=1}^{N} x_i x_i - \bar{x}\bar{x} & \frac{1}{N}\sum_{i=1}^{N} x_i y_i - \bar{x}\bar{y} \\ \frac{1}{N}\sum_{i=1}^{N} y_i x_i - \bar{y}\bar{x} & \frac{1}{N}\sum_{i=1}^{N} y_i y_i - \bar{y}\bar{y} \end{bmatrix} \\ &= \frac{1}{N}\begin{bmatrix} \sum_{i=1}^{N} x_i x_i & \sum_{i=1}^{N} x_i y_i \\ \sum_{i=1}^{N} y_i x_i & \sum_{i=1}^{N} y_i y_i \end{bmatrix} - \begin{bmatrix} \bar{x}\bar{x} & \bar{x}\bar{y} \\ \bar{y}\bar{x} & \bar{y}\bar{y} \end{bmatrix} \end{aligned} \quad (4.37)$$

where the upper case Roman letter is used to denote a matrix of sample statistics. While we will develop the matrix notation more completely in Chapter 10, the sample covariance matrix can then be written as

$$S = \frac{1}{N} \begin{bmatrix} x_1 & \cdots & x_N \\ y_1 & \cdots & y_N \end{bmatrix} \begin{bmatrix} x_1 & y_1 \\ \cdots & \cdots \\ x_N & y_N \end{bmatrix} - \begin{bmatrix} \bar{x} \\ \bar{y} \end{bmatrix} \begin{bmatrix} \bar{x} & \bar{y} \end{bmatrix}. \tag{4.38}$$

Next, we need to develop a couple of theorems regarding the variance of linear combinations of random variables. First consider the linear sum or difference of two random variables X and Y.

Theorem 4.16. $\mathrm{V}(X \pm Y) = \mathrm{V}(X) + \mathrm{V}(Y) \pm \mathrm{Cov}(X,Y)$.

Proof.

$$\begin{aligned} \mathrm{V}[X \pm Y] &= \mathrm{E}[(X \pm Y)(X \pm Y)] \\ &= \mathrm{E}[XX \pm 2XY + YY] \\ &= \mathrm{E}[XX] + \mathrm{E}[YY] \pm \mathrm{E}[XY] \\ &= \mathrm{V}(X) + \mathrm{V}(Y) \pm 2\mathrm{Cov}(X,Y). \end{aligned} \tag{4.39}$$

□

Note that this result can be obtained from the variance matrix. Specifically, $X+Y$ can be written as a vector operation:

$$\begin{bmatrix} X & Y \end{bmatrix} \begin{bmatrix} 1 \\ 1 \end{bmatrix} = X + Y. \tag{4.40}$$

Given this vectorization of the problem, we can define the variance of the sum as

$$\begin{aligned} \begin{bmatrix} 1 & 1 \end{bmatrix} \begin{bmatrix} \sigma_{xx} & \sigma_{xy} \\ \sigma_{xy} & \sigma_{yy} \end{bmatrix} \begin{bmatrix} 1 \\ 1 \end{bmatrix} &= \begin{bmatrix} \sigma_{xx} + \sigma_{xy} & \sigma_{xy} + \sigma_{yy} \end{bmatrix} \begin{bmatrix} 1 \\ 1 \end{bmatrix} \\ &= \sigma_{xx} + 2\sigma_{xy} + \sigma_{yy}. \end{aligned} \tag{4.41}$$

Next, consider the variance for the sum of a collection of random variables X_i where $i = 1, \cdots N$.

Theorem 4.17. *Let X_i, $i = 1, 2, \cdots$ be pairwise independent (i.e., where $\sigma_{ij} = 0$ for $i \neq j$, $i, j = 1, 2, \cdots N$). Then*

$$\mathrm{V}\left(\sum_{i=1}^{N} X_i\right) = \sum_{i=1}^{N} \mathrm{V}(X_i). \tag{4.42}$$

Proof. The simplest proof to this theorem is to use the variance matrix. Note in the preceding example, if X and Y are independent, we have

$$\begin{bmatrix} 1 & 1 \end{bmatrix} \begin{bmatrix} \sigma_{xx} & \sigma_{xy} \\ \sigma_{xy} & \sigma_{yy} \end{bmatrix} \begin{bmatrix} 1 \\ 1 \end{bmatrix} = \sigma_{xx} + \sigma_{yy} \tag{4.43}$$

if $\sigma_{xy} = 0$. Extending this result to three variables implies

$$\begin{bmatrix} 1 \\ 1 \\ 1 \end{bmatrix}' \begin{bmatrix} \sigma_{11} & \sigma_{12} & \sigma_{13} \\ \sigma_{21} & \sigma_{22} & \sigma_{23} \\ \sigma_{31} & \sigma_{32} & \sigma_{33} \end{bmatrix} \begin{bmatrix} 1 \\ 1 \\ 1 \end{bmatrix} = \sigma_{11} + 2\sigma_{12} + 2\sigma_{13} + \sigma_{22} + 2\sigma_{23} + \sigma_{33} \tag{4.44}$$

if the xs are independent, the covariance terms are zero and this expression simply becomes the sum of the variances.

□

One of the difficulties with the covariance coefficient for making intuitive judgments about the strength of the relationship between two variables is that it is dependent on the magnitude of the variance of each variable. Hence, we often compute a normalized version of the covariance coefficient called the **correlation coefficient**.

Definition 4.18. The correlation coefficient for two variables is defined as

$$\text{Corr}(X, Y) = \frac{\text{Cov}(X, Y)}{\sqrt{\sigma_{xx}}\sqrt{\sigma_{yy}}}. \tag{4.45}$$

Note that the covariance between any random variable and a constant is equal to zero. Letting Y equal zero, we have

$$\text{E}\left[(X - \text{E}[X])(Y - \text{E}[Y])\right] = \text{E}\left[(X - \text{E}[X])(0)\right] = 0. \tag{4.46}$$

It stands to reason the correlation coefficient between a random variable and a constant is also zero.

It is now possible to derive the ordinary least squares estimator for a linear regression equation.

Definition 4.19. We define the ordinary least squares estimator as that set of parameters that minimizes the squared error of the estimate.

$$\min_{\alpha,\beta} = \text{E}\left[(Y - \alpha - \beta X)^2\right] = \min_{\alpha,\beta} \text{E}\left[Y^2 - 2\alpha Y - 2\beta XY + \alpha^2 + 2\alpha\beta X + \beta^2 X^2\right]. \tag{4.47}$$

The first order conditions for this minimization problem then become

$$\begin{aligned} \frac{\partial S}{\partial \alpha} &= -2\text{E}[Y] + 2\alpha + 2\beta\text{E}[X] = 0 \\ \frac{\partial S}{\partial \beta} &= -2\text{E}[XY] + 2\alpha\text{E}[X] + 2\beta\text{E}\left[X^2\right] = 0. \end{aligned} \tag{4.48}$$

Solving the first equation for α yields

$$\alpha = \mathrm{E}[Y] - \beta \mathrm{E}[X]. \tag{4.49}$$

Substituting this expression into the second first order condition yields

$$\begin{aligned}
-\mathrm{E}[XY] + (\mathrm{E}[Y] - \beta\mathrm{E}[X])\,\mathrm{E}[X] + \beta\mathrm{E}\left[X^2\right] &= 0 \\
-\mathrm{E}[XY] + \mathrm{E}[Y]\,\mathrm{E}[X] + \beta\left(\mathrm{E}\left[X^2\right] - (\mathrm{E}[X])^2\right) &= 0 \\
-\mathrm{Cov}(X,Y) + \beta\mathrm{V}(X) &= 0 \\
\Rightarrow \beta &= \frac{\mathrm{Cov}(X,Y)}{V(X)}.
\end{aligned} \tag{4.50}$$

Theorem 4.20. *The best linear predictor (or more exactly, the minimum mean-squared-error linear predictor) of Y based on X is given by $\alpha^* + \beta^* X$, where α^* and β^* are the least square estimates where α^* and β^* are defined by Equations 4.49 and 4.50, respectively.*

4.4 Conditional Mean and Variance

Next,, we consider the formulation where we are given some information about the bivariate random variable and wish to compute the implications of this knowledge for the other random variable. Specifically, assume that we are given the value of X and want to compute the expectation of $\phi(X,Y)$ given that information. For the case of the discrete random variable, we can define this expectation, called the **conditional mean**, using Definition 4.21.

Definition 4.21. Let (X,Y) be a bivariate discrete random variable taking values (x_i, y_j) $i,j = 1,2,\cdots$. Let $\mathrm{P}(y_j|X)$ be the conditional probability of $Y = y_j$ given X. Let $\phi(x_i, y_j)$ be an arbitrary function. Then the conditional mean of $\phi(X,Y)$ given X, denoted $\mathrm{E}[\phi(X,Y)|X]$ or by $\mathrm{E}_{Y|X}[\phi(X,Y)]$, is defined by

$$\mathrm{E}_{Y|X}[\phi(X,Y)] = \sum_{i=1}^{\infty} \phi(X, y_i)\,\mathrm{P}(y_i|X). \tag{4.51}$$

Similarly, Definition 4.22 presents the conditional mean for the continuous random variable.

Definition 4.22. Let (X,Y) be a bivariate continuous random variable with conditional density $f(y|x)$. Let $\phi(x,y)$ be an arbitrary function. Then the conditional mean of $\phi(X,Y)$ given X is defined by

$$\mathrm{E}_{Y|X}[\phi(X,Y)] = \int_{-\infty}^{\infty} \phi(X,y)\, f(y|X)\, dy. \tag{4.52}$$

Building on these definitions, we can demonstrate the **Law of Iterated Means** given in Theorem 4.23.

Theorem 4.23 (Law of Iterated Means). $\mathrm{E}\left[\phi\left(X,Y\right)\right]=\mathrm{E}_X\mathrm{E}_{Y|X}\left[\phi\left(X,Y\right)\right]$. *(where the symbol* E_x *denotes the expectation with respect to* X*).*

Proof. Consider the general expectation of $\phi\left(X,Y\right)$ assuming a continuous random variable.

$$\mathrm{E}\left[\phi\left(X,Y\right)\right]\equiv\int_{x_a}^{x_b}\int_{y_a}^{y_b}\phi\left(x,y\right)f\left(x,y\right)dydx \tag{4.53}$$

as developed in Theorem 4.9. Next we can group without changing the result, yielding

$$\mathrm{E}\left[\phi\left(X,Y\right)\right]=\int_{x_a}^{x_b}\left[\int_{y_a}^{y_b}\phi\left(x,y\right)f\left(x,y\right)dy\right]dx. \tag{4.54}$$

Notice that the term in brackets is the $\mathrm{E}_{Y|X}\left[\phi\left(X,Y\right)\right]$ by Definition 4.22. To complete the proof, we rewrite Equation 4.54 slightly.

$$\mathrm{E}\left[\phi\left(X,Y\right)\right]=\int_{x_a}^{x_b}\mathrm{E}_{Y|X}\left[\phi\left(X,Y\right)\right]f\left(x\right)dx \tag{4.55}$$

basically rewriting $f\left(x,y\right)=f\left(y|\,x\right)f\left(x\right)$. □

Building on the conditional means, the conditional variance for $\phi\left(X,Y\right)$ is presented in Theorem 4.24.

Theorem 4.24.

$$\mathrm{V}(\phi(X,Y))=\mathrm{E}_X[\mathrm{V}_{Y|X}[\phi(X,Y)]]+\mathrm{V}_X[\mathrm{E}_{Y|X}[\phi(X,Y)]]. \tag{4.56}$$

Proof.

$$\mathrm{V}_{Y|X}[\phi]=\mathrm{E}_{Y|X}[\phi^2]-(\mathrm{E}_{Y|X}[\phi])^2 \tag{4.57}$$

implies

$$\mathrm{E}_X[\mathrm{V}_{Y|X}(\phi)]=\mathrm{E}[\phi^2]-\mathrm{E}_X[\mathrm{E}_{Y|X}[\phi]]. \tag{4.58}$$

By the definition of conditional variance,

$$\mathrm{V}_X\left(\mathrm{E}_{Y|X}\left[\phi\right]\right)=\mathrm{E}_X\left[\mathrm{E}_{Y|X}\left[\phi\right]\right]^2-\left(\mathrm{E}\left[\phi\right]\right)^2. \tag{4.59}$$

Adding these expressions yields

$$\mathrm{E}_X\left[\mathrm{V}_{Y|X}\left(\phi\right)\right]+\mathrm{V}_X\left(\mathrm{E}_{Y|X}\left[\phi\right]\right)=\mathrm{E}\left[\phi^2\right]-\left(\mathrm{E}\left[\phi\right]\right)=\mathrm{V}\left(\phi\right). \tag{4.60}$$

□

Finally, we can link the least squares estimator to the projected variance in Theorem 4.20 using Theorem 4.25.

Theorem 4.25. *The best predictor (or the minimum mean-squared-error predictor) of* Y *based on* X *is given by* $\mathrm{E}\left[Y|X\right]$.

4.5 Moment-Generating Functions

The preceding sections of this chapter have approached the moments of random variables one at a time. One alternative to this approach is to define a function called a **Moment Generating Function** that systematically generates the moments for random variables.

Definition 4.26. Let X be a random variable with a cumulative distribution function $F(X)$. The moment generating function of X (or $F(X)$), denoted $M_X(t)$, is

$$M_X(t) = \mathrm{E}\left[e^{tX}\right] \tag{4.61}$$

provided that the expectation exists for t in some neighborhood of 0. That is, there is an $h > 0$ such that, for all t in $-h < t < h$, $\mathrm{E}\left[e^{tX}\right]$ exists.

If the expectation does not exist in a neighborhood of 0, we say that the moment generating function does not exist. More explicitly, the moment generating function can be defined as

$$\begin{aligned} M_X(t) &= \int_{-\infty}^{\infty} e^{tx} f(x)\,dx \text{ for continuous random variables, and} \\ M_x(t) &= \sum_x e^{tx}\mathrm{P}[X = x] \text{ for discrete random variables.} \end{aligned} \tag{4.62}$$

Theorem 4.27. *If X has a moment generating function $M_X(t)$, then*

$$\mathrm{E}[X^n] = M_x^{(n)}(0) \tag{4.63}$$

where we define

$$M_X^{(n)}(0) = \frac{d^n}{dt^n} M_X(t)|_{t\to 0}\,. \tag{4.64}$$

First note that e^{tX} can be approximated around zero using a Taylor series expansion.

$$\begin{aligned} M_X(t) = \mathrm{E}\left[e^{tx}\right] = \mathrm{E}\Big[e^0 + te^{t0}(x-0) + \\ \frac{1}{2}t^2e^{t0}(x-0)^2 + \frac{1}{6}t^3e^{t0}(x-0)^3 + \cdots\Big] \\ = 1 + \mathrm{E}[x]t + \mathrm{E}\left[x^2\right]\frac{t^2}{2} + \mathrm{E}\left[x^3\right]\frac{t^3}{6} + \cdots. \end{aligned} \tag{4.65}$$

Note for any moment n

$$M_x^{(n)}(t) = \frac{d^n}{dt^n}M_X(t) = \mathrm{E}[x^n] + \mathrm{E}\left[x^{n+1}\right]t + \mathrm{E}\left[x^{n+2}\right]t^2 + \cdots. \tag{4.66}$$

Thus, as $t \to 0$,

$$M_x^{(n)}(0) = \mathrm{E}[x^n]. \tag{4.67}$$

Definition 4.28. Leibnitz's Rule: If $f(x,\theta)$, $a(\theta)$, and $b(\theta)$ are differentiable with respect to θ, then

$$\frac{d}{d\theta}\int_{a(\theta)}^{b(\theta)} f(x,\theta)\,dx = -f(b(\theta),\theta)\frac{d}{d\theta}a(\theta) + f(a(\theta))\frac{d}{d\theta}b(\theta) \int_{a(\theta)}^{b(\theta)} \frac{\partial}{\partial\theta} f(x,\theta)\,dx. \tag{4.68}$$

Lemma 4.29. Casella and Berger's proof: *Assume that we can differentiate under the integral using Leibnitz's rule; we have*

$$\begin{aligned}\frac{d}{dt}M_X(t) &= \frac{d}{dt}\int_{-\infty}^{\infty} e^{tx} f(x)\,dx \\ &= \int_{-\infty}^{\infty}\left(\frac{d}{dt}e^{tx}\right) f(x)\,dx \\ &= \int_{-\infty}^{\infty} x e^{tx} f(x)\,dx.\end{aligned} \tag{4.69}$$

Letting $t \to 0$, this integral simply becomes

$$\int_{-\infty}^{\infty} x f(x)\,dx = \mathrm{E}[x] \tag{4.70}$$

[7].

This proof can be extended for any moment of the distribution function.

4.5.1 Moment-Generating Functions for Specific Distributions

The moment generating function for the uniform distribution is

$$M_X(t) = \int_a^b \frac{e^{tx}}{b-a}dx = \frac{1}{b-a}\frac{1}{t}\left(e^{tx}\right|_a^b = \frac{e^{bt}-e^{at}}{t(b-a)}. \tag{4.71}$$

Following the expansion developed earlier, we have

$$\begin{aligned}M_X(t) &= \frac{(1-1)+(b-a)t+\frac{1}{2}\left(b^2-a^2\right)t^2+\frac{1}{6}\left(b^3-a^3\right)t^3+\cdots}{(b-a)t} \\ &= 1+\frac{\left(b^2-a^2\right)t^2}{2(b-a)t}+\frac{\left(b^3-a^3\right)t^3}{6(b-a)t}+\cdots \\ &= 1+\frac{1}{2}\frac{(b-a)(b+a)}{(b-a)}\frac{t^2}{t}+\frac{1}{6}\frac{(b-a)\left(b^2+ab+a^2\right)}{(b-a)}\frac{t^3}{t}+\cdots \\ &= 1+\tfrac{1}{2}(a+b)t+\tfrac{1}{6}\left(a^2+ab+b^2\right)t^2+\cdots.\end{aligned} \tag{4.72}$$

Letting $b = 1$ and $a = 0$, the last expression becomes

$$M_X(t) = 1 + \frac{1}{2}t + \frac{1}{6}t^2 + \frac{1}{24}t^3 + \cdots. \tag{4.73}$$

The first three moments of the uniform distribution are then

$$\begin{aligned} M_X^{(1)} &= \tfrac{1}{2} \\ M_X^{(2)} &= \tfrac{1}{6}2 = \tfrac{1}{3} \\ M_X^{(3)} &= \tfrac{1}{24}6 = \tfrac{1}{4}. \end{aligned} \tag{4.74}$$

The moment generating function for the univariate normal distribution

$$\begin{aligned} M_X(t) &= \frac{1}{\sigma\sqrt{2\pi}} \int_{-\infty}^{\infty} e^{tx} e^{-\frac{1}{2}\frac{(x-\mu)^2}{\sigma^2}} dx \\ &= \frac{1}{\sigma\sqrt{2\pi}} \int_{-\infty}^{\infty} \exp\left[tx - \frac{1}{2}\frac{(x-\mu)^2}{\sigma^2}\right] dx. \end{aligned} \tag{4.75}$$

Focusing on the term in the exponent, we have

$$\begin{aligned} tx - \frac{1}{2}\frac{(x-\mu)^2}{\sigma^2} &= -\frac{1}{2}\frac{(x-\mu)^2 - 2tx\sigma^2}{\sigma^2} \\ &= -\frac{1}{2}\frac{x^2 - 2x\mu + \mu^2 - 2tx\sigma^2}{\sigma^2} \\ &= -\frac{1}{2}\frac{x^2 - 2\left(x\mu + tx\sigma^2\right) + \mu^2}{\sigma^2} \\ &= -\frac{1}{2}\frac{x^2 - 2x\left(\mu + t\sigma^2\right) + \mu^2}{\sigma^2}. \end{aligned} \tag{4.76}$$

The next step is to complete the square in the numerator.

$$\begin{aligned} x^2 - 2x\left(\mu + t\sigma^2\right) + \mu^2 + c &= 0 \\ \left(x - \left(\mu + t\sigma^2\right)\right)^2 &= 0 \\ x^2 - 2x\left(\mu + t\sigma\right) + \mu^2 + 2t\sigma^2\mu + t^2\sigma^4 &= 0 \\ c &= 2t\sigma^2\mu + t^2\sigma^4. \end{aligned} \tag{4.77}$$

The complete expression then becomes

$$\begin{aligned} tx - \frac{1}{2}\frac{(x-\mu)^2}{\sigma^2} &= -\frac{1}{2}\frac{\left(x-\mu-t\sigma^2\right) - 2\mu\sigma^2 t - \sigma^4 t^2}{\sigma^2} \\ &= -\tfrac{1}{2}\frac{\left(x-\mu-t\sigma^2\right)}{\sigma^2} + \mu t + \tfrac{1}{2}\sigma^2 t^2. \end{aligned} \tag{4.78}$$

The moment generating function then becomes

$$\begin{aligned} M_X(t) &= \exp\left(\mu t + \frac{1}{2}\sigma^2 t^2\right)\frac{1}{\sigma\sqrt{2\pi}}\int_{-\infty}^{\infty}\exp\left(-\frac{1}{2}\frac{\left(x-\mu-t\sigma^2\right)}{\sigma^2}\right)dx \\ &= \exp\left(\mu t + \frac{1}{2}\sigma^2 t^2\right). \end{aligned} \tag{4.79}$$

Taking the first derivative with respect to t, we get

$$M_X^{(1)}(t) = \left(\mu + \sigma^2 t\right)\exp\left(\mu t + \frac{1}{2}\sigma^2 t^2\right). \tag{4.80}$$

Letting $t \to 0$, this becomes

$$M_X^{(0)} = \mu. \tag{4.81}$$

The second derivative of the moment generating function with respect to t yields

$$\begin{aligned} M_X^{(2)}(t) = \sigma^2 \exp\left(\mu t + \frac{1}{2}\sigma^2 t^2\right) + \\ \left(\mu + \sigma^2 t\right)\left(\mu + \sigma^2 t\right)\exp\left(\mu t + \frac{1}{2}\sigma^2 t^2\right). \end{aligned} \tag{4.82}$$

Again, letting $t \to 0$ yields

$$M_X^{(2)}(0) = \sigma^2 + \mu^2. \tag{4.83}$$

Let X and Y be independent random variables with moment generating functions $M_X(t)$ and $M_Y(t)$. Consider their sum $Z = X + Y$ and its moment generating function.

$$\begin{aligned} M_Z(t) = \mathrm{E}\left[e^{tz}\right] = \mathrm{E}\left[e^{t(x+y)}\right] = \mathrm{E}\left[e^{tx}e^{ty}\right] = \\ \mathrm{E}\left[e^{tx}\right]\mathrm{E}\left[e^{ty}\right] = M_X(t)\,M_Y(t). \end{aligned} \tag{4.84}$$

We conclude that the moment generating function for two independent random variables is equal to the product of the moment generating functions of each variable. Skipping ahead slightly, the multivariate normal distribution function can be written as

$$f(x) = \frac{1}{\sqrt{2\pi}} |\Sigma|^{-1/2} \exp\left(-\frac{1}{2}(x-\mu)' \Sigma^{-1} (x-\mu)\right). \tag{4.85}$$

In order to derive the moment generating function, we now need a vector $\tilde{t}$. The moment generating function can then be defined as

$$M_{\tilde{X}}(\tilde{t}) = \exp\left(\mu'\tilde{t} + \frac{1}{2}\tilde{t}'\Sigma\tilde{t}\right). \tag{4.86}$$

Normal variables are independent if the variance matrix is a diagonal matrix. Note that if the variance matrix is diagonal, the moment generating function for the normal can be written as

$$\begin{aligned}
M_{\tilde{X}}(\tilde{x}) &= \exp\left(\mu'\tilde{t} + \frac{1}{2}\tilde{t}'\left(\begin{array}{ccc} \sigma_1^2 & 0 & 0 \\ 0 & \sigma_2^2 & 0 \\ 0 & 0 & \sigma_3^2 \end{array}\right)\tilde{t}\right) \\
&= \exp\left(\mu_1 t_1 + \mu_2 t_2 + \mu_3 t_3 + \frac{1}{2}\left(t_1^2\sigma_1^2 + t_2^2\sigma_2^2 + t_3^2\sigma_3^2\right)\right) \\
&= \exp\left(\left(\mu_1 t_1 + \frac{1}{2}\sigma_1^2 t_1^2\right) + \left(\mu_2 t_2 + \frac{1}{2}\sigma_2^2 t_2^2\right) + \left(\mu_3 t_3 + \frac{1}{2}\sigma_3^2 t_3^2\right)\right) \\
&= M_{X_1}(t)\, M_{X_2}(t)\, M_{X_3}(t).
\end{aligned} \tag{4.87}$$

4.6 Chapter Summary

- The moments of the distribution are the kth power of the random variable.
  - The first moment is its expected value or the mean of the distribution. It is largely a measure of the central tendancy of the random variable.
  - The second central moment of the distribution (i.e., the expected value of $(x-\mu)^2$) is a measure of the expected distance between the value of the random variable and its mean.
- The existence of an expectation is dependent on the boundedness of each side of the distribution.
  - Several random variables have an infinite range (i.e., the normal distribution's range is $x \in (-\infty, \infty)$).

  - The value of the expectation depends on whether the value of the distribution function converges to zero faster than the value of the random variable.
  - Some distribution functions such as the Cauchy distribution have no mean.

- The variance of the distribution can also be defined as the expected squared value of the random variable minus the expected value squared.
  - This squared value specification is useful in defining the covariance between two random variables (i.e., $\mathrm{Cov}(X,Y) = \mathrm{E}[XY] - \mathrm{E}[X]\,\mathrm{E}[Y]$.
  - The correlation coefficient is a normalized form of the covariance $\rho_{XY} = \mathrm{Cov}[X,Y] / \sqrt{\mathrm{V}[X]\,\mathrm{V}[Y]}$.

- This chapter also introduces the concepts of sample means and variances versus theoretical means and variances.

- Moment generating functions are functions whose derivatives give the moments of a particular function. Two random variables with the same moment generating function have the same distribution.

4.7 Review Questions

4-1R. What are the implications of autocorrelation (i.e., $y_t = \alpha_0 + \alpha_1 x_t + \epsilon_t + \rho\epsilon_{t-1}$) for the covariance between t and $t+1$?

4-2R. What does this mean for the variance of the sum y_t and y_{t-1}?

4-3R. Derive the normal equations (see Definition 4.19) of a regression with two independent variables (i.e., $y_i = \alpha_0 + \alpha_1 x_{1i} + \alpha_2 x_{2i} + \epsilon_i$).

4-4R. Use the moment generating function for the standard normal distribution to demonstrate that the third central moment of the normal distribution is zero and the fourth central moment of the standard normal is 3.0.

4.8 Numerical Exercises

4-1E. What is the expected value of a two-die roll of a standard six-sided die?

4-2E. What is the expected value of the roll of an eight-sided die and a six-sided die?

4-3E. Compute the expectation and variance of a random variable with the distribution function

$$f(x) = \frac{3}{4}\left(x^2 - 1\right)\ x \in (-1, 1)\,. \tag{4.88}$$

4-4E. What is the expected value of the negative exponential distribution

$$f(x|\,\lambda) = \lambda \exp(-\lambda x)\ x \in (0, \infty)? \tag{4.89}$$

4-5E. Compute the correlation coefficient for the rainfall in August–September and October–December using the data presented in Table 2.1.

5

Binomial and Normal Random Variables

CONTENTS

At several points in this textbook, we have encountered the normal distribution. I have noted that the distribution is frequently encountered in applied econometrics. Its centrality is due in part to its importance in sampling theory. As we will develop in Chapter 6, the normal distribution is typically the limiting distribution of sums of averages. In the vernacular of statistics, this result is referred to as the **Central Limit Theorem**. However, normality is important apart from its asymptotic nature. Specifically, since the sum of a normal distribution is also a normal distribution, we sometimes invoke the normal distribution function in small samples. For example, since a typical sample average is simply the sum of observations divided by a constant (i.e., the sample size), the mean of a collection of normal random variables is also normal. Extending this notion, since ordinary least squares is simply a weighted sum, then the normal distribution provides small sample statistics for most regression applications.

The reasons for the importance of the normal distribution developed above are truly arguments that only a statistician or an applied econometrician could love. The normal distribution in the guise of the **bell curve** has spawned a plethora of books with titles such as *Bell Curve: Intelligence and Class Structure in American Life*, *Intelligence, Genes, and Success: Scientists Respond to the Bell Curve*, and *Poisoned Apple: The Bell-Curve Crisis and How Our Schools Create Mediocrity and Failure.* Whether or not the authors of these books have a firm understanding of the mathematics of the normal distribution, the topic appeals to readers outside of a narrow band of specialists. The popular appeal is probably due to the belief that the distribution of observable characteristics is concentrated around some point, with extreme characteristics being relatively rare. For example, an individual drawn randomly from a sample of people would agree that height would follow a bell curve regardless of their knowledge of statistics.

In this chapter we will develop the **binomial distribution** based on simple Bernoulli distributions. The binomial distribution provides a bridge between rather non-normal random variables (i.e., the $X = 0$ and $X = 1$ Bernoulli random variable) and the normal distribution as the sample size becomes large. In addition, we formally develop the bivariate and multivariate forms of the normal distribution.

5.1 Bernoulli and Binomial Random Variables

In Chapter 2, we developed the Bernoulli distribution to characterize a random variable with two possible outcomes (i.e., whether a coin toss was a heads ($X = 1$) or a tails ($X = 0$) or whether or not it would rain tomorrow). In these cases the probability distribution function $\mathrm{P}[X]$ can be written as

$$\mathrm{P}[X] = p^x (1-p)^x \tag{5.1}$$

where p is the probability of the event occurring. Extending this basic formulation slightly, consider the development of the probability of two independent Bernoulli events. Suppose that we are interested in the outcome of two coin tosses or whether it will rain two days in a row. Mathematically we can specify this random variable as $Z = X + Y$. If X and Y are identically distributed (both Bernoulli) and independent, the probability becomes

$$\begin{aligned} \mathrm{P}[X,Y] = \mathrm{P}[X]\,\mathrm{P}[Y] &= p^x p^y (1-p)^{1-x} (1-p)^{1-y} \\ &= p^{x+y} (1-p)^{2-x-y}. \end{aligned} \tag{5.2}$$

This density function is only concerned with three outcomes, $Z = X + Y = \{0, 1, 2\}$. Notice that there is only one way each for $Z = 0$ or $Z = 2$. Specifically, $Z = 0$, $X = 0$, and $Y = 0$. Similarly, $Z = 2$, $X = 1$, and $Y = 1$. However, for $Z = 1$, either $X = 1$ and $Y = 0$ or $X = 0$ and $Y = 1$. Thus, we can derive

$$\begin{aligned} \mathrm{P}[Z=0] &= p^0 (1-p)^{2-0} \\ \mathrm{P}[Z=1] &= \mathrm{P}[X=1, Y=0] + \mathrm{P}[X=0, Y=1] \\ &= p^{1+0} (1-p)^{2-1-0} + p^{0+1} (1-p)^{2-0-1} \\ &= 2p^1 (1-p)^1 \\ \mathrm{P}[Z=2] &= p^2 (1-p)^0. \end{aligned} \tag{5.3}$$

Next we expand the distribution to three independent Bernoulli events where $Z = W + X + Y = \{0, 1, 2, 3\}$.

$$\begin{aligned} \mathrm{P}[Z] &= \mathrm{P}[W, X, Y] \\ &= \mathrm{P}[W]\,\mathrm{P}[X]\,\mathrm{P}[Y] \\ &= p^w p^x p^y (1-p)^{1-w} (1-p)^{1-x} (1-p)^{1-y} \\ &= p^{w+x+y} (1-p)^{3-w-x-y} \\ &= p^z (1-p)^{3-z}. \end{aligned} \tag{5.4}$$

Again, there is only one way for $Z = 0$ and $Z = 3$. However, there are now three ways for $Z = 1$ or $Z = 2$. Specifically, $Z = 1$ if $W = 1$, $X = 1$, or $Y = 1$. In addition, $Z = 2$ if $W = 1$, $X = 1$, and $Y = 0$, or if $W = 1$, $X = 0$, and $Y = 1$, or if $W = 0$, $X = 1$, and $Y = 1$. Thus the general distribution function for Z can now be written as

$$\begin{aligned} \mathrm{P}[Z = 0] &= p^0 (1-p)^{3-0} \\ \mathrm{P}[Z = 1] &= p^{1+0+0} (1-p)^{3-1-0-0} + p^{0+1+0} (1-p)^{3-0-1-0} + \\ & \quad p^{0+0+1} (1-p)^{3-0-0-1} = 3p^1 (1-p)^2 \\ \mathrm{P}[Z = 2] &= p^{1+1+0} (1-p)^{3-1-1-0} + p^{1+0+1} (1-p)^{3-1-0-1} + \\ & \quad p^{0+1+1} (1-p)^{3-0-1-1} = 3p^2 (1-p)^1 \\ \mathrm{P}[Z = 3] &= p^3 (1-p)^0. \end{aligned} \tag{5.5}$$

Based on our discussion of the Bernoulli distribution, the binomial distribution can be generalized as the sum of n Bernoulli events.

$$\mathrm{P}[Z = r] = \begin{pmatrix} n \\ r \end{pmatrix} p^r (1-p)^{n-r} \tag{5.6}$$

where C_r^n is the combinatorial of n and r developed in Defintion 2.14. Graphically, the combinatorial can be depicted as Pascal's triangle in Figure 5.1. The relationship between the combinatorial and the binomial function can be developed through the general form of the polynomial. Consider the first four polynomial forms:

$$\begin{aligned} (a+b)^1 &= a + b \\ (a+b)^2 &= a^2 + 2ab + b^2 \\ (a+b)^3 &= a^3 + 3a^2b + 3ab^2 + b^3 \\ (a+b)^4 &= a^4 + 4a^3b + 6a^2b^2 + 4ab^3 + b^4. \end{aligned} \tag{5.7}$$

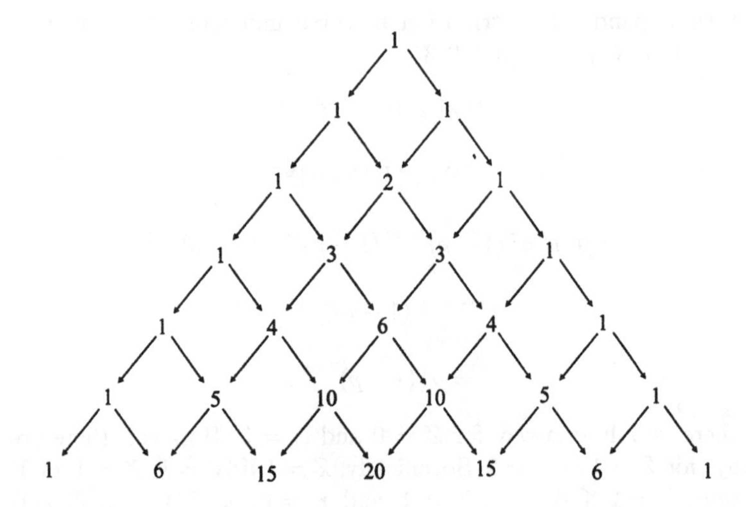

FIGURE 5.1
Pascal's Triangle.

Based on this sequence, the general form of the polynomial form can then be written as

$$(a+b)^n = \sum_{r=1}^{n} C_r^n a^r b^{n-r}. \tag{5.8}$$

As a first step, consider changing the polynomial from $a+b$ to $a-b$. Hence, $(a-b)^n$ can be written as

$$\begin{aligned}(a-b)^n &= (a+(-1)\,b)^n \\ &= \sum_{r=1}^{n} C_r^n a^r (-1)^{n-4} b^{n-r} \\ &= \sum_{r=1}^{n} (-1)^{n-r} C_r^n a^r b^{n-r}.\end{aligned} \tag{5.9}$$

This sequence can be linked to our discussion of the Bernoulli system by letting $a = p$ and $b = 1-p$. Thus, the binomial distribution $X \sim B(n,p)$ can be written as

$$\mathrm{P}(X=k) = C_k^n p^k (1-p)^{n-k}. \tag{5.10}$$

In Equation 5.5 above, $n = 3$. The distribution function can be written as

$$\begin{aligned}\mathrm{P}[Z=0] &= \begin{pmatrix}3\\0\end{pmatrix} p^0 (1-p)^{3-0} \\ \mathrm{P}[Z=1] &= \begin{pmatrix}3\\1\end{pmatrix} p^1 (1-p)^{3-1} \\ \mathrm{P}[Z=2] &= \begin{pmatrix}3\\2\end{pmatrix} p^2 (1-p)^{3-2} \\ \mathrm{P}[Z=3] &= \begin{pmatrix}3\\3\end{pmatrix} p^3 (1-p)^{3-3}.\end{aligned} \tag{5.11}$$

Next, recalling Theorem 4.11, ($\mathrm{E}[aX+bY] = a\mathrm{E}[X] + b\mathrm{E}[Y]$), the expectation of the binomial distribution function can be recovered from the Bernoulli distributions.

$$\begin{aligned}
\mathrm{E}[X] &= \sum_{k=0}^{n} C_k^n p^k (1-p)^{n-k} k \\
&= \sum_{i=1}^{n} \left[\sum_{X_i} p^{X_i} (1-p)^{1-X_i} X_i \right] \\
&= \sum_{i=1}^{n} \left[p^1 (1-p)^0 (1) + p^0 (1-p)^1 (0) \right] \\
&= \sum_{i=1}^{n} p = np.
\end{aligned} \tag{5.12}$$

In addition, by Theorem 4.17

$$\mathrm{V}\left(\sum_{i=1}^{n} X_i \right) = \sum_{i=1}^{n} \mathrm{V}(X_i). \tag{5.13}$$

Thus, the variance of the binomial is simply the sum of the variances of the Bernoulli distributions or n times the variance of a single Bernoulli distribution.

$$\begin{aligned}
\mathrm{V}(X) &= \mathrm{E}\left[X^2\right] - (\mathrm{E}[X])^2 \\
&= \left[p^1 (1-p)^0 (1^2) + p^0 (1-p)^1 (0^2) \right] - p^2 \\
&= p - p^2 = p(1-p) \\
\mathrm{V}\left(\sum_{i=1}^{n} \right) &= np(1-p).
\end{aligned} \tag{5.14}$$

5.2 Univariate Normal Distribution

In Section 3.5 we introduced the normal distribution function as

$$f(x) = \frac{1}{\sigma\sqrt{2\pi}} \exp\left[-\frac{1}{2} \frac{(x-\mu)^2}{\sigma^2} \right] \quad -\infty < x < \infty,\ \sigma > 0. \tag{5.15}$$

In Section 3.7 we demonstrated that the standard normal form of this distribution written as

$$f(x) = \frac{1}{\sqrt{2\pi}} e^{-\frac{x^2}{2}} \tag{5.16}$$

integrated to one. In addition, Equations 3.116 and 3.117 demonstrate how the more general form of the normal distribution in Equation 5.15 can be derived from Equation 5.16 by the change in variables technique.

Theorem 5.1. *Let X be* $\mathrm{N}\left(\mu, \sigma^2\right)$ *as defined in Equation 5.15; then* $\mathrm{E}[X] = \mu$ *and* $\mathrm{V}[X] = \sigma^2$.

Proof. Starting with the expectation of the normal function as presented in Equation 5.15,

$$\mathrm{E}[X] = \int_{-\infty}^{\infty} \frac{1}{\sigma\sqrt{2\pi}} x \exp\left[-\frac{1}{2}\frac{(x-\mu)^2}{\sigma^2}\right] dx. \tag{5.17}$$

Using the change in variables technique, we create a new random variable z such that

$$\begin{gathered} z = \frac{x-\mu}{\sigma} \Rightarrow x = z\sigma + \mu \\ dx = \sigma dz. \end{gathered} \tag{5.18}$$

Substituting into the original integral yields

$$\begin{aligned} \mathrm{E}[X] &= \int_{-\infty}^{\infty} \frac{1}{\sigma\sqrt{2\pi}} (z\sigma + \mu) \exp\left[-\frac{1}{2}z^2\right] dz \\ &= \int_{-\infty}^{\infty} \frac{1}{\sqrt{2\pi}} z \exp\left[-\frac{1}{2}z^2\right] + \mu \int_{-\infty}^{\infty} \frac{1}{\sqrt{2\pi}} \exp\left[-\frac{1}{2}z^2\right] dz. \end{aligned} \tag{5.19}$$

Taking the integral of the first term first, we have

$$\begin{aligned} \int_{-\infty}^{\infty} \frac{1}{\sqrt{2\pi}} z \exp\left[-\frac{1}{2}z^2\right] dx &= C \int_{-\infty}^{\infty} z \exp\left[-\frac{1}{2}z^2\right] dz \\ &= C \left(-\exp\left[-\frac{1}{2}z^2\right]\right|_{-\infty}^{\infty} = 0. \end{aligned} \tag{5.20}$$

The value of the second integral becomes μ by polar integration (see Section 3.7).

The variance of the normal is similarly defined except that the initial integral now becomes

$$\begin{aligned} \mathrm{V}[X] &= \frac{1}{\sigma\sqrt{2\pi}} \int_{-\infty}^{\infty} (x-\mu)^2 \exp\left[-\frac{1}{2}\frac{(x-\mu)^2}{\sigma^2}\right] dx \\ &= \frac{1}{\sigma\sqrt{2\pi}} \int_{-\infty}^{\infty} (z\sigma + \mu - \mu)^2 \exp\left[-\frac{1}{2}z^2\right] \sigma dz \\ &= \frac{1}{\sqrt{2\pi}} \int_{-\infty}^{\infty} z^2\sigma^2 \exp\left[-\frac{1}{2}z^2\right] dz. \end{aligned} \tag{5.21}$$

This formulation is then completed using integration by parts.

$$u = -z \Rightarrow d = -1$$
$$dv = -z \exp\left[-\frac{1}{2}z^2\right] \Rightarrow v = \exp\left[-\frac{1}{2}z^2\right]$$
$$\int_{-\infty}^{\infty} zz \exp\left[-\frac{1}{2}z^2\right] dz = -\left(z \exp\left[-\frac{1}{2}z^2\right]\right|_{-\infty}^{\infty} + \int_{-\infty}^{\infty} \exp\left[-\frac{1}{2}z^2\right] dz. \tag{5.22}$$

The first term of the integration by parts is clearly zero, while the second is defined by polar integral. Thus,

$$\mathrm{V}[X] = 0 + \sigma^2 \int_{-\infty}^{\infty} \frac{1}{\sqrt{2\pi}} \exp\left[-\frac{1}{2}z^2\right] = \sigma^2. \tag{5.23}$$

□

Theorem 5.2. *Let X be distributed $\mathrm{N}(\mu, \sigma^2)$ and let $Y = \alpha + \beta Y$. Then $Y \sim \mathrm{N}(\alpha + \beta\mu, \beta^2\sigma^2)$.*

Proof. This theorem can be demonstrated using Theorem 3.33 (the theorem on changes in variables).

$$g(y) = f\left[\phi^{-1}(y)\right] \left|\frac{d\phi^{-1}(y)}{dy}\right|. \tag{5.24}$$

In this case

$$\phi(x) = \alpha + \beta x \Leftrightarrow \phi^{-1}(y) = \frac{y - \alpha}{\beta}$$
$$\frac{d\phi^{-1}(y)}{dy} = \frac{1}{\beta}. \tag{5.25}$$

The transformed normal then becomes

$$g(y) = \frac{1}{\sigma|\beta|\sqrt{2\pi}} \exp\left[-\frac{1}{2}\frac{\left(\frac{y-\alpha}{\beta} - \mu\right)^2}{\sigma^2}\right]$$
$$= \frac{1}{\sigma|\beta|\sqrt{2\pi}} \exp\left[-\frac{1}{2}\frac{(y - \alpha - \beta\mu)^2}{\sigma^2\beta^2}\right]. \tag{5.26}$$

□

Note that probabilities can be derived for any normal based on the standard normal integral. Specifically, in order to find the probability that $X \sim \mathrm{N}(10, 4)$ lies between 4 and 8 ($\mathrm{P}[4 < X < 8]$) implies

$$\mathrm{P}[4 < X < 8] = \mathrm{P}[X < 8] - \mathrm{P}[X < 4]. \tag{5.27}$$

Transforming each boundary to standard normal space,

$$\begin{aligned} z_1 &= \frac{x_1 - 10}{2} = \frac{-6}{2} = -3 \\ z_2 &= \frac{x_2 - 10}{2} = \frac{-2}{2} = -1. \end{aligned} \tag{5.28}$$

Thus, the equivalent boundary becomes

$$\mathrm{P}\left[-3 < z < -1\right] = \mathrm{P}\left[z < -1\right] - \mathrm{P}\left[z < -3\right] \tag{5.29}$$

where z is a standard normal variable. These values can be found in a standard normal table as $\mathrm{P}\left[z < -1\right] = 0.1587$ and $\mathrm{P}\left[z < -3\right] = 0.0013$.

5.3 Linking the Normal Distribution to the Binomial

To develop the linkage between the normal and binomial distributions, consider the probabilities for binomial distributions with 6, 9, and 12 draws presented in columns two, three, and four in Table 5.1. To compare the shape of these distributions with the normal distribution, we construct normalized

TABLE 5.1
Binomial Probabilities and Normalized Binomial Outcomes

Random	Probability			Normalized Outcome		
Variable	6	9	12	6	9	12
0	0.016	0.002	0.000	−2.449	−3.000	−3.464
1	0.094	0.018	0.003	−1.633	−2.333	−2.887
2	0.234	0.070	0.016	−0.816	−1.667	−2.309
3	0.313	0.164	0.054	0.000	−1.000	−1.732
4	0.234	0.246	0.121	0.816	−0.333	−1.155
5	0.094	0.246	0.193	1.633	0.333	−0.577
6	0.016	0.164	0.226	2.449	1.000	0.000
7		0.070	0.193		1.667	0.577
8		0.018	0.121		2.333	1.155
9		0.002	0.054		3.000	1.732
10			0.016			2.309
11			0.003			2.887
12			0.000			3.464
Mean	3.00	4.50	6.00			
Variance	1.50	2.25	3.00			

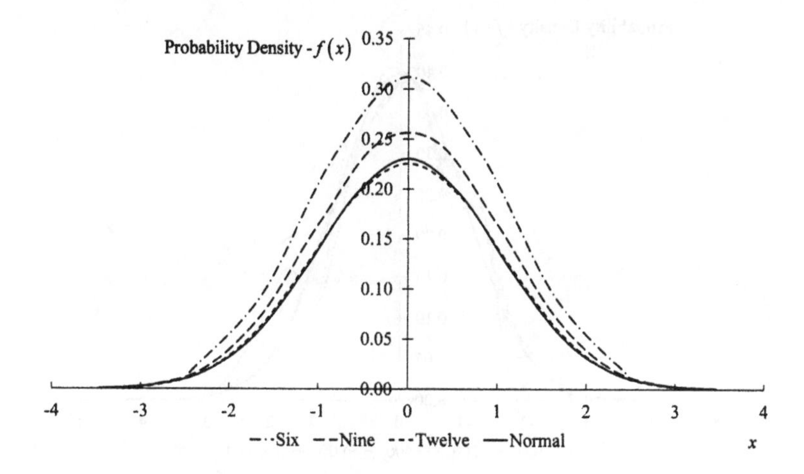

FIGURE 5.2
Comparison of Binomial and Normal Distribution.

outcomes for each set of outcomes.

$$\tilde{z}_i = \frac{x_i - \mu^1_{x_i}}{\sqrt{\mu^2_{x_i} - \left(\mu^1_{x_i}\right)^2}} \tag{5.30}$$

where $\mu^1_{x_i}$ and $\mu^2_{x_i}$ are the first and second moments of the binomial distribution. Basically, Equation 5.30 is the level of the random variable minus the theoretical mean divided by the square root of the theoretical variance. Thus, taking the outcome of $x = 2$ for six draws as an example,

$$\tilde{z}_i = \frac{2 - 3}{\sqrt{1.5}} = -0.816. \tag{5.31}$$

We frequently use this procedure to normalize random variables to be consistent with the standard normal. The sample equivalent to Equation 5.30 can be expressed as

$$\tilde{z} = \frac{x - \bar{x}}{\sqrt{s^2_x}}. \tag{5.32}$$

Figure 5.2 presents graphs of each distribution function and the standard normal distribution (i.e., $x \sim \mathrm{N}\left[0, 1\right]$). The results in Figure 5.2 demonstrate the distribution functions for normalized binomial variables converge rapidly to the normal density function. For 12 draws, the two distributions are almost identical.

This convergence to normality extends to other random variables. Table 5.2 presents 15 samples of eight uniform draws with the sum of the sample and

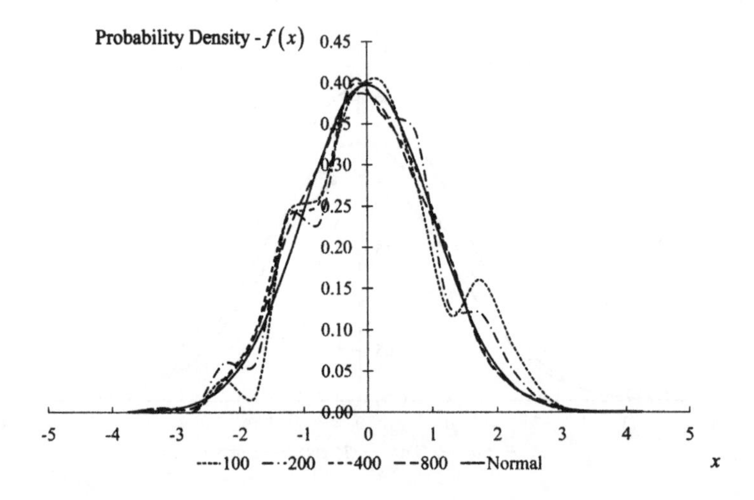

FIGURE 5.3
Limit of the Sum of Uniform Random Variables.

the normalized sum using the definition in Equation 5.30. Figure 5.3 presents the distribution of the normalized random variable in Table 5.2 for sample sizes of 100, 200, 400, and 800. These results demonstrate that as the sample size increases, the empirical distribution of the normalized sum approaches the standard normal distribution.

It is important to notice that the limiting results behind Figures 5.2 and 5.3 are similar, but somewhat different. The limiting result for the binomial distribution presented in Figure 5.2 is due to a direct relationship between the binomial and the normal distribution. The limiting result for the sum of uniform random variables in Figure 5.3 is the result of the central limit theorem developed in Section 6.5.

5.4 Bivariate and Multivariate Normal Random Variables

In addition to its limiting characteristic, the normal density function provides for a simple way to conceptualize correlation between random variables. To develop this specification, we start with the bivariate form of the normal distribution and then expand the bivariate formulation to the general multivariate form.

TABLE 5.2
Samples of Uniform Random Draws

	1	2	3	4	5	6	7	8	Sum	Normalized
1	0.702	0.361	0.914	0.875	0.227	0.011	0.038	0.844	3.972	−0.035
2	0.804	0.560	0.114	0.003	0.652	0.961	0.407	0.779	4.280	0.343
3	0.078	0.781	0.128	0.263	0.875	0.847	0.065	0.656	3.693	−0.376
4	0.774	0.953	0.488	0.179	0.864	0.153	0.974	0.222	4.606	0.743
5	0.465	0.464	0.723	0.222	0.570	0.042	0.892	0.932	4.310	0.379
6	0.850	0.175	0.123	0.490	0.890	0.901	0.948	0.001	4.377	0.462
7	0.930	0.522	0.551	0.202	0.171	0.591	0.767	0.957	4.691	0.846
8	0.730	0.490	0.133	0.905	0.745	0.326	0.177	0.162	3.670	−0.405
9	0.432	0.743	0.104	0.082	0.832	0.740	0.068	0.735	3.735	−0.325
10	0.037	0.954	0.242	0.774	0.689	0.102	0.582	0.123	3.503	−0.608
11	0.545	0.497	0.981	0.957	0.875	0.285	0.242	0.483	4.866	1.061
12	0.656	0.321	0.179	0.309	0.342	0.274	0.907	0.915	3.902	−0.119
13	0.561	0.646	0.435	0.654	0.816	0.225	0.689	0.216	4.243	0.297
14	0.134	0.121	0.740	0.430	0.232	0.546	0.035	0.175	2.414	−1.943
15	0.273	0.747	0.014	0.103	0.028	0.647	0.950	0.395	3.157	−1.033

5.4.1 Bivariate Normal Random Variables

In order to develop the general form of the bivariate normal distribution, consider a slightly expanded form of a normal random variable:

$$f\left(x|\,\mu,\sigma^2\right)=\frac{1}{\sigma\sqrt{2\pi}}\exp\left[-\frac{\left(x-\mu\right)^2}{\sigma^2}\right]. \tag{5.33}$$

The left-hand side of Equation 5.33 explicitly recognizes that the general form of the normal distribution function is conditioned on two parameters – the mean of the distribution μ and the variance of the distribution σ^2. Given this specification, consider a bivariate normal distribution function where the two random variables are independent.

$$f\left(x,y|\,\mu_x,\mu_y,\sigma_x^2,\sigma_y^2\right)=\frac{1}{\sigma_x\sigma_y 2\pi}\exp\left[-\frac{\left(x-\mu_x\right)^2}{2\sigma_x^2}-\frac{\left(y-\mu_y\right)^2}{2\sigma_y^2}\right]. \tag{5.34}$$

Equation 5.34 can be easily factored into forms similar to Equation 5.33.

Definition 5.3 builds on Equation 5.34 by introducing a coefficient that controls the correlation between the two random variables (ρ).

Definition 5.3. The bivariate normal density is defined by

$$f\left(x,y|\,\mu_x,\mu_y,\sigma_x,\sigma_y,\rho\right)=\frac{1}{2\pi\sigma_x\sigma_y\sqrt{1-\rho^2}}$$

$$\exp\left\{-\frac{1}{2\left(1-\rho^2\right)}\left[\left(\frac{x-\mu_x}{\sigma_x}\right)^2+\left(\frac{y-\mu_y}{\sigma_y}\right)^2-2\rho\left(\frac{x-\mu_x}{\sigma_x}\right)\left(\frac{y-\mu_y}{\sigma_y}\right)\right]\right\}. \tag{5.35}$$

Theorem 5.4 develops some of the conditional moments of the bivariate normal distribution presented in Equation 5.35.

Theorem 5.4. *Let (X,Y) have the bivariate normal density. Then the marginal densities $f_X(X)$ and $f_Y(Y)$ and the conditional densities $f(Y|X)$ and $f(X|Y)$ are univariate normal densities, and we have* $\mathrm{E}[X]=\mu_X$, $\mathrm{E}[Y]=\mu_Y$, $\mathrm{V}[X]=\sigma_X^2$, $\mathrm{V}[Y]=\sigma_Y^2$, $\mathrm{Corr}(X,Y)=\rho$, *and*

$$\begin{aligned}\mathrm{E}[Y|X]&=\mu_Y+\rho\frac{\sigma_Y}{\sigma_X}\left(X-\mu_X\right)\\ \mathrm{V}[Y|X]&=\sigma_Y^2\left(1-\rho^2\right).\end{aligned} \tag{5.36}$$

Proof. Let us start by factoring $\left(x-\mu_x\right)/\sigma_x$ out of the exponent term in Equation 5.35. Specifically, suppose that we solve for K such that

$$\frac{1}{2\left(1-\rho^2\right)}\left(\frac{x-\mu_x}{\sigma_x}\right)^2=\frac{1}{2}\left(\frac{x-\mu_x}{\sigma_x}\right)^2+K\left(\frac{x-\mu_x}{\sigma_x}\right)^2. \tag{5.37}$$

Technically, this solution technique is referred to as the method of unknown coefficients – we introduce an unknown coefficient, in this case K, and attempt to solve for the original expression. Notice that $(x - \mu_x)/\sigma_x$ is irrelevant. The problem becomes to solve for K such that

$$\begin{aligned} \frac{1}{2(1-\rho^2)} &= \frac{1}{2} + K \\ K = \frac{1}{2(1-\rho^2)} - \frac{(1-\rho^2)}{2(1-\rho^2)} &= \frac{\rho^2}{2(1-\rho^2)}. \end{aligned} \tag{5.38}$$

Therefore we can rewrite Equation 5.37 as

$$\frac{1}{2(1-\rho^2)}\left(\frac{x-\mu_x}{\sigma_x}\right)^2 = \frac{1}{2}\left(\frac{x-\mu_x}{\sigma_x}\right)^2 + \frac{\rho^2}{2(1-\rho^2)}\left(\frac{x-\mu_x}{\sigma_x}\right)^2. \tag{5.39}$$

Substituting the result of Equation 5.39 into the exponent term in Equation 5.35 yields

$$\begin{aligned} -\frac{1}{2(1-\rho^2)}&\left[\rho^2\left(\frac{x-\mu_x}{\sigma_x}\right)^2 + \left(\frac{y-\mu_y}{\sigma_y}\right)^2 - 2\rho\left(\frac{x-\mu_x}{\sigma_x}\right)\left(\frac{y-\mu_y}{\sigma_y}\right)^2\right] \\ &-\frac{1}{2}\left(\frac{x-\mu_x}{\sigma_x}\right)^2. \end{aligned} \tag{5.40}$$

We can rewrite the first term (in brackets) of Equation 5.40 as a square.

$$\begin{aligned} \rho^2\left(\frac{x-\mu_x}{\sigma_x}\right)^2 &+ \left(\frac{y-\mu_y}{\sigma_y}\right)^2 - 2\rho\left(\frac{x-\mu_x}{\sigma_x}\right)\left(\frac{y-\mu_y}{\sigma_y}\right)^2 \\ &= \left[\left(\frac{y-\mu_y}{\sigma_y}\right) - \rho\left(\frac{x-\mu_x}{\sigma_x}\right)\right]^2. \end{aligned} \tag{5.41}$$

Multiplying the last result in Equation 5.41 by σ_y^2/σ_y^2 yields

$$\frac{1}{\sigma_y^2}\left[y - \mu_y - \rho\frac{\sigma_y}{\sigma_x}(x-\mu_x)\right]. \tag{5.42}$$

Hence the density function in Equation 5.35 can be rewritten as

$$\begin{aligned} f = \frac{1}{\sigma_Y\sqrt{2\pi}\sqrt{1-\rho^2}} &\exp\left\{-\frac{1}{2\sigma_y^2(1-\rho^2)}\left[y-\mu_y-\rho\frac{\sigma_y}{\sigma_x}(x-\mu_x)\right]^2\right\} \\ &\times\frac{1}{\sigma_x\sqrt{2\pi}}\exp\left[-\frac{1}{2\sigma_x}(x-\mu_x)^2\right] \\ &= f_{2|1}f_1 \end{aligned} \tag{5.43}$$

where f_1 is the density of $\mathrm{N}\left(\mu_x, \sigma_x^2\right)$ and $f_{2|1}$ is the conditional density function of $\mathrm{N}\left(\mu_y + \rho\sigma_y\sigma_x^{-1}\left(x - \mu_x\right), \sigma_Y^2\left(1 - \rho^2\right)\right)$. To complete the proof, start by taking the expectation with respect to Y.

$$\begin{aligned} f(x) &= \int_{-\infty}^{\infty} f_1 f_{2|1} dy \\ &= f_1 \int_{-\infty}^{\infty} f_{2|1} dy \\ &= f_1. \end{aligned} \tag{5.44}$$

This gives us $x \sim \mathrm{N}\left(\mu_X, \sigma_X^2\right)$. Next, we have

$$f(y|x) = \frac{f(x,y)}{f(x)} = \frac{f_{2|1} f_1}{f_1} = f_{2|1} \tag{5.45}$$

which proves the conditional relationship.

By Theorem 4.23 (Law of Iterated Means), $\mathrm{E}\left[\phi(X,Y)\right] = \mathrm{E}_X \mathrm{E}_{Y|X}\left[\phi(X,Y)\right]$ where E_X denotes the expectation with respect to X.

$$\begin{aligned} \mathrm{E}[XY] &= \mathrm{E}_X \mathrm{E}[XY|X] = \mathrm{E}_X\left[X \mathrm{E}[Y|X]\right] \\ &= \mathrm{E}_X\left[X\mu_Y + \rho\frac{\sigma_Y}{\sigma_X} X\left(X - \mu_X\right)\right] \\ &= \mu_X\mu_Y + \rho\sigma_X\sigma_Y. \end{aligned} \tag{5.46}$$

□

Also notice that if the random variables are uncorrelated then $\rho \to 0$ and by Equation 5.46 $\mathrm{E}[XY] = \mu_X\mu_Y$.

Following some of the results introduced in Chapter 4, a linear sum of normal random variables is normal, as presented in Theorem 5.5.

Theorem 5.5. *If X and Y are bivariate normal and α and β are constants, then $\alpha X + \beta Y$ is normal.*

The mean of a sample of pairwise independent normal random variables is also normally distributed with a variance of σ^2/N, as depicted in Theorem 5.6.

Theorem 5.6. *Let $\left\{\bar{X}_i\right\}$, $i = 1, 2, \ldots N$ be pairwise independent and identically distributed sample means where the original random variables are distributed $\mathrm{N}\left(\mu, \sigma^2\right)$. Then $\bar{x} = 1/N \sum_{i=1}^{N} \bar{X}_i$ is $\mathrm{N}\left(\mu, \sigma^2/N\right)$.*

And, finally, if two normally distributed random variables are uncorrelated, then they are independent.

Theorem 5.7. *If X and Y are bivariate normal and $\mathrm{Cov}(X,Y) = 0$, then X and Y are independent.*

5.4.2 Multivariate Normal Distribution

Expansion of the normal distribution function to more than two variables requires some matrix concepts introduced in Chapter 10. However, we start with a basic introduction of the multivariate normal distribution to build on the basic concepts introduced in our discussion of the bivariate distribution.

Definition 5.8. We say X is multivariate normal with mean μ and variance-covariance matrix Σ, denoted $\mathrm{N}(\mu, \Sigma)$, if its density is given by

$$f(x) = (2\pi)^{-n/2} |\Sigma|^{-1/2} \exp\left[-\frac{1}{2}(x-\mu)' \Sigma^{-1} (x-\mu)\right]. \tag{5.47}$$

Note first that $|\Sigma|$ denotes the determinant of the variance matrix (discussed in Section 10.1.1.5). For our current purposes, we simply define the determinant of the 2×2 matrix as

$$|\Sigma| = \begin{vmatrix} \sigma_{11} & \sigma_{12} \\ \sigma_{21} & \sigma_{22} \end{vmatrix} = \sigma_{11}\sigma_{22} - \sigma_{12}\sigma_{21}. \tag{5.48}$$

Given that the variance matrix is symmetric (i.e., $\sigma_{12} = \sigma_{21}$), we could write $|\Sigma| = \sigma_{11}\sigma_{22} - \sigma_{12}^2$.

The inverse of the variance matrix (i.e., Σ^{-1} in Equation 5.47) is a little more complex. We will first invert the matrix where the coefficients are unknown scalars (i.e., single numbers) by row reduction.

$$\begin{aligned}
&\begin{bmatrix} 1 & 0 \\ -\frac{\sigma_{21}}{\sigma_{11}} & 1 \end{bmatrix} \left[\begin{array}{cc|cc} \sigma_{11} & \sigma_{12} & 1 & 0 \\ \sigma_{21} & \sigma_{22} & 0 & 1 \end{array}\right] \\
&= \left[\begin{array}{cc|cc} \sigma_{11} & \sigma_{12} & 1 & 0 \\ 0 & \sigma_{22} - \frac{\sigma_{12}\sigma_{21}}{\sigma_{11}} & -\frac{\sigma_{21}}{\sigma_{11}} & 1 \end{array}\right] \\
&\begin{bmatrix} 1 & 0 \\ 0 & \frac{\sigma_{11}}{\sigma_{11}\sigma_{22} - \sigma_{12}\sigma_{21}} \end{bmatrix} \left[\begin{array}{cc|cc} \sigma_{11} & \sigma_{12} & 1 & 0 \\ 0 & \frac{\sigma_{12}\sigma_{22} - \sigma_{12}\sigma_{21}}{\sigma_{11}} & -\frac{\sigma_{21}}{\sigma_{11}} & 1 \end{array}\right] \\
&= \left[\begin{array}{cc|cc} \sigma_{11} & \sigma_{12} & 1 & 0 \\ 0 & 1 & -\frac{\sigma_{21}}{\sigma_{11}\sigma_{22} - \sigma_{12}\sigma_{21}} & \frac{\sigma_{11}}{\sigma_{11}\sigma_{22} - \sigma_{12}\sigma_{21}} \end{array}\right] \\
&\begin{bmatrix} 1 & -\frac{\sigma_{12}}{\sigma_{11}} \\ 0 & 1 \end{bmatrix} \left[\begin{array}{cc|cc} 1 & \frac{\sigma_{12}}{\sigma_{11}} & \frac{1}{\sigma_{11}} & 0 \\ 0 & 1 & -\frac{\sigma_{21}}{\sigma_{11}\sigma_{22} - \sigma_{12}\sigma_{21}} & \frac{\sigma_{11}}{\sigma_{11}\sigma_{22} - \sigma_{12}\sigma_{21}} \end{array}\right] \\
&= \left[\begin{array}{cc|cc} 1 & 0 & \frac{1}{\sigma_{11}} + \frac{\sigma_{12}\sigma_{21}}{\sigma_{11}(\sigma_{11}\sigma_{22} - \sigma_{12}\sigma_{21})} & -\frac{\sigma_{12}}{\sigma_{11}\sigma_{22} - \sigma_{12}\sigma_{21}} \\ 0 & 1 & -\frac{\sigma_{21}}{\sigma_{11}\sigma_{22} - \sigma_{12}\sigma_{21}} & \frac{\sigma_{11}}{\sigma_{11}\sigma_{22} - \sigma_{12}\sigma_{21}} \end{array}\right].
\end{aligned} \tag{5.49}$$

Focusing on the first term in the inverse,

$$\begin{aligned}\frac{1}{\sigma_{11}}+\frac{\sigma_{12}\sigma_{21}}{\sigma_{11}\left(\sigma_{11}\sigma_{22}-\sigma_{12}\sigma_{21}\right)}&=\frac{\sigma_{11}\sigma_{22}-\sigma_{12}\sigma_{21}+\sigma_{12}\sigma_{21}}{\sigma_{11}\left(\sigma_{11}\sigma_{22}-\sigma_{12}\sigma_{21}\right)}\\&=\frac{\sigma_{11}\sigma_{22}}{\sigma_{11}\left(\sigma_{11}\sigma_{22}-\sigma_{12}\sigma_{21}\right)}=\frac{\sigma_{11}}{\sigma_{11}\sigma_{22}-\sigma_{12}\sigma_{22}}.\end{aligned} \tag{5.50}$$

Next, consider inverting a matrix of matrices by row reduction following the same approach used in inverting the matrix of scalars.

$$\begin{gathered}\left[\begin{array}{cc|cc}\Sigma_{XX} & \Sigma_{XY} & I & 0\\ \Sigma_{YX} & \Sigma_{YY} & 0 & I\end{array}\right]\ \Sigma_{XX}^{-1}R_1\\[2ex]\left[\begin{array}{cc|cc}I & \Sigma_{XX}^{-1}\Sigma_{XY} & \Sigma_{XX}^{-1} & 0\\ \Sigma_{YX} & \Sigma_{YY} & 0 & I\end{array}\right]\ R_2-\Sigma_{YX}R_1\\[2ex]\left[\begin{array}{cc|cc}I & \Sigma_{XX}^{-1}\Sigma_{XY} & \Sigma_{XX}^{-1} & 0\\ 0 & \Sigma_{YY}-\Sigma_{YX}\Sigma_{XX}^{-1}\Sigma_{XY} & -\Sigma_{YX}\Sigma_{XX}^{-1} & I\end{array}\right].\end{gathered} \tag{5.51}$$

Multiplying the last row of Equation 5.51 by $\left(\Sigma_{YY}-\Sigma_{YX}\Sigma_{XX}^{-1}\Sigma_{XY}\right)^{-1}$ and then subtracting $\Sigma_{XX}^{-1}\Sigma_{XY}$ times the second row from the first row yields

$$\left[\begin{array}{c}\Sigma_{XX}^{-1}+\Sigma_{XX}^{-1}\Sigma_{XY}\left(\Sigma_{YY}-\Sigma_{YX}\Sigma_{XX}^{-1}\Sigma_{XY}\right)^{-1}\Sigma_{YX}\Sigma_{XX}^{-1}\\ -\left(\Sigma_{YY}-\Sigma_{YX}\Sigma_{XX}^{-1}\Sigma XY\right)^{-1}\Sigma_{YX}\Sigma_{XX}^{-1}\end{array}\right.$$
$$\left.\begin{array}{c}-\Sigma_{XX}^{-1}\Sigma_{XY}\left(\Sigma_{YY}-\Sigma_{YX}\Sigma_{XX}^{-1}\Sigma_{XY}\right)^{-1}\\ \left(\Sigma_{YY}-\Sigma_{YX}\Sigma_{XX}^{-1}\Sigma_{XY}\right)^{-1}\end{array}\right]. \tag{5.52}$$

As unwieldy as the result in Equation 5.52 appears, it yields several useful implications. For example, suppose we were interested in the matrix relationship

$$y=Bx\Rightarrow\begin{bmatrix}y_1\\ y_2\end{bmatrix}=\begin{bmatrix}I & B\\ 0 & I\end{bmatrix}\begin{bmatrix}x_1\\ x_2\end{bmatrix} \tag{5.53}$$

so that $y_1=x_1+Bx_2$ and $y_2=x_2$. Notice that the equation leaves open the possibility that both y_1 and y_2 are vectors and B is a matrix.

Next, assume that we want to derive the value of B so that y_1 is uncorrelated with y_2 (i.e., $\mathrm{E}\left[(y_1-\mu_1)'(y_2-\mu_2)\right]=0$). Therefore,

$$\begin{gathered}\mathrm{E}\left[(x_1+Bx_2-\mu_1-B\mu_2)'(x_2-\mu_2)\right]=0\\[1ex]\mathrm{E}\left[\{(x_1-\mu_1)+B(x_2-\mu_2)\}'(x_2-\mu_2)\right]=0\\[1ex]\mathrm{E}\left[(x_1-\mu_1)'(x_2-\mu_2)+B(x_2-\mu_2)'(x_2-\mu_2)\right]=0\\[1ex]\Sigma_{12}+B\Sigma_{22}=0\\[1ex]\Rightarrow B=-\Sigma_{12}\Sigma_{22}^{-1}\text{ and }y_1=x_1-\Sigma_{12}\Sigma_{22}^{-1}x_2.\end{gathered} \tag{5.54}$$

Not to get too far ahead of ourselves, but $\Sigma_{12}\Sigma_{22}^{-1}$ is the general form of the regression coefficient presented in Equation 4.50. Substituting the result from Equation 5.54 yields a general form for the conditional expectation and conditional variance of the normal distribution.

$$\begin{pmatrix} Y_1 \\ Y_2 \end{pmatrix} = \begin{pmatrix} I & -\Sigma_{12}\Sigma_{22}^{-1} \\ 0 & I \end{pmatrix} \begin{pmatrix} X_1 \\ X_2 \end{pmatrix}$$

$$\mathrm{E}\begin{pmatrix} Y_1 \\ Y_2 \end{pmatrix} = \begin{pmatrix} I & -\Sigma_{12}\Sigma_{22}^{-1} \\ 0 & I \end{pmatrix} \begin{pmatrix} \mu_1 \\ \mu_2 \end{pmatrix} = \begin{pmatrix} \mu_1 - \Sigma_{12}\Sigma_{22}^{-1}\mu_2 \\ \mu_2 \end{pmatrix} \quad (5.55)$$

$$V\begin{pmatrix} Y_1 \\ Y_2 \end{pmatrix} = \begin{pmatrix} \Sigma_{11} - \Sigma_{12}\Sigma_{22}^{-1}\Sigma_{21} & 0 \\ 0 & \Sigma_{22} \end{pmatrix}.$$

5.5 Chapter Summary

- The normal distribution is the foundation of a wide variety of econometric applications.
  - The Central Limit Theorem (developed in Chapter 6) depicts how the sum of random variables (and hence their sample averages) will be normally distributed regardless of the original distribution of random variables.
  - If we assume a random variable is normally distributed in small samples, we know the small sample distribution of a variety of sample statistics such as the mean, variance, and regression coefficients. As will be developed in Chapter 7, the mean and regression coefficients follow the Student's t distribution while the variance follows a chi-squared distribution.
- The multivariate normal distribution provides for the analysis of relationships between random variables within the distribution function.

5.6 Review Questions

5-1R. Derive the general form of the normal variable from the standard normal using a change in variables technique.

5-2R. In Theorem 5.4, rewrite the conditional expectation in terms of the regression β (i.e., $\beta = \mathrm{Cov}[XY]/\mathrm{V}[X]$).

5-3R. Prove the variance portion of Theorem 5.6.

5.7 Numerical Exercises

5-1E. Derive the absolute approximation error for 6, 9, 12, and 15. Draw binomial variables and compute the percent absolute deviation between the binomial variable and a standard normal random variable. Does this absolute deviation decline?

5-2E. Given that a random variable is distributed normally with a mean of 5 and a variance of 6, what is the probability that the outcome will be less than zero?

5-3E. Construct a histogram for 10 sums of 10 Bernoulli draws (normalized by subtracting their means and dividing by their theoretical standard deviations). Compare this histogram with a histogram of 20 sums of 10 Bernoulli draws normalized in the same manner. Compare each histogram with the standard normal distribution. Does the large sample approach the standard normal?

5-4E. Compute the sample covariance matrix for the farm interest rate for Alabama and the Baa Corporate bond rate using the data presented in Appendix D. What is the correlation coefficient between these two series?

5-4E. Appendix D presents the interest rate and the change in debt to asset ratio in the southeastern United States for 1960 through 2003 as well as the interest rate on Baa Corporate bonds from the St. Louis Federal Reserve Bank. The covariance matrix for interest rates in Florida, the change in debt to asset ratio for Florida, and the Baa Corporate bond rate is

$$S = \begin{bmatrix} 0.0002466 & -0.0002140 & 0.00031270 \\ -0.0002140 & 0.0032195 & -0.00014738 \\ 0.0003127 & -0.00014738 & 0.00065745 \end{bmatrix}. \tag{5.56}$$

Compute the projected variance of the interest rate for Florida conditional on Florida's change in debt to asset ratio and the Baa Corporate bond rate.

Part II

Estimation

6

Large Sample Theory

CONTENTS

In this chapter we want to develop the foundations of sample theory. First assume that we want to make an inference, either estimation or some test, based on a sample. We are interested in how well parameters or statistics based on that sample represent the parameters or statistics of the whole population.

6.1 Convergence of Statistics

In statistical terms, we want to develop the concept of convergence. Specifically, we are interested in whether or not the statistics calculated on the sample converge toward the population values. Let $\{X_n\}$ be a sequence of samples. We want to demonstrate that statistics based on $\{X_n\}$ converge toward the population statistics for X.

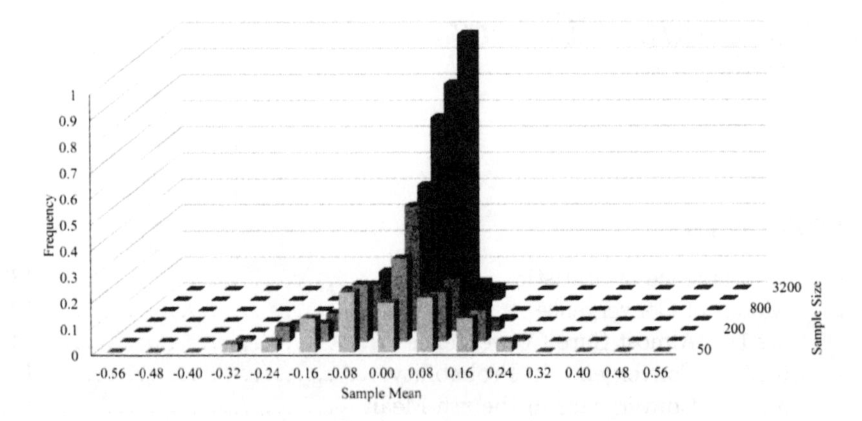

FIGURE 6.1
Probability Density Function for the Sample Mean.

As a starting point, consider a simple estimator of the first four central moments of the standard normal distribution.

$$\begin{aligned}
\hat{M}_1(x) &= \frac{1}{N}\sum_{i=1}^{N} x_i \\
\hat{M}_2(x) &= \frac{1}{N}\sum_{i=1}^{N} \left(x_i - \hat{M}_1(x)\right)^2 \\
\hat{M}_3(x) &= \frac{1}{N}\sum_{i=1}^{N} \left(x_i - \hat{M}_1(x)\right)^3 \\
\hat{M}_4(x) &= \frac{1}{N}\sum_{i=1}^{N} \left(x_i - \hat{M}_1(x)\right)^4 .
\end{aligned} \tag{6.1}$$

Given the expression for the standard normal distribution developed in Chapter 5, the theoretical first four moments are $\mu_1(x) = 0$, $\mu_2(x) = 1$, $\mu_3(x) = 0$, and $\mu_4(x) = 3$. Figure 6.1 presents the empirical probability density functions for the first empirical moment (the sample mean) defined in Equation 6.1 for samples sizes of 50, 100, 200, 400, 800, 1600, and 3200. As depicted in Figure 6.1, the probability density function of the sample mean concentrates (or converges) around its theoretical value as the sample size increases.

Another way of looking at the concept of convergence involves examining the quantiles of the distribution of the sample moments. The quantile is a measure that asks what level of random variable X yields a probability of a value less than p. Mathematically, the quantile is defined as

$$x^*(p) \Rightarrow \int_{-\infty}^{x^*} f(z)\,dz \tag{6.2}$$

where $f(z)$ is a valid density function. If $p = \{0, 0.25, 0.50, 0.75, 1.0\}$, this measure defines the quartiles of the distribution. Table 6.1 presents the

TABLE 6.1
Quartiles for Sample Moments of the Standard Normal

	Quartiles					Interquartile Range	
	0%	25%	50%	75%	100%	Raw	Normalized
			Mean – First Moment				
50	−0.3552	−0.0979	−0.0128	0.0843	0.2630	0.1821	
100	−0.3485	−0.0586	0.0012	0.0622	0.2028	0.1208	
200	−0.2264	−0.0510	−0.0028	0.0377	0.1412	0.0888	
400	−0.1330	−0.0323	0.0010	0.0439	0.1283	0.0762	
800	−0.0991	−0.0172	0.0052	0.0259	0.0701	0.0431	
1600	−0.0622	−0.0220	0.0029	0.0213	0.0620	0.0433	
3200	−0.0425	−0.0164	−0.0003	0.0110	0.0397	0.0274	
			Variance – Second Central Moment				
50	0.5105	0.8676	0.9481	1.0918	1.5562	0.2242	
100	0.7115	0.8947	0.9862	1.0927	1.3715	0.1980	
200	0.8024	0.9248	0.9971	1.0857	1.3140	0.1609	
400	0.8695	0.9491	0.9966	1.0419	1.2266	0.0928	
800	0.9160	0.9540	0.9943	1.0287	1.1443	0.0747	
1600	0.9321	0.9771	1.0047	1.0267	1.0933	0.0495	
3200	0.9414	0.9856	0.9978	1.0169	1.0674	0.0313	
			Skewness – Third Central Moment				
50	−0.7295	−0.1412	−0.0333	0.1582	0.8134	0.2994	
100	−0.6869	−0.1674	−0.0008	0.1646	0.7121	0.3320	
200	−0.4252	−0.0968	−0.0116	0.1392	0.4867	0.2361	
400	−0.3163	−0.0797	−0.0036	0.0703	0.3953	0.1500	
800	−0.2185	−0.0601	−0.0133	0.0605	0.1841	0.1205	
1600	−0.1397	−0.0428	−0.0011	0.0448	0.1356	0.0877	
3200	−0.1288	−0.0208	0.0058	0.0285	0.1068	0.0493	
			Kurtosis – Fourth Central Moment				
50	0.7746	2.0197	2.6229	3.3221	7.9826	1.3025	0.4342
100	1.3950	2.1279	2.7092	3.4366	6.1016	1.3086	0.4362
200	1.5611	2.4127	2.8683	3.5025	5.0498	1.0898	0.3633
400	1.9865	2.6823	2.8981	3.3730	4.7347	0.6907	0.2302
800	2.4249	2.7026	3.0089	3.2071	3.9311	0.5045	0.1682
1600	2.5575	2.8189	2.9881	3.2241	4.0558	0.4052	0.1351
3200	2.6229	2.9042	3.0351	3.1240	3.6167	0.2198	0.0733

quartiles for each sample statistics and sample size. Next, consider the interquartile range defined as

$$\mathrm{R}(p) = x^{*}(0.75) - x^{*}(0.25). \tag{6.3}$$

Examining the quartile results in Table 6.1, notice that the true value of the moment is contained in the quartile range. In addition, the quartile range declines as the sample size increases. Notice that the quartile range for the fourth moment is normalized by its theoretical value. However, even with this adjustment, the values in Table 6.1 indicate that higher order moments converge less rapidily than lower order moments (i.e., kurtosis converges more slowly than the mean).

Taking a slightly different tack, the classical assumptions for ordinary least squares (OLS) as presented in White [53] are identified by Theorem 6.1.

Theorem 6.1. *The following are the assumptions of the classical linear model.*

(i) The model is known to be $y = X\beta + \epsilon$, $\beta < \infty$.

(ii) X *is a nonstochastic and finite* $n \times k$ *matrix.*

(iii) $X'X$ *is nonsingular for all* $n \geq k$.

(iv) $\mathrm{E}(\epsilon) = 0$.

(v) $\epsilon \sim \mathrm{N}(0, \sigma_0^2 I)$, $\sigma_0^2 < \infty$.

Given these assumptions, we can conclude that

a) *Existence* given (i) – (iii) β_n exists for all $n \geq k$ and is unique.

b) *Unibiasedness* given (i) – (v) $\mathrm{E}[\beta_n] = \beta_0$.

c) *Normality* given (i) – (v) $\beta_n \sim \mathrm{N}\left(\beta_0, \sigma^2 (X'X)^{-1}\right)$.

d) *Efficiency* given (i) – (v) β_n is the maximum likelihood estimator and the best unbiased estimator in the sense that the variance of any other unbiased estimator exceeds that of β_n by a positive semi-definite matrix regardless of the value of β_0.

Existence, unbiasedness, normality, and efficiency are small sample analogs of asymptotic theory. Unbiased implies that the distribution of β_n is centered around β_0. Normality allows us to construct t-distribution or F-distribution tests for restrictions. Efficiency guarantees that the ordinary least squares estimates have the greatest possible precision.

Asymptotic theory involves the behavior of the estimator under the failure of certain assumptions – specifically, assumptions (ii) or (v). The possible failure of assumption (ii) depends on the ability of the econometrician to control the sample. Specifically, the question is whether to control the experiment by designing or selecting the levels of exogenous variables. The alternative is the conjecture that the econometrician observed a **sample of convenience**. In other words, the economy generated the sample through its normal operation. Under this scenario, the X matrix was in fact stochastic or random and the researcher simply observes one possible outcome.

Failure of assumption (v) also pervades economic applications. Specifically, the fact that residual or error is normally distributed may be one of the primary testable hypotheses in the study. For example, Moss and Shonkwiler [33] found that corn yields were non-normally distributed. Further, the fact that the yields were non-normally distributed was an important finding of the study because it affected the pricing of crop insurance contracts.

The potential non-normality of the error term is important for the classical linear model because normality of the error term is required to strictly apply t-distributions or F-distributions. However, the central limit theorem can be used if n is large enough to guarantee that β_n is approximately normal.

Given that the data collected by the researcher conforms to the classical linear model, the estimated coefficients are unbiased and hypothesis tests on the results are correct (i.e., the parameters are distributed t and linear combinations of the paramters are distributed F). However, if the data fails to meet the classical assumptions, the estimated parameters converge to their true value as the sample size becomes large. In addition, we are interested in modifying the statistical test of significance for the model's parameters.

6.2 Modes of Convergence

As a starting point, consider a simple mathematical model of convergence for a nonrandom variable. Most students have seen the basic proof that

$$\lim_{x \to \infty} \frac{1}{x} = 0 \tag{6.4}$$

which basically implies that as x becomes infinitely large, the function value $f(x) = 1/x$ can be made arbitrarily close to zero. To put a little more mathematical rigor on the concept, let us define δ and ϵ such that

$$f(x+\delta) = \left| \frac{1}{x+\delta} - 0 \right| \leq \epsilon. \tag{6.5}$$

Convergence implies that for any ϵ there exists a δ that meets the criteria in Equation 6.5. The basic concept in Equation 6.5 is typically used in calculus courses to define derivatives. In this textbook we are interested in a slightly different formulation – the limit of a sequence of numbers as defined in Definition 6.2.

Definition 6.2. A sequence of real numbers $\{\alpha_n\}$, $n = 1, 2, \cdots$ is said to converge to a real number α if for any $\epsilon > 0$ there exists an integer N such that for all $n > N$ we have

$$|\alpha_n - \alpha| < \epsilon. \tag{6.6}$$

This convergence is expressed $\alpha_n \to \alpha$ as $n \to \infty$ or $\lim_{n\to\infty} \alpha_n = \alpha$.

This definition must be changed for random variables because we cannot require a random variable to approach a specific value. Instead, we require the probability of the variable to approach a given value. Specifically, we want the probability of the event to equal 1 or zero as n goes to infinity. This concept defines three different concepts or modes of convergence. First, the **convergence in probability** implies that the distance between the sample value and the true value (i.e., the absolute difference) can be made small.

Definition 6.3 (Convergence in Probability). A sequence of random variables $\{X_n\}$, $n = 1, 2, \cdots$ is said to converge to a random variable X in probability if for any $\epsilon > 0$ and $\delta > 0$ there exists an integer N such that for all $n > N$ we have $\mathrm{P}(|X_n - X| < \epsilon) > 1 - \delta$. We write

$$X_n \xrightarrow{\mathrm{P}} X \tag{6.7}$$

$\text{plim}_{n\to\infty} X_n = X$. The last equality reads – the probability limit of X_n is X. (Alternatively, the clause may be paraphrased as $\lim \mathrm{P}(|X_n - X| < \epsilon) = 1$ for any $\epsilon > 0$).

Intuitively, this convergence in probability is demonstrated in Table 6.1. As the sample size expands, the quartile range declines and $|X_n - X|$ becomes small.

A slightly different mode of convergence is **convergence in mean square.**

Definition 6.4 (Convergence in Mean Square). A sequence $\{X_n\}$ is said to *converge* to X in *mean square* if $\lim_{n\to\infty} \mathrm{E}(X_n - X)^2 = 0$. We write

$$X_n \xrightarrow{\mathrm{M}} X. \tag{6.8}$$

Table 6.2 presents the expected mean squared errors for the sample statistics for the standard normal distribution. The results indicate that the sample statistics converge in mean squared error to their theoretical values – that is, the mean squared error becomes progressively smaller as the sample size increases.

A final type of convergence, **convergence in distribution**, implies that the whole distribution of X_n approaches the distribution function of X.

Definition 6.5 (Convergence in Distribution). A sequence $\{X_n\}$ is said to *converge* to X *in distribution* if the distribution function F_n of X_n converges

TABLE 6.2
Convergence in Mean Square for Standard Normal

Statistic	Sample Size						
	50	100	200	400	800	1600	3200
Mean	0.03410	0.01148	0.00307	0.00073	0.00018	0.00005	0.00001
Variance	0.07681	0.01933	0.00535	0.00130	0.00029	0.00007	0.00002
Skewness	0.17569	0.06556	0.01614	0.00379	0.00092	0.00021	0.00006
Kurtosis	3.40879	0.98780	0.25063	0.06972	0.01580	0.00506	0.00110

to the distribution function F of X at every continuity point of F. We write

$$X_n \xrightarrow{\text{d}} X \tag{6.9}$$

and call F the limit distribution of $\{X_n\}$. If $\{X_n\}$ and $\{Y_n\}$ have the same limit distribution, we write

$$X_n \overset{\text{LD}}{=} Y_n. \tag{6.10}$$

Figures 5.2 and 5.3 demonstrate convergence in probability. Figure 5.2 demonstrates that the normalized binomial converges to the standard normal, while Figure 5.3 shows that the normalized sum of uniform random variables converges in probability to the standard normal.

The differences in types of convergence are related. Comparing Definition 6.3 with Definition 6.4, $\text{E}\left(X_n - X\right)^2$ will tend to zero faster than $|X_n - X|$. Basically, the squared convergence is faster than the linear convergence implied by the absolute value. The result is Chebyshev's theorem.

Theorem 6.6 (Chebyshev).

$$X_n \xrightarrow{\text{M}} X \Rightarrow X_n \xrightarrow{\text{P}} X. \tag{6.11}$$

Next, following Definition 6.5, assume that $f_n(x) \to f(x)$, or that the sample distribution converges to a limiting distribution.

$$\begin{aligned} &\int_{-\infty}^{\infty} (z - \bar{z})^k f_n(z)\, dz - \int_{-\infty}^{\infty} (z - \bar{z})^k f(z)\, dz \\ &\int_{-\infty}^{\infty} (z - \bar{z})^k \left[f_n(z) - f(z)\right] dz = 0. \end{aligned} \tag{6.12}$$

Therefore convergence in distribution implies convergence in mean square.

Theorem 6.7.

$$X_n \xrightarrow{\text{P}} X \Rightarrow X_n \xrightarrow{\text{d}} X. \tag{6.13}$$

The convergence results give rise to **Chebyshev's inequality,**

$$P\left[g\left(X_n\right) \geq \epsilon^2\right] \leq \frac{E\left[g\left(X_n\right)\right]}{\epsilon^2} \tag{6.14}$$

which can be used to make probabilistic statements about a variety of statistics. For example, replacing $g\left(X_n\right)$ with the sample mean yields a form of the confidence interval for the sample mean. Extending the probability statements to general functions yields

Theorem 6.8. *Let X_n be a vector of random variables with a fixed finite number of elements. Let g be a function continuous at a constant vector point α. Then*

$$X_n \xrightarrow{P} \alpha \Rightarrow g\left(X_n\right) \xrightarrow{P} g\left(\alpha\right). \tag{6.15}$$

Similar results are depicted in **Slutsky's theorem.**

Theorem 6.9 (Slutsky). *If $X_n \xrightarrow{d} X$ and $Y_n \xrightarrow{d} \alpha$, then*

$$\begin{gathered} X_n + Y_n \xrightarrow{d} X + \alpha \\ X_n Y_n \xrightarrow{d} \alpha X \\ \left(\frac{X_n}{Y_n}\right) \xrightarrow{d} \frac{X}{\alpha} \text{ if } \alpha \neq 0. \end{gathered} \tag{6.16}$$

6.2.1 Almost Sure Convergence

Let ω represent the entire random sequence $\{Z_t\}$. As before, our interest typically centers around the averages of this sequence.

$$b_n\left(\omega\right) = \frac{1}{n}\sum_{t=1}^{n} Z_t. \tag{6.17}$$

Definition 6.10. Let $\{b_n\left(\omega\right)\}$ be a sequence of real-valued random variables. We say that $b_n\left(\omega\right)$ converges *almost surely* to b, written

$$b_n\left(\omega\right) \xrightarrow{a.s.} b \tag{6.18}$$

if and only if there exists a real number b such that

$$P\left[\omega : b_n\left(\omega\right) \rightarrow b\right] = 1. \tag{6.19}$$

The probability measure P describes the distribution of ω and determines the joint distribution function for the entire sequence $\{Z_t\}$. Other common terminology is that $b_n\left(\omega\right)$ converges to b with probability 1 or that $b_n\left(\omega\right)$ is strongly consistent for b.

Example 6.11. Let

$$\bar{Z}_n = \frac{1}{n}\sum_{i=1}^{n} Z_t \tag{6.20}$$

where $\{Z_t\}$ is a sequence of independently and identically distributed (i.i.d.) random variables with $\mathrm{E}[Z_t] = \mu < \infty$. Then

$$\bar{Z}_n \xrightarrow{\text{a.s.}} \mu \tag{6.21}$$

by Kolmogorov's strong law of large numbers, Proposition 6.22 [8, p. 3].

Proposition 6.12. *Given $g : R^k \to R^l$ $(k, l < \infty)$ and any sequence $\{b_n\}$ such that*

$$b_n \xrightarrow{\text{a.s.}} b \tag{6.22}$$

where b_n and b are $k \times 1$ vectors, if g is continuous at b, then

$$g(b_n) \xrightarrow{\text{a.s.}} g(b). \tag{6.23}$$

This result then allows us to extend our results to include the matrix operations used to define the ordinary least squares estimators.

Theorem 6.13. *Suppose*

a) $y = X\beta_0 + \epsilon$;

b) $\frac{X'\epsilon}{n} \xrightarrow{\text{a.s.}} 0$;

c) $\frac{X'X}{n} \xrightarrow{\text{a.s.}} M$ *is finite and positive definite.*

Then β_n exists a.s. *for all n sufficiently large, and $\beta_n \xrightarrow{\text{a.s.}} \beta_0$.*

Proof. Since $X'X/n \xrightarrow{\text{a.s.}} M$, it follows from Proposition 2.11 that $\det(X'X/n) \xrightarrow{\text{a.s.}} \det(M)$. Because M is positive definite by (c), $\det(M) > 0$. It follows that $\det(X'X/n) > 0$ a.s. for all n sufficiently large, so $(X'X)^{-1}$ exists for all n sufficiently large. Hence

$$\hat{\beta}_n \equiv \left(\frac{X'X}{n}\right)^{-1} \frac{X'y}{n} \tag{6.24}$$

exists for all n sufficiently large. In addition,

$$\hat{\beta}_n = \beta_0 + \left(\frac{X'X}{n}\right)^{-1} \frac{X'\epsilon}{n}. \tag{6.25}$$

It follows from Proposition 6.12 that

$$\hat{\beta}_n \xrightarrow{\text{a.s.}} \beta_0 + M^{-1}0 = \beta_0. \tag{6.26}$$

□

The point of Theorem 6.13 is that the ordinary least squares estimator $\hat{\beta}_n$ converges almost surely to the true value of the parameters β. The concept is very powerful. Using the small sample properties, we are sure that the ordinary least squares estimator is unbiased. However, what if we cannot guarantee the small sample properties? What if the error is not normal, or what if the values of the independent variables are random? Given Theorem 6.13, we know that the estimator still converges almost surely to the true value.

6.2.2 Convergence in Probability

A weaker stochastic convergence concept is that of convergence in probability.

Definition 6.14. Let $\{b_n(\omega)\}$ be a sequence of real-valued random variables. If there exists a real number b such that for every $\epsilon > 0$,

$$P[\omega : |b_n(\omega) - b| < \epsilon] \to 1 \tag{6.27}$$

as $n \to \infty$, then $b_n(\omega)$ converges in probability to b.

The almost sure measure of probability takes into account the joint distribution of the entire sequence $\{Z_t\}$, but with convergence in probability, we only need to be concerned with the joint distribution of those elements that appear in $b_n(\omega)$. Convergence in probability is also referred to as *weak consistency*.

Theorem 6.15. *Let $\{b_n(\omega)\}$ be a sequence of random variables. If*

$$b_n \xrightarrow{a.s.} b \text{ , then } b_n \xrightarrow{P} b. \tag{6.28}$$

If b_n converges in probability to b, then there exists a subsequence $\{b_{n_j}\}$ such that

$$b_{n_j} \xrightarrow{a.s.} b. \tag{6.29}$$

Given that we know Theorem 6.13, what is the point of Theorem 6.15? Theorem 6.13 states that the ordinary least squares estimator converges almost surely to the true value. Theorem 6.15 states that given Theorem 6.13, then b_n converges in probability to b. The point is that we can make fewer assumptions to guarantee Theorem 6.15. Making fewer assumptions is always preferred to making more assumptions – the results are more robust.

6.2.3 Convergence in the rth Mean

Earlier in this chapter we developed the concept of mean squared convergence. Generalizing the concept we consider the rth power of the expectation yielding

Definition 6.16. Let $\{b_n(\omega)\}$ be a sequence of real-valued random variables. If there exists a real number b such that

$$E[|b_n(\omega) - b|^r] \to 0 \tag{6.30}$$

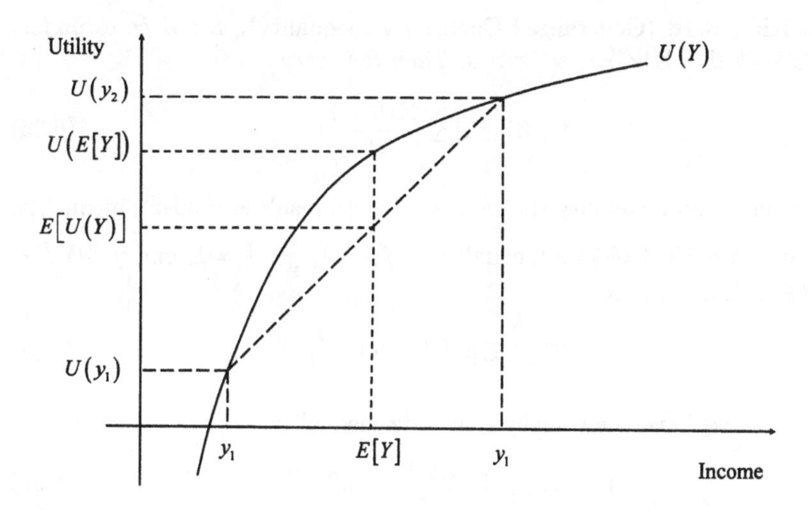

FIGURE 6.2
Expected Utility.

as $n \to \infty$ for some $r > 0$, then $b_n(\omega)$ *converges in the* rth *mean* to b, written as

$$b_n(\omega) \xrightarrow{\text{r.m.}} b. \tag{6.31}$$

Next, we consider a proposition that has applications for both estimation and economic applications – Jensen's inequality. Following the development of expected utility in Moss [32, pp. 57–82], economic decision makers choose among alternatives to maximize their expected utility. As depicted in Figure 6.2, the utility function for risk averse decision makers is concave in income. Hence, the expected utility is less than the utility at the expected level of income.

$$\mathrm{E}[U(Y)] = pU(y_1) + (1-p)U(y_2) < U[p \times y_1 + (1-p) \times y_2] = U[\mathrm{E}(Y)]. \tag{6.32}$$

Hence, risk averse decision makers are worse off under risk. Jensen's inequality provides a general statement of this concept.

Proposition 6.17 (Jensen's inequality). *Let $g : R^1 \to R^1$ be a convex function on the interval $B \subset R^1$ and Z be a random variable such that $P[Z \in B] = 1$. Then $g[\mathrm{E}(Z)] \leq \mathrm{E}[g(Z)]$. If g is concave on B, then $g[\mathrm{E}(Z)] \geq \mathrm{E}[g(Z)]$.*

In addition, the development of the convergence of the rth mean allows for the statement of a generalized version of Chebychev's inequality presented in Equation 6.14.

Proposition 6.18 (Generalized Chebyshev Inequality). *Let Z be a random variable such that $\mathrm{E}|Z|^r < \infty$, $r > 0$. Then for every $\epsilon > 0$*

$$\mathrm{P}[|Z| \geq \epsilon] \leq \frac{\mathrm{E}(|Z|^r)}{\epsilon^r}. \tag{6.33}$$

Another useful inequality that follows the rth result is Holder's inequality.

Proposition 6.19 (Holder's Inequality). *If $p > 1$, $\frac{1}{p} + \frac{1}{q} = 1$, and if $\mathrm{E}|Y|^p < \infty$ and $\mathrm{E}|Z|^q < \infty$, then*

$$\mathrm{E}|YX| \leq [\mathrm{E}|y|^p]^{1/p} [\mathrm{E}|X|^q]^{1/q}. \tag{6.34}$$

If $p = q = 2$, we have the Cauchy–Schwartz inequality

$$\mathrm{E}[|YZ|] \leq \mathrm{E}\left[Y^2\right]^{1/2} \mathrm{E}\left[Z^2\right]^{1/2} \tag{6.35}$$

The rth convergence is also useful in demonstrating the ordering of convergence presented in Theorem 6.6.

Theorem 6.20. *If $b_n(\omega) \xrightarrow{\text{r.m.}} b$ for some $r > 0$, then $b_n(\omega) \xrightarrow{\text{P}} b$.*

The real point to Theorem 6.21 is that we will select the estimator that minimizes the mean squared error (i.e., $r = 2$ in the root mean squared error). Hence, Theorem 6.21 states that an estimator that converges in mean squared error also converges in probability.

6.3 Laws of Large Numbers

Given the above convergence results, we can show that as the size of the sample increases, the sample statistic converges to the underlying population statistic. Taking our initial problem as an example,

$$\begin{aligned} \lim_{n\to\infty} \hat{M}_1(x_n) = \mu_1(x) &\Leftarrow \lim_{n\to\infty} M_1(x_n) - \mu_1(x) = 0 \\ \lim_{n\to\infty} \hat{M}_2(x_n) = \mu_2(x) &\Leftarrow \lim_{n\to\infty} M_2(x_n) - \mu_2(x) = 0. \end{aligned} \tag{6.36}$$

Proposition 6.21. *Given restrictions on the dependence, heterogeneity, and moments of a sequence of random variables $\{X_t\}$,*

$$\bar{X}_n - \bar{\mu}_n \xrightarrow{\text{a.s.}} 0 \tag{6.37}$$

where

$$\bar{X}_n = \frac{1}{n}\sum_{i=1}^{n} X_t \text{ and } \bar{\mu}_n = \mathrm{E}\left(\bar{X}_n\right). \tag{6.38}$$

Specifically, if we can assume that the random variables in the sample (X_t) are independently and identically distributed then

Theorem 6.22 (Kolmogorov). *Let $\{X_t\}$ be a sequence of indepedently and identically distributed (i.i.d.) random variables. Then*

$$\bar{X}_n \xrightarrow{\text{a.s.}} \mu \tag{6.39}$$

if and only if $\mathrm{E}|X_t| < \infty$ *and* $\mathrm{E}(X_t) = \mu$.

Thus, the sample mean converges almost surely to the population mean. In addition, letting $\{X_i\}$ be independent and identically distributed with $\mathrm{E}[X_i] = \mu$,

$$\bar{X}_n \xrightarrow{\text{P}} \mu. \tag{6.40}$$

This result is known as Khintchine's law of large numbers [8, p. 2].

6.4 Asymptotic Normality

Under the traditional assumptions of the linear model (fixed regressors and normally distributed error terms), β_n is distributed multivariate normal with

$$\begin{aligned} \mathrm{E}\left[\hat{\beta}_n\right] &= \beta_0 \\ V\left[\hat{\beta}_n\right] &= \sigma_0^2 (X'X)^{-1} \end{aligned} \tag{6.41}$$

for any sample size n. However, when the sample size becomes large the distribution of β_n is approximately normal under some general conditions.

Definition 6.23. Let $\{b_n\}$ be a sequence of random finite-dimensional vectors with joint distribution functions $\{F_n\}$. If $F_n(z) \to F(z)$ as $n \to \infty$ for every continuity point z, where F is the distribution function of a random variable Z, then b_n *converges in distribution* to the random variable Z, denoted

$$b_n \xrightarrow{\text{d}} Z. \tag{6.42}$$

Intuitively, the distribution of b_n becomes closer and closer to the distribution function of the random variable Z. Hence, the distribution F can be used as an approximation of the distribution function of b_n. Other ways of stating this concept are that b_n *converges in law* to Z.

$$b_n \xrightarrow{\text{L}} Z. \tag{6.43}$$

Or, b_n is *asymptotically distributed* as F

$$b_n \overset{A}{\sim} F. \tag{6.44}$$

In this case, F is called the *limiting distribution* of b_n.

Example 6.24. Let $\{X_t\}$ be an i.i.d. sequence of random variables with mean μ and variance $\sigma^2 < \infty$. Define

$$b_n = \frac{\bar{X}_n - \mathrm{E}\left[\bar{X}_n\right]}{\left(V\left[\bar{X}_n\right]\right)^{1/2}} = \left(\frac{1}{n}\right)^{1/2} \sum_{t=1}^{n} \frac{(X_t - \mu)}{\sigma}. \tag{6.45}$$

Then by the Lindeberg–Levy central limit theorem (Theorem 6.24),

$$b_n \overset{A}{\sim} \mathrm{N}(0,1). \tag{6.46}$$

Theorem 6.25 (Lindeberg–Levy). *Let $\{X_i\}$ be i.i.d. with $\mathrm{E}[X_i] = \mu$ and $\mathrm{V}(X_i) = \sigma^2$. Then, defining Z_n as above, $Z_n \to \mathrm{N}(0,1)$.*

In this textbook, we will justify the Lindeberg–Levy theory using a general **characteristic function** for a sequence of random variables. Our demonstration stops a little short of a formal proof but contains the essential points necessary to justify the result. The characteristic function of a random variable X is defined as

$$\begin{aligned} \phi_X(t) &= \mathrm{E}\left[e^{itX}\right] = \mathrm{E}\left[\cos(tX) + i\sin(tX)\right] \\ &= \mathrm{E}\left[\cos(tX)\right] + i\mathrm{E}\left[\sin(tX)\right]. \end{aligned} \tag{6.47}$$

This function may appear intimidating, but recalling our development of the normal distribution in Section 3.7, we can rewrite the function $f(x) = (x-5)^2/5$ as

$$\phi(\theta) = r(\theta)\cos(\theta) \tag{6.48}$$

where the imaginary term is equal to zero. Hence, we start by defining the characteristic function of a random variable in Definition 6.26.

Definition 6.26. Let Z be a $k \times 1$ random vector with distribution function F. The characteristic function of F is defined as

$$f(\lambda) = \mathrm{E}\left[\exp(i\lambda' Z)\right] \tag{6.49}$$

where $i^2 = -1$ and λ is a $k \times 1$ real vector.

Notice the similarity between the definition of the moment generating function and the characteristic function.

$$\begin{aligned} M_X(t) &= \mathrm{E}\left[\exp(tx)\right] \\ f(\lambda) &= \mathrm{E}\left[\exp(i\lambda' z)\right]. \end{aligned} \tag{6.50}$$

In fact, a weaker form of the central limit theorem can be demonstrated using the moment generating function instead of the characteristic function.

Next, we define the characteristic function for the standard normal distribution as

Definition 6.27. Let $Z \sim \mathrm{N}\left(\mu, \sigma^2\right)$. Then

$$f(\lambda) = \exp\left(i\lambda\mu - \frac{\lambda^2\sigma^2}{2}\right). \tag{6.51}$$

It is important that the characteristic function is unique given the density function for any random variable. Two random variables with the same characteristic function also have the same density function.

Theorem 6.28 (Uniqueness Theorem). *Two distribution functions are identical if and only if their characteristic functions are identical.*

This result also holds for moment generating functions. Hence, the point of the Lindeberg–Levy proof is to demonstrate that the characteristic function for Z_t (the standardized mean) approaches the distribution function for the standard normal distribution as the sample size becomes large.

Lindeberg–Levy. First define $f(\lambda)$ as the characteristic function for $(Z_t - \mu)/\sigma$ and let $f_n(\lambda)$ be the characteristic function of

$$\frac{\sqrt{n}\left(\bar{Z}_n - \bar{\mu}_n\right)}{\bar{\sigma}_n} = \left(\frac{1}{n}\right)^{1/2} \sum_{t=1}^{n} \left(\frac{Z_t - \mu}{\sigma}\right). \tag{6.52}$$

By the structure of the characteristic function we have

$$\begin{aligned} f_n(\lambda) &= f\left(\frac{\lambda}{\sigma\sqrt{n}}\right) \\ \ln\left(f_n(\lambda)\right) &= n \ln\left(f\left(\frac{\lambda}{\sigma\sqrt{n}}\right)\right). \end{aligned} \tag{6.53}$$

Taking a second order Taylor series expansion of $f(\lambda)$ around $\lambda = 0$ gives

$$f(\lambda) = 1 - \frac{\sigma^2\lambda^2}{2} + o\left(\lambda^2\right). \tag{6.54}$$

Thus,

$$\ln\left(f_n(\lambda)\right) = n \ln\left[1 - \frac{\lambda^2}{2n} + o\left(\frac{\lambda^2}{n}\right)\right] \to \frac{\lambda^2}{2} \text{ as } n \to \infty. \tag{6.55}$$

□

Another proof of the central limit theorem involves taking a Taylor series expansion of the characteristic function around the point $t = 0$, yielding

$$\begin{gathered}\phi_z(t) = \phi_z(0) + \frac{1}{1!}\phi_z'(0)\,t + \frac{1}{2!}\phi_z''(0)\,t^2 + o\left(t^2\right) \\ \text{s.t.} Z = \frac{X-\mu}{\sigma\sqrt{n}}.\end{gathered} \tag{6.56}$$

To work on this expression we note that

$$\phi_X(0) = 1 \tag{6.57}$$

for any random variable X, and

$$\phi_X^{(k)}(0) = i^k \mathrm{E}\left(X^k\right). \tag{6.58}$$

Putting these two results into the second-order Taylor series expansion,

$$\begin{gathered}1 + \frac{\mathrm{E}(Z)}{i}t + \frac{\mathrm{E}\left(Z^2\right)}{i^2}\frac{t^2}{2} + o\left(t^2\right) = 1 - \frac{t^2}{2} + o\left(t^2\right) \\ \ni: \mathrm{E}(Z) = 0,\ \mathrm{E}\left(Z^2\right) = 1.\end{gathered} \tag{6.59}$$

Thus,

$$\begin{aligned}\phi_z(t) &= \phi_z(0) + \frac{1}{1!}\phi_z'(0)\,t + \frac{1}{2!}\phi_z''(0)\,t^2 + o\left(t^2\right) \\ &= 1 + \frac{\mathrm{E}(Z)}{i}t + \frac{\mathrm{E}(Z)}{i^2}\frac{t^2}{2} + o\left(t^2\right)\end{aligned} \tag{6.60}$$

$$= 1 - \frac{t^2}{2} + o\left(t^2\right) \Leftrightarrow 1 - \frac{t^2}{2} = \mathrm{E}\left(e^{iy}\right) \ni: y \sim \mathrm{N}(0,1)$$

or Z is normally distributed. This approach can be used to prove the central limit theorem from the moment generating function.

For completeness, there are other characteristic functions that the limit does not approach. The characteristic function of the uniform distribution function is

$$\phi_X(t) = e^{it} - 1. \tag{6.61}$$

The gamma distribution's characteristic function is

$$\phi_X(t) = \frac{1}{\left(1 - \dfrac{it}{\alpha}\right)^r}. \tag{6.62}$$

Thus, our development of the Lindeberg–Levy theorem does not assume the result. The fact that the sample means converge to the true population mean and the variance of the means converges to a fixed value drives the result. Essentially, only the first two moments matter asymptotically.

6.5 Wrapping Up Loose Ends

Finally, we can use some of the implications of convergence to infer results about the correlation coefficient between any two random variables, bound the general probabilities, and examine the convergence of the binomial distribution to the normal.

6.5.1 Application of Holder's Inequality

Using Holder's inequality it is possible to place a general bound on the correlation coefficient regardless of the distribution.

Example 6.29. If X and Y have means μ_X, μ_Y and variances σ_X^2, σ_Y^2, respectively, we can apply the Cauchy–Schwartz Inequality (Holder's inequality with $p = q = 1/2$) to get

$$\mathrm{E}\left|(X-\mu_X)(Y-\mu_Y)\right| \leq \left\{\mathrm{E}\left[(X-\mu_x)^2\right]\right\}^{1/2}\left\{\mathrm{E}\left[(Y-\mu_y)^2\right]\right\}^{1/2}. \tag{6.63}$$

Squaring both sides and substituting for variances and covariances yields

$$(\mathrm{Cov}(X,Y))^2 \leq \sigma_X^2\sigma_Y^2 \tag{6.64}$$

which implies that the absolute value of the correlation coefficient is less than one.

6.5.2 Application of Chebychev's Inequality

Using Chebychev's inequality we can bound the probability of an outcome of a random variable (x) being different from the mean for any distribution.

Example 6.30. The most widespread use of Chebychev's inequality involves means and variances. Let $g(x) = (x-\mu)^2/\sigma^2$, where $\mu = \mathrm{E}[X]$ and $\sigma^2 = V(X)$. Let $r = t^2$.

$$\mathrm{P}\left(\frac{(X-\mu)^2}{\sigma^2} \geq t^2\right) \leq \frac{1}{t^2}\mathrm{E}\left[\frac{(X-\mu)^2}{\sigma^2}\right] = \frac{1}{t^2}. \tag{6.65}$$

Since

$$\begin{gathered}\frac{(X-\mu)^2}{\sigma^2} \geq t^2 \Rightarrow (X-\mu)^2 \geq \sigma^2t^2 \Rightarrow |X-\mu| \geq \sigma t \\ \mathrm{P}(|X-\mu| \geq \sigma t) \leq \frac{1}{t^2}\end{gathered} \tag{6.66}$$

letting $t = 2$, this becomes

$$\mathrm{P}(|X-\mu| \geq 2\sigma) \leq \frac{1}{4} = 0.25. \tag{6.67}$$

However, this inequality may not say much, since for the normal distribution

$$P(|X - \mu| \geq 2\sigma) = \left[1 - \int_{-\infty}^{2\sigma} \frac{1}{\sqrt{2\pi}\sigma} \exp\left[-\frac{(x-\mu)^2}{2\sigma^2}\right] dx\right] \tag{6.68}$$

$$= 2 \times (0.0227) = 0.0455.$$

Thus, the actual probability under the standard normal distribution is much smaller than the Chebychev inequality. Put slightly differently, the Chebychev method gives a very loose probability bound.

6.5.3 Normal Approximation of the Binomial

Starting from the binomial distribution function

$$b(n, r, p) = \frac{n!}{(n-r)!r!} p^r (1-p)^{n-r} \tag{6.69}$$

first assume that $n = 10$ and $p = 0.5$. The probability of $r \leq 3$ is

$$P(r \leq 3) = b(10, 0, 0.5) + b(10, 1, 0.5) + b(10, 3, 0.5) = 0.1719. \tag{6.70}$$

Note that this distribution has a mean of 5 and a variance of 2.5. Given this we can compute

$$z^* = \frac{3-5}{\sqrt{2.5}} = -1.265. \tag{6.71}$$

Integrating the standard normal distribution function from negative infinity to -1.265 yields

$$P(z^* \leq -1.265) = \int_{-\infty}^{-1.265} \frac{1}{\sqrt{2\pi}} \exp\left[-\frac{z^2}{2}\right] dz = 0.1030. \tag{6.72}$$

Expanding the sample size to 20 and examining the probability that $r \leq 6$ yields

$$P(r \leq 6) = \sum_{i=0}^{6} b(20, i, 5) = 0.0577. \tag{6.73}$$

This time the mean of the distribution is 10 and the variance is 5. The resulting $z^* = -1.7889$. The integral of the normal distribution function from negative infinity to -1.7889 is 0.0368.

As the sample size increases, the binomial probability approaches the normal probability. Hence, the binomial converges in probability to the normal distribution, as depicted in Figure 5.2.

6.6 Chapter Summary

- Small sample assumptions such as the assumption that the set of independent variables is fixed by the choice of the experimenter and the normality of residuals yield very powerful identification conditions for estimation.

- Econometricians very seldom are able to make these assumptions. Hence, we are interested in the limiting behavior of estimators as the sample size becomes very large.
- Of the modes of convergence, we are typically most interested in convergence in distribution. Several distributions of interest converge in probability to the normal distribution such as ordinary least squares and maximum likelihood estimators.
- Jensen's inequality has implications for economic concepts such as expected utility.

6.7 Review Questions

6-1R. A couple of production functions are known for their numerical limits. For example, the Spillman production function

$$f(x_1) = \alpha_0 \left(1 - \exp\left(\alpha_1 - \alpha_2 x_1\right)\right) \tag{6.74}$$

and its generalization called the Mitcherlich–Baule

$$f(x_1, x_2) = \beta_0 \left(1 - \exp\left(\beta_1 - \beta_2 x_1\right)\right) \left(1 - \exp\left(\beta_3 - \beta_4 x_2\right)\right) \tag{6.75}$$

have limiting levels. Demonstrate the limit of each of these functions as $x_1, x_2 \to \infty$.

6-2R. Following the discussion of Sandmo [40], demonstrate that the expected value of a concave production function lies below the production function at its expected value.

6.8 Numerical Exercises

6-1E. Construct 10 samples of a Bernoulli distribution for 25, 50, 75, and 100 draws. Compute the mean and standard deviations for each draw. Demonstrate that the mean and variance of the samples converge numerically.

6-2E. Compare the probability that fewer than 10 heads will be tossed out of 50 with the comparable event under a normal distribution.

7

Point Estimation

CONTENTS

We will divide the discussion into the estimation of a single number such as a mean or standard deviation, or the estimation of a range such as a confidence interval. At the most basic level, the definition of an estimator involves the distinction between a sample and a population. In general, we assume that we have a random variable (X) with some distribution function. Next, we assume that we want to estimate something about that population, for example, we may be interested in estimating the mean of the population or probability that the outcome will lie between two numbers. In a farm-planning model, we may be interested in estimating the expected return for a particular crop. In a regression context, we may be interested in estimating the average effect of price or income on the quantity of goods consumed. This estimation is typically based on a sample of outcomes drawn from the population instead of the population itself.

7.1 Sampling and Sample Image

Focusing on the sample versus population dichotomy for a moment, the sample image of X, denoted X^*, and the empirical distribution function for $f(X)$ can be depicted as a discrete distribution function with probability $1/n$.

Consider an example from production economics. Suppose that we observe data on the level of production for a group of firms and their inputs (e.g., the capital (K), labor (L), energy (E), and material (M) data from Dale Jorgenson's KLEM dataset [22] for a group of industries $i = 1, \cdots N$). Next, assume that we are interested in measuring the inefficiency given an estimate of the efficient amount of production associated with each input ($\hat{y}_i(k_i, l_i, e_i, m_i)$).

$$\epsilon_i = y_i - \hat{y}_i(k_i, l_i, e_i, m_i). \tag{7.1}$$

For the moment assume that the efficient level of production is known without error. One possible assumption is that $-\epsilon_i \sim \Gamma(\alpha, \beta)$, or all firms are at most efficient ($y_i - \hat{y}_i(k_i, l_i, e_i, m_i) \leq 0$). An example of the gamma distribution is presented in Figure 7.1.

Given this specification, we could be interested in estimating the characteristics of the inefficiency for a firm in a specific industry – say the average

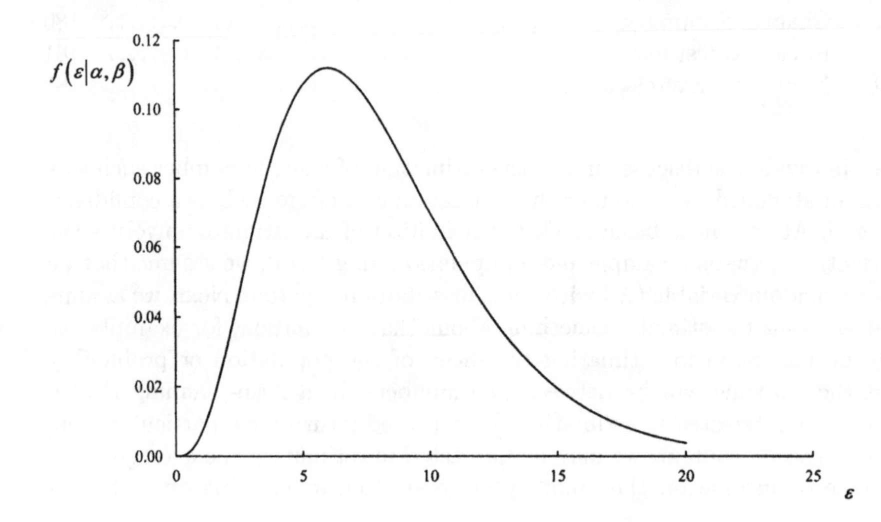

FIGURE 7.1
Density Function for a Gamma Distribution.

TABLE 7.1
Small Sample of Gamma Random Variates

Obs.	ϵ	$\hat{F}(\epsilon_i)$	$F(\epsilon_i)$	Obs.	ϵ	$\hat{F}(\epsilon_i)$	$F(\epsilon_i)$
1	0.4704	0.02	0.0156	26	1.7103	0.52	0.4461
2	0.4717	0.04	0.0157	27	1.7424	0.54	0.4601
3	0.5493	0.06	0.0256	28	1.8291	0.56	0.4971
4	0.6324	0.08	0.0397	29	1.9420	0.58	0.5436
5	0.6978	0.10	0.0532	30	1.9559	0.60	0.5491
6	0.7579	0.12	0.0676	31	1.9640	0.62	0.5524
7	0.9646	0.14	0.1303	32	2.1041	0.64	0.6061
8	0.9849	0.16	0.1375	33	2.2862	0.66	0.6698
9	0.9998	0.18	0.1428	34	2.3390	0.68	0.6868
10	1.0667	0.20	0.1677	35	2.3564	0.70	0.6923
11	1.0927	0.22	0.1778	36	2.5629	0.72	0.7522
12	1.1193	0.24	0.1883	37	2.6581	0.74	0.7766
13	1.1895	0.26	0.2169	38	2.8669	0.76	0.8234
14	1.2258	0.28	0.2321	39	2.9415	0.78	0.8381
15	1.3933	0.30	0.3051	40	3.0448	0.80	0.8566
16	1.4133	0.32	0.3140	41	3.0500	0.82	0.8575
17	1.4354	0.34	0.3238	42	3.0869	0.84	0.8637
18	1.5034	0.36	0.3543	43	3.1295	0.86	0.8705
19	1.5074	0.38	0.3561	44	3.1841	0.88	0.8788
20	1.5074	0.40	0.3561	45	4.0159	0.90	0.9585
21	1.5459	0.42	0.3733	46	4.1773	0.92	0.9667
22	1.5639	0.44	0.3814	47	4.2499	0.94	0.9699
23	1.5823	0.46	0.3896	48	4.4428	0.96	0.9770
24	1.5827	0.48	0.3898	49	4.4562	0.98	0.9774
25	1.6533	0.50	0.4211	50	4.6468	1.00	0.9828

technical inefficiency of firms in the Food and Fiber Sector. Table 7.1 presents one such sample for 50 firms in ascending order (i.e., this is not the order the sample was drawn in). In this table we define the empirical cumulative distribution as

$$\hat{F}(\epsilon_i) = \frac{i}{N} \tag{7.2}$$

where $N = 50$ (the number of oberservations). The next column gives the theoretical cumulative density function ($F(\epsilon_i)$) – integrating the gamma density function from 0 to ϵ_i. The relationship between the empirical and theoretical cumulative distribution functions is presented in Figure 7.2. From this graphical depiction, we conclude that the sample image (i.e., the empirical cumulative distribution) approaches the theoretical distibution. Given the data presented in Table 7.1, the sample mean is 2.03 and the sample variance is 1.27.

Next, we extend the sample to $N = 200$ observations. The empirical and theoretical cumulative density functions for this sample are presented in Figure 7.3. Intuitively, the sample image for the larger sample is closer to the

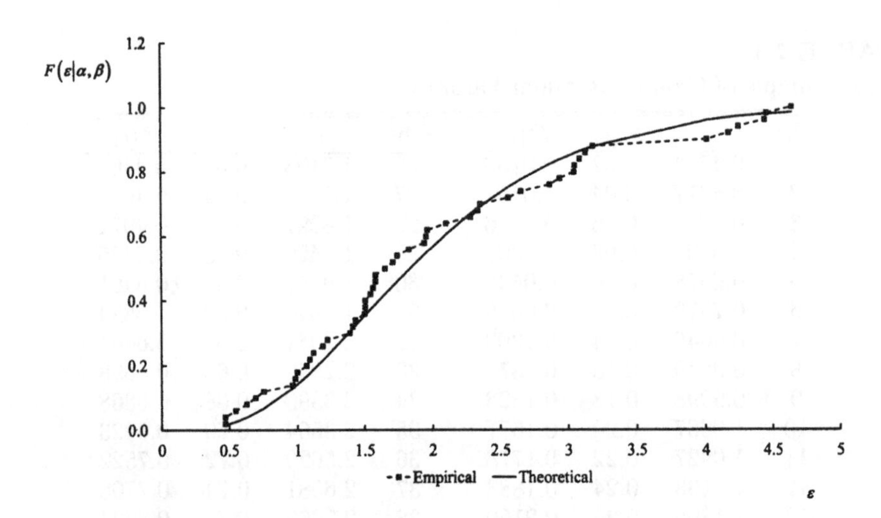

FIGURE 7.2
Empirical versus Theoretical Cumulative Distribution Functions — Small Sample.

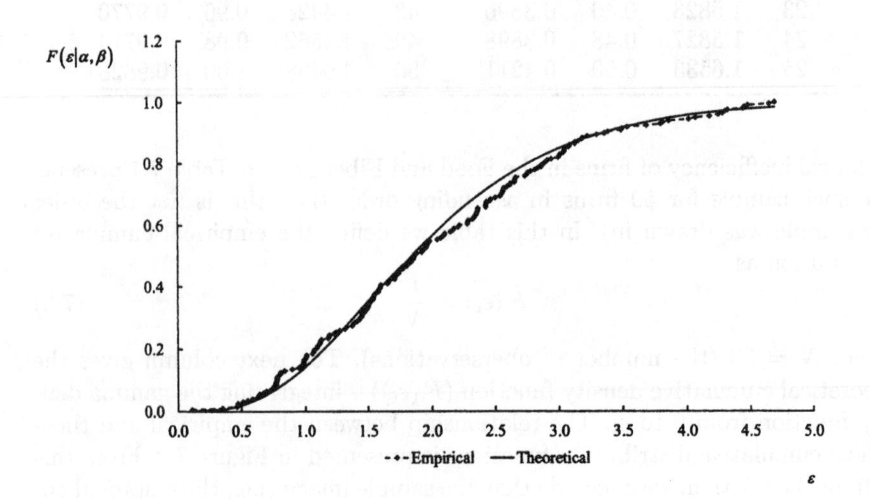

FIGURE 7.3
Empirical versus Theoretical Cumulative Distribution Functions — Large Sample.

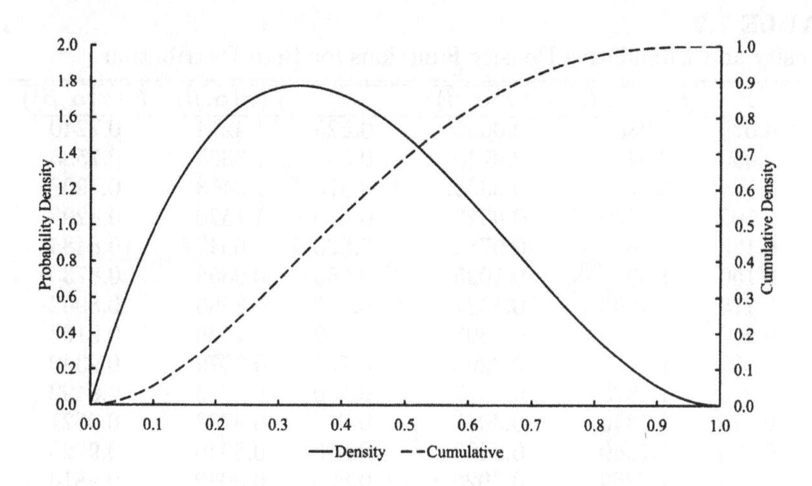

FIGURE 7.4
Probability and Cumulative Beta Distributions.

underlying distribution function than the smaller sample. Empirically, the mean of the larger sample is 2.03 and the variance is 1.02. Given that the true underlying distribution is a $\Gamma(\alpha = 4, \beta = 2)$, the theoretical mean is 2 and the variance is 1. Hence, while there is little improvement in the estimate of the mean from the larger sample, the estimate of the variance for the larger sample is much closer to its true value.

To develop the concept of sampling from a distribution, assume that we are interested in estimating the share of a household's income spent on housing. One possibility for this effort is the beta distribution, which is a two parameter distribution for a continuous random variable with values between zero and one (depicted in Figure 7.4). Assume that our population is the set of 40 faculty of some academic department. Further assume that the true underlying beta distribution is the one depicted in Table 7.2. Assume that it is too costly to sample all 40 faculty members for some reason and that we will only be able to collect a sample of 8 faculty (i.e., there are two days remaining in the spring semester so the best you can hope for is to contact 8 faculty). The question is how does our sample of eight faculty relate to the true beta distribution?

First, assume that we rank the faculty by the percent of their income spent on housing from the lowest to the highest. Next, assume that we draw a sample of eight faculty members from this list (or sample) at random.

$$s = \{34, 27, 19, 29, 33, 12, 23, 35\}. \tag{7.3}$$

Taking the first point, 34/40 is equivalent to a uniform outcome of 0.850. Graphically, we can map this draw from a uniform random outcome into a

TABLE 7.2
Density and Cumulative Density Functions for Beta Distribution

x	$f(x\|\alpha,\beta)$	$F(x\|\alpha,\beta)$	x	$f(x\|\alpha,\beta)$	$F(x\|\alpha,\beta)$
0.025	0.2852	0.0036	0.525	1.4214	0.7240
0.050	0.5415	0.0140	0.550	1.3365	0.7585
0.075	0.7701	0.0305	0.575	1.2463	0.7908
0.100	0.9720	0.0523	0.600	1.1520	0.8208
0.125	1.1484	0.0789	0.625	1.0547	0.8484
0.150	1.3005	0.1095	0.650	0.9555	0.8735
0.175	1.4293	0.1437	0.675	0.8556	0.8962
0.200	1.5360	0.1808	0.700	0.7560	0.9163
0.225	1.6217	0.2203	0.725	0.6579	0.9340
0.250	1.6875	0.2617	0.750	0.5625	0.9492
0.275	1.7346	0.3045	0.775	0.4708	0.9621
0.300	1.7640	0.3483	0.800	0.3840	0.9728
0.325	1.7769	0.3926	0.825	0.3032	0.9814
0.350	1.7745	0.4370	0.850	0.2295	0.9880
0.375	1.7578	0.4812	0.875	0.1641	0.9929
0.400	1.7280	0.5248	0.900	0.1080	0.9963
0.425	1.6862	0.5675	0.925	0.0624	0.9984
0.450	1.6335	0.6090	0.950	0.0285	0.9995
0.475	1.5711	0.6491	0.975	0.0073	0.9999
0.500	1.5000	0.6875	1.000	0.0000	1.0000

beta outcome, as depicted in Figure 7.5, yielding a value of the beta random variable of 0.6266. This value requires a linear interpolation. The uniform value (i.e., the value of the cumulative distribution for beta) lies between 0.8484 ($x = 0.625$) and 0.8735 ($x = 0.650$).

$$x = 0.625 + (0.8500 - 0.8484) \times \frac{0.650 - 0.625}{0.8735 - 0.8484} = 0.6266. \qquad (7.4)$$

Thus, if our distribution is true ($B(\alpha = 3, \beta = 2)$), the 34th individual in the sample will spend 0.6266 of their income on housing. The sample of house shares for these individuals are then

$$t = \{0.6266, 0.4919, 0.3715, 0.5257, 0.6038, 0.2724, 0.4295, 0.6516\}. \qquad (7.5)$$

Table 7.3 presents a larger sample of random variables drawn according to the theoretical distribution. Figure 7.6 presents the sample and theoretical cumulative density functions for the data presented in Table 7.3.

The point of the discussion is that a sample drawn at random from a population that obeys any specific distribution function will replicate that distribution function (the sample converges in probability to the theoretical distribution). The uniform distribution is simply the collection of all individuals in the population. We assume that each individual is equally likely to be

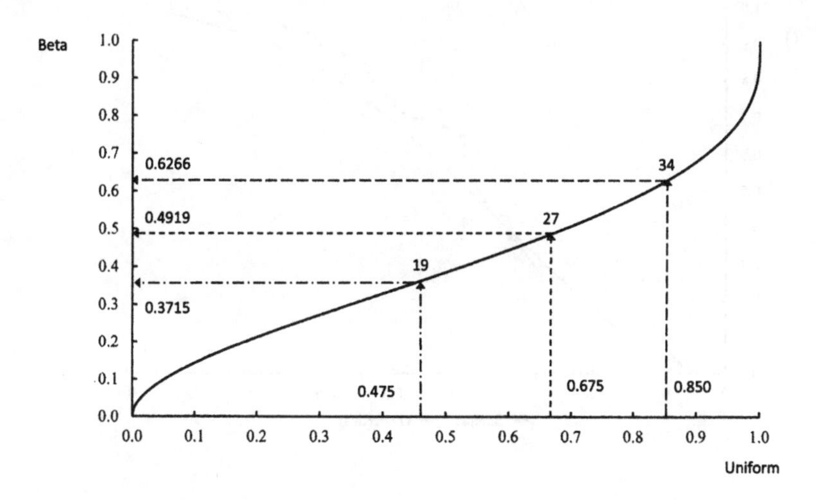

FIGURE 7.5
Inverse Beta Distribution.

TABLE 7.3
Random Sample of Betas

Obs.	$U[0,1]$	$B(\alpha,\beta)$	Obs.	$U[0,1]$	$B(\alpha,\beta)$
1	0.3900	0.3235	21	0.3944	0.3260
2	0.8403	0.6177	22	0.0503	0.0977
3	0.3312	0.2902	23	0.5190	0.3967
4	0.5652	0.4236	24	0.4487	0.3566
5	0.7302	0.5295	25	0.7912	0.5753
6	0.4944	0.3826	26	0.4874	0.3785
7	0.3041	0.2748	27	0.7320	0.5307
8	0.3884	0.3227	28	0.4588	0.3623
9	0.2189	0.2241	29	0.1510	0.1799
10	0.9842	0.8357	30	0.9094	0.6915
11	0.8840	0.6616	31	0.6834	0.4973
12	0.0244	0.0657	32	0.6400	0.4694
13	0.0354	0.0806	33	0.6833	0.4973
14	0.0381	0.0837	34	0.3476	0.2996
15	0.8324	0.6105	35	0.3600	0.3066
16	0.0853	0.1302	36	0.0993	0.1417
17	0.5128	0.3931	37	0.5149	0.3943
18	0.7460	0.5409	38	0.7397	0.5364
19	0.4754	0.3717	39	0.0593	0.1066
20	0.0630	0.1101	40	0.4849	0.3771

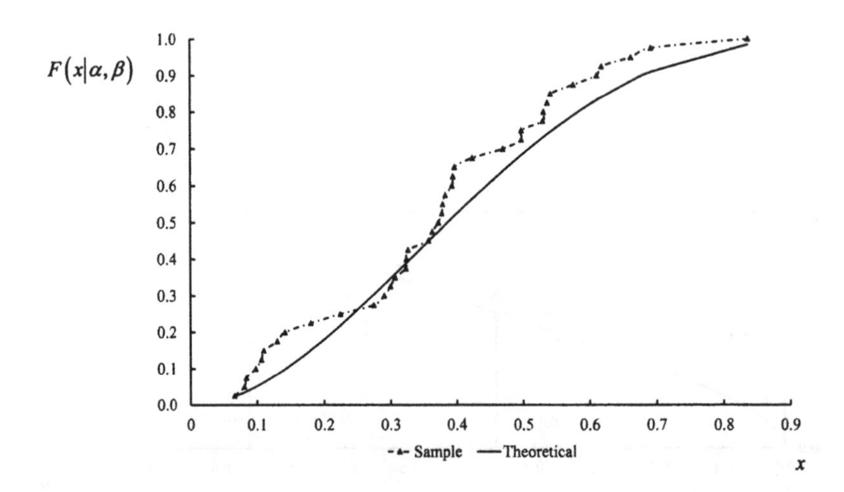

FIGURE 7.6
Sample and Theoretical Beta Distributions.

drawn for the sample. We order the underlying uniform distribution in our discussion as a matter of convenience. However, given that we draw the sample population randomly, no assumption about knowing the underlying ordering of the population is actually used.

7.2 Familiar Estimators

As a starting point, let us consider a variety of estimators that students have seen in introductory statistics courses. For example, we start by considering the sample mean

$$\bar{X} = \frac{1}{n}\sum_{i=1}^{n} X_i. \tag{7.6}$$

Based on this accepted definition, we ask the question – what do we know about the properties of the mean? Using Theorem 4.23, we know that

$$\mathrm{E}\left[\bar{X}\right] = \mathrm{E}\left[X\right] \tag{7.7}$$

which means that the population mean is close to a "center" of the distribution of the sample mean. Suppose $\mathrm{V}(X) = \sigma^2$ is finite. Then, using Theorem 4.17, we know that

$$\mathrm{V}(X) = \frac{\sigma^2}{n} \tag{7.8}$$

which shows that the degree of dispersion of the distribution of the sample mean around the population mean is inversely related to the sample size n. Using Theorem 6.22 (Khinchine's law of large numbers), we know that

$$\text{plim}_{n\to\infty}\bar{X} = \text{E}[X]. \tag{7.9}$$

If $V(X)$ is finite, the same result follows from Equations 7.7 and 7.8 above because of Chebychev's inequality.

Other familiar statistics include the sample variance.

$$S_X^2 = \frac{1}{n}\sum_{i=1}^{n}(X_t - \bar{X})^2 = \frac{1}{n}\sum_{i=1}^{n}X_t^2 - (\bar{X})^2. \tag{7.10}$$

Another familiar statistic is the kth sample moment around zero

$$M^k = \frac{1}{n}\sum_{i=1}^{n}X_i^k. \tag{7.11}$$

and the kth moment around the mean

$$\tilde{M}^k = \frac{1}{n}\sum_{i=1}^{n}(X_i - \bar{X})^k. \tag{7.12}$$

As discussed in a previous example, the kth moment around the mean is used to draw conclusions regarding the skewness and kurtosis of the sample. In addition, most students have been introduced to the sample covariance

$$\text{Cov}(X,Y) = S_{xy} = \frac{1}{n}\sum_{i=1}^{n}(X_i - \bar{X})(Y_i - \bar{Y}) \tag{7.13}$$

and the sample correlation

$$\rho_{xy} = \frac{S_{xy}}{\sqrt{S_{xx}S_{yy}}}. \tag{7.14}$$

In each case the student is typically introduced to an intuitive meaning of each statistic. For example, the mean is related to the expected value of a random variable, the variance provides a measure of the dispersion of the random variables, and the covariance provides a measure of the tendency of two random variables to vary together (either directly or indirectly). This chapter attempts to link estimators with parameters of underlying distributions.

7.2.1 Estimators in General

In general, an estimator is a function of the sample, not based on population parameters. First, the estimator is a known function of random variables.

$$\hat{\theta} = \phi(X_1, X_2, X_3, \cdots X_n). \tag{7.15}$$

The value of an estimator is then a random variable. As with any other random variable, it is possible to define the distribution of the estimator based on distribution of the random variables in the sample. These distributions will be used in the next chapter to define confidence intervals. Any function of the sample is referred to as a statistic. Most of the time in econometrics, we focus on the moments as sample statistics. Specifically, we may be interested in the sample means, or may use the sample covariances with the sample variances to define the least squares estimator.

Other statistics may be important. For example, we may be interested in the probability of a given die role (for example, the probability of a three). If we define a new set of variables, Y, such that $Y = 1$ if $X = 3$ and $Y = 0$ otherwise, the probability of a three becomes

$$\hat{p}_3 = \frac{1}{n} \sum_{i=1}^{n} Y_i. \tag{7.16}$$

Amemiya [1, p. 115] suggests that this probability could also be estimated from the moments of the distribution. Assume that you have a sample of 50 die rolls. Compute the sample distribution for each moment $k = 0, 1, 2, 3, 4, 5$

$$m_k = \frac{1}{50} \sum_{i=1}^{n} X_i^k \tag{7.17}$$

where X_i is the value of the die roll. The method of moments estimate of each probability p_i is defined by the solution of the five equation system

$$\begin{aligned}
M_\theta^{(0)}(0) &= \sum_{j=1}^{6} \theta_j \\
M_\theta^{(1)}(0) &= \sum_{j=1}^{6} j\theta_j \\
M_\theta^{(2)}(0) &= \sum_{j=1}^{6} j^2\theta_j \\
M_\theta^{(3)}(0) &= \sum_{j=1}^{6} j^3\theta_j \\
M_\theta^{(4)}(0) &= \sum_{j=1}^{6} j^4\theta_j \\
M_\theta^{(5)}(0) &= \sum_{j=1}^{6} j^5\theta_j
\end{aligned} \tag{7.18}$$

TABLE 7.4
Sample of Die Rolls

Observation	Outcome	Observation	Outcome	Observation	Outcome
1	1	18	3	35	6
2	5	19	5	36	4
3	3	20	5	37	6
4	2	21	3	38	2
5	6	22	2	39	1
6	5	23	1	40	3
7	4	24	5	41	3
8	2	25	1	42	5
9	1	26	2	43	4
10	3	27	1	44	2
11	1	28	6	45	1
12	5	29	2	46	1
13	6	30	1	47	3
14	1	31	6	48	3
15	6	32	2	49	3
16	5	33	2	50	3
17	1	34	2		

where $\theta = (\theta_1, \theta_2, \theta_3, \theta_4, \theta_5, \theta_6)$ are the probabilities of rolling a $1, 2, \cdots 6$, respectively. Consider the sample of 50 observations presented in Table 7.4. We can solve for the method of moments estimator for the parameters in θ by solving

$$\begin{aligned}
M_\theta^{(0)}(0) &= \sum_{j=1}^{6} \theta_j = \hat{m}_0 = \frac{1}{50}\sum_{i=1}^{50} 1 = 1 \\
M_\theta^{(1)}(0) &= \sum_{j=1}^{6} j\theta_j = \hat{m}_1 = \frac{1}{50}\sum_{i=1}^{50} x_i = 3.12 \\
M_\theta^{(2)}(0) &= \sum_{j=1}^{6} j^2\theta_j = \hat{m}_2 = \frac{1}{50}\sum_{i=1}^{50} x_i^2 = 12.84 \\
M_\theta^{(3)}(0) &= \sum_{j=1}^{6} j^3\theta_j = \hat{m}_3 = \frac{1}{50}\sum_{i=1}^{50} x_1^3 = 61.32 \\
M_\theta^{(4)}(0) &= \sum_{j=1}^{6} j^4\theta_j = \hat{m}_4 = \frac{1}{50}\sum_{i=1}^{50} x_i^4 = 316.44 \\
M_\theta^{(5)}(0) &= \sum_{j=1}^{6} j^5\theta_j = \hat{m}_5 = \frac{1}{50}\sum_{i=1}^{50} x_i^5 = 1705.32.
\end{aligned} \tag{7.19}$$

The method of moments estimator can then be written as

$$\begin{bmatrix} 1 & 1 & 1 & 1 & 1 & 1 \\ 1 & 2 & 3 & 4 & 5 & 6 \\ 1 & 8 & 27 & 64 & 125 & 216 \\ 1 & 16 & 81 & 256 & 625 & 1296 \\ 1 & 32 & 243 & 1024 & 3125 & 7776 \end{bmatrix} \begin{bmatrix} \hat{\theta}_1 \\ \hat{\theta}_2 \\ \hat{\theta}_3 \\ \hat{\theta}_4 \\ \hat{\theta}_5 \\ \hat{\theta}_6 \end{bmatrix} = \begin{bmatrix} 1 \\ 3.12 \\ 12.84 \\ 61.32 \\ 316.44 \\ 1705.32 \end{bmatrix}$$

$$\Rightarrow \begin{bmatrix} \hat{\theta}_1 \\ \hat{\theta}_2 \\ \hat{\theta}_3 \\ \hat{\theta}_4 \\ \hat{\theta}_5 \\ \hat{\theta}_6 \end{bmatrix} = \begin{bmatrix} 0.24 \\ 0.20 \\ 0.20 \\ 0.06 \\ 0.16 \\ 0.14 \end{bmatrix}. \tag{7.20}$$

Thus, we can estimate the parameters of the distribution by solving for that set of parameters that equates the theoretical moments of the distribution with the empirical moments.

For another example of a method of moments estimator, consider the gamma distribution. The theoretical moments for this distribution are

$$M^{(1)}_{\alpha,\beta} = \alpha\beta \; \tilde{M}^{(2)}_{\alpha,\beta} = \alpha\beta^2 \tag{7.21}$$

where $\tilde{M}^{(2)}_{\alpha,\beta}$ is the central moment. Using the data from Table 7.1, we have

$$\left.\begin{array}{r} \alpha\beta = 2.0331 \\ \alpha\beta^2 = 1.2625 \end{array}\right\} \Rightarrow \hat{\beta}_{MOM} = \frac{1.2625}{2.0331} = 0.6210. \tag{7.22}$$

Next, returning to the theoretical first moment

$$\hat{\alpha}_{MOM} = \frac{2.0311}{0.6210}. \tag{7.23}$$

Each of these estimators relies on sample information in the guise of the sample moments. Further, the traditional estimator of the mean and variance of the normal distribution can be justified using a method of moments estimator.

7.2.2 Nonparametric Estimation

At the most general level, we can divide estimation procedures into distribution specific estimators and nonparameteric or distribution free estimators. Intuitively, distribution specific methods have two sources of information – information from the sample and information based on distributional assumptions. For example, we can assume that the underlying random variable obeys a normal distribution. Given this assumption, the characteristics of the distribution are based on two parameters – the mean and the variance. Hence, the estimators focus on the first two moments of the sample. The estimation procedure can be tailored to the estimation of specific distribution parameters.

Extending our intuitive discussion, if the distributional assumption is correct, tailoring the estimation to parameters of the distribution improves our ability to describe the random variable. However, if our assumption about the distribution form is incorrect, the distribution specific characteristics of the estimator could add noise or confuse our ability to describe the distribution. For example, suppose that we hypothesized the random variable was normally distributed but the true underlying distribution was negative exponential

$$f(x|\lambda) = \lambda e^{-\lambda x} \; x \geq 0 \text{ and } 0 \text{ otherwise.} \tag{7.24}$$

The negative exponential distribution has a theoretical mean of $1/\lambda$ and a variance of $1/\lambda^2$. Hence, the negative exponential distribution provides more restrictions than the normal distribution.

Nonparametric or distribution free methods are estimators that are not based on specific distributional assumptions. These estimators are less efficient in that they cannot take advantage of assumptions such as the relationship between the moments of the distribution. However, they are not fragile to distributional assumptions (i.e., assuming that the distribution is a normal when it is in fact a gamma distribution could significantly affect the estimated parameters).

7.3 Properties of Estimators

In general, any parameter such as a population mean or variance (i.e., the μ and σ^2 parameters of the normal distribution) may have several different estimators. For example, we could estimate a simple linear model

$$y_i = \alpha_0 + \alpha_1 x_{1i} + \alpha_2 x_{2i} + \nu_i \tag{7.25}$$

where y_i, x_{1i}, and x_{2i} are observed and α_0, α_1, and α_2 are parameters using ordinary least squares, maximum likelihood, or a method of moments estimator. Many of the estimators are mathematically similar. For example, if we assume that the error in Equation 7.25 is normally distributed, the least squares estimator is also the maximum likelihood estimator. In the cases where the estimators are different, we need to develop criteria for comparing the goodness of each estimator.

7.3.1 Measures of Closeness

As a starting point for our discussion, consider a relatively innocuous criteria — suppose that we choose the parameter that is close to its true value. For example, suppose that we want to estimate the probability of a Bernoulli event being 1 (i.e., the probability that the coin toss results in a head). The general form of the distribution becomes

$$f(Z|\theta) = \theta^{Z}(1-\theta)^{(1-Z)}. \tag{7.26}$$

Next, assume that we develop an estimator

$$X = \frac{1}{N}\sum_{i=1}^{N} z_i \tag{7.27}$$

where z_is are observed outcomes where $z_i = 1$ denotes a head and $z_i = 0$ denotes a tail. Next, suppose that we had a different estimator

$$Y = \sum_{i=1}^{N} w_i z_i \tag{7.28}$$

where w_i is a weighting function different from $1/N$. One question is whether X produces an estimate closer to the true θ than Y.

Unfortunately, there are several different possible measures of closeness:

1. $\mathrm{P}(|X-\theta| \leq |Y-\theta|) = 1$.
2. $\mathrm{E}[g(X-\theta)] \leq \mathrm{E}[g(Y-\theta)]$ for every continuous function $g(.)$ which is nonincreasing for $x < 1$ and nondecreasing for $x > 0$.
3. $\mathrm{E}[g(|X-\theta|)] \leq \mathrm{E}[g(|Y-\theta|)]$ for every continuous function and nondecreasing $g(.)$.
4. $\mathrm{P}(|X-\theta| > \epsilon) \leq \mathrm{P}(|Y-\theta| > \epsilon)$ for every ϵ.
5. $\mathrm{E}(X-\theta)^2 \leq \mathrm{E}(Y-\theta)^2$.
6. $\mathrm{P}(|X-\theta| < |Y-\theta|) \geq \mathrm{P}(|X-\theta| > |Y-\theta|)$.

Of these possibilities, the most widely used in econometrics are minimize mean error squared (5) and a likelihood comparison (akin to 2).

7.3.2 Mean Squared Error

To develop the mean squared error comparison, we will develop an example presented by Amemiya [1, p. 123]. Following our Bernoulli example, suppose that we have a sample of two outcomes and want to estimate θ using one of three estimators,

$$\begin{aligned} T &= \frac{(z_1 + z_2)}{2} \\ S &= z_1 \\ W &= \frac{1}{2}. \end{aligned} \tag{7.29}$$

Note that the first estimator corresponds with the estimator presented in Equation 7.27, while the second estimator corresponds with the estimator in Equation 7.28 with $w_1 = 1$ and $w_2 = 0$. The third estimator appears ridiculous – no matter what the outcome, I think that $\theta = 1/2$.

As a starting point, consider constructing a general form of the mean squared error for each estimator. Notice that the probability of a single event z_1 in the Bernoulli formulation becomes

$$f(z_1|\theta) = \theta^{z_1}(1-\theta)^{(1-z_1)} \Rightarrow \begin{cases} f(z_1 = 1|\theta) = \theta \\ f(z_1 = 0|\theta) = (1-\theta). \end{cases} \tag{7.30}$$

Given the probability function presented in Equation 7.30, we can express the expected value of estimator S as

$$\mathrm{E}[S(z_1)] = (z_1 = 1) \times \theta + (z_1 = 0)(1-\theta) = \theta. \tag{7.31}$$

Thus, even though the estimator always estimates either a zero ($z_1 = 0$) or a one ($z_1 = 1$), on average it is correct. However, the estimate may not be very close to the true value using the mean squared error measure. The mean squared error of the estimate for this estimator can then be written as

$$\begin{aligned} \mathrm{MSE}_S(\theta) &= \sum_{z_1} f(z_1|\theta)(S(z_1) - \theta)^2 \\ &= f(z_1 = 0|\theta)(0-\theta)^2 + f(z_1 = 1|\theta)(1-\theta)^2 \\ &= (1-\theta) \times \theta^2 + \theta \times (1-\theta)^2. \end{aligned} \tag{7.32}$$

Next, consider the same logic for estimator T. In the case of T, there are three outcomes: $z_1 + z_2 = 0$, $z_1 + z_2 = 1$, and $z_1 + z_2 = 2$. In the case of $z_1 + z_2 = 1$, either $z_1 = 1$ and $z_2 = 0$ or $z_1 = 0$ and $z_2 = 1$. In other words, there are two ways to generate this event. Following our approach in Equation 7.30, we write the distribution function as

$$f(z_1, z_2|\theta) = \theta^{(z_1+z_2)}(1-\theta)^{(1-z_1-z_2)} \Rightarrow \begin{cases} f(z_1 = 1, z_2 = 1|\theta) = \theta^2 \\ f(z_1 = 1, z_2 = 0|\theta) = \theta(1-\theta) \\ f(z_1 = 0, z_2 = 1|\theta) = \theta(1-\theta) \\ f(z_1 = 0, z_2 = 0|\theta) = (1-\theta)^2. \end{cases} \tag{7.33}$$

The expected value of the estimator T can then be derived as

$$\begin{aligned} \mathrm{E}\left[T\left(z_1,z_2\right)\right] = & \frac{z_1+z_2=0}{2}\left(1-\theta\right)^2 + 2\frac{z_1+z_2=1}{2}\theta\left(1-\theta\right) + \\ & \frac{z_1+z_2=2}{2}\theta^2 = \left(\theta-\theta^2\right)+\theta^2 = \theta. \end{aligned} \tag{7.34}$$

Again, the expected value of the estimator is correct, but the estimator only yields three possible values – $T(z_1,z_2)=0$, $T(z_1,z_2)=0.5$, or $T(z_1,z_2)=1$. We derive the mean squared error as a measure of closeness.

$$\begin{aligned} \mathrm{MSE}_T\left(\theta\right) = & \sum_{z_1+z_2} f\left(z_1+z_2|\,\theta\right)\left(T\left(z_1,z_2\right)-\theta\right)^2 \\ = & \left(1-\theta\right)^2\left(0-\theta\right)^2 + 2\theta\left(1-\theta\right)\left(\frac{1}{2}-\theta\right)^2 + \theta^2\left(1-\theta\right)^2 \\ = & 2\left(\left(1-\theta\right)^2\theta^2 + \theta\left(1-\theta\right)\left(\frac{1}{2}-\theta\right)^2\right). \end{aligned} \tag{7.35}$$

Finally, for completeness we define the mean squared error of the W estimator as

$$\mathrm{MSE}_W\left(\theta\right) = \left(1-\theta\right)\left(\frac{1}{2}-\theta\right)^2 + \theta\left(\frac{1}{2}-\theta\right)^2. \tag{7.36}$$

The mean squared error for each estimator is presented in Figure 7.7. The question (loosely phrased) is then which is the best estimator of θ? In answering this question, however, two kinds of ambiguities occur. For a particular value of the parameter, say $\theta = 3/4$, it is not clear which of the three estimators is preferred. T dominates W for $\theta = 0$, but W dominates T for $\theta = 1/2$.

Definition 7.1. Let X and Y be two estimators of θ. We say that X is better (or more efficient) than Y if $\mathrm{E}\left(X-\theta\right)^2 \leq \mathrm{E}\left(Y-\theta\right)^2$ for all $\theta \in \Theta$ and strictly less than for at least one $\theta \in \Theta$.

When an estimator is dominated by another estimator, the dominated estimator is inadmissable.

Definition 7.2. Let $\hat{\theta}$ be an estimator of θ. We say that $\hat{\theta}$ is inadmissible if there is another estimator which is better in the sense that it produces a lower mean square error of the estimate. An estimator that is not inadmissible is admissible.

Thus, we assume that at least one of these estimators is inadmissible and in fact T always performs better than S, so S is dominated and inadmissible. However, this criterion does not allow us to rank S and W.

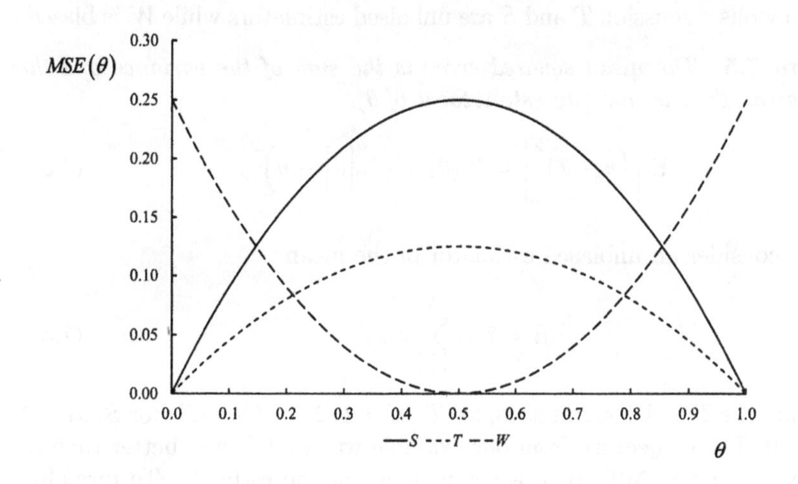

FIGURE 7.7
Comparison of MSE for Various Estimators.

7.3.3 Strategies for Choosing an Estimator

Subjective strategy: This strategy considers the likely outcome of θ and selects the estimator that is best in that likely neighborhood. *Minimax Strategy*: According to the minimax strategy, we choose the estimator for which the largest possible value of the mean squared error is the smallest.

Definition 7.3. Let $\hat{\theta}$ be an estimator of θ. It is a minimax estimator if, for any other estimator of $\tilde{\theta}$, we have

$$\max_{\theta} \mathrm{E}\left[\left(\hat{\theta}-\theta\right)^2\right] \leq \max_{\theta} \mathrm{E}\left[\left(\tilde{\theta}-\theta\right)^2\right]. \tag{7.37}$$

Returning to our previous example, T is chosen over W according to the minimax strategy because the maximum MSE for T is 0.10 while the maximum MSE for W is 0.25.

7.3.4 Best Linear Unbiased Estimator

To begin our development of the best linear unbiased estimator, we need to develop the concept of an unbiased estimator in Definition 7.4.

Definition 7.4. $\hat{\theta}$ is said to be an unbiased estimator of θ if $\mathrm{E}\left[\hat{\theta}\right]$ for all $\theta \in \Theta$. We call $\mathrm{E}\left[\hat{\theta}-\theta\right]$ the bias.

In our previous discussion T and S are unbaised estimators while W is biased.

Theorem 7.5. *The mean squared error is the sum of the variance and the bias squared. That is, for any estimator* $\hat{\theta}$ *of* θ,

$$E\left[\left(\hat{\theta}-\theta\right)^2\right] = V(\theta) + \left(E\left[\hat{\theta}\right]-\theta\right)^2. \tag{7.38}$$

Next, consider an unbiased estimator of the mean

$$\hat{\mu} = \bar{x} = \sum_{i=1}^{N} a_i x_i. \tag{7.39}$$

Comparing the T and S estimators, for T, $a_i = 1/2 = 1/N$ while for S, $a_1 = 1$ and $a_2 = 0$. The conjecture from our example was that T was better than S. It produced a lower MSE or a lower variance of the estimate. To formalize this conjecture, consider Theorem 7.6.

Theorem 7.6. *Let* $\{X_i\}$, $i = 1, 2, \cdots, N$ *be independent and have a common mean* μ *and variance* σ^2. *Consider the class of linear estimators of* μ *which can be written in the form*

$$\hat{\mu} = \bar{X} = \sum_{i=1}^{N} a_i X_i \tag{7.40}$$

and impose the unbiasedness condition

$$E\left[\sum_{i=1}^{N} a_i X_i\right] = \mu. \tag{7.41}$$

Then

$$V\left(\bar{X}\right) \leq V\left(\sum_{i=1}^{N} a_i X_i\right) \tag{7.42}$$

for all a_i, *satisfying the unbiasedness condition. Further, this condition holds with equality only for* $a_i = 1/N$.

To prove these points, note that the a_is must sum to one for unbiasedness.

$$E\left[\sum_{i=1}^{N} a_i X_i\right] = \sum_{i=1}^{N} a_i E\left[X_i\right] = \sum_{i=1}^{N} a_i \mu = \mu \sum_{i=1}^{N} a_i. \tag{7.43}$$

Therefore, $\sum_{i=1}^{N} a_i = 1$ results in an unbiased estimator. The final condition can be demonstrated through the identity

$$\sum_{i=1}^{N}\left(a_i - \frac{1}{N}\right)^2 = \sum_{i=1}^{N} a_i^2 - \frac{2}{N}\sum_{i=1}^{N} a_i + \frac{1}{N}. \tag{7.44}$$

If $a_i = 1/N$ then $\sum_{i=1}^{N}(a_i - 1/N)^2 = 0$. Thus, any other a_i must yield a higher variance $\left(\sum_{i=1}^{N}(a_i - 1/N)^2 \geq 0\right)$

$$\sum_{i=1}^{N}\left(a_i - \frac{1}{N}\right)^2 \sigma^2 \geq 0 \Rightarrow \sum_{i=1}^{N} a_i^2 \sigma^2 \geq \left(\frac{1}{N}\right)\sigma^2. \tag{7.45}$$

Theorem 7.7. *Consider the problem of minimizing*

$$\sum_{i=1}^{N} a_i^2 \tag{7.46}$$

with respect to $\{a_i\}$ *subject to the condition*

$$\sum_{i=1}^{N} a_i b_i = 1. \tag{7.47}$$

The solution to this problem is given by

$$a_i = \frac{b_i}{\sum_{j=1}^{N} b_j^2}. \tag{7.48}$$

Proof. Consider the Lagrange formulation for this minimization problem

$$L = \sum_{i=1}^{N} a_i^2 + \lambda\left(1 - \sum_{i=1}^{N} a_i b_i\right) \tag{7.49}$$

yielding the general first order condition

$$\frac{\partial L}{\partial a_i} = 2a_i - \lambda b_i = 0 \Rightarrow a_i = \frac{\lambda}{2} b_i. \tag{7.50}$$

Substituting this result back into the constraint

$$1 - \frac{\lambda}{2}\sum_{i=1}^{N} b_i^2 = 0 \Rightarrow \lambda = \frac{2}{\sum_{i=1}^{N} b_i^2}. \tag{7.51}$$

Substituting Equation 7.51 into Equation 7.50 yields Equation 7.48. Holding

$$\lambda = \frac{2a_i}{b_i} = \frac{2a_j}{b_j}\ \forall i, j \tag{7.52}$$

implying that $b_i = b_j = 1 \Rightarrow a_i = 1/N$. □

Thus, equally weighting the observations yields the minimum variance estimator.

7.3.5 Asymptotic Properties

Unbiasedness works well for certain classes of estimators such as the mean. Other estimators are somewhat more complicated. For example, the maximum likelihood estimator of the variance can be written as

$$\hat{\sigma}_x^2 = \frac{1}{N}\sum_{i=1}^{N}(x_i - \bar{x})^2 . \tag{7.53}$$

However, this estimator is biased. As we will develop in our discussion of the χ^2 distribution, the unbiased estimator of the variance is

$$\tilde{\sigma}_x^2 = \frac{1}{N-1}\sum_{i=1}^{N}(x_i - \bar{x})^2 . \tag{7.54}$$

Notice that as the sample size becomes large the maximum likelihood estimator of the variance converges to the unbiased estimator of the variance. Rephrasing the discussion slightly, the maximum likelihood estimator of the variance is a consistent estimator of the underlying variance, as decribed in Definition 7.8.

Definition 7.8. We say that $\hat{\theta}$ is a consistent estimator of θ if

$$\text{plim}_{n\rightarrow\infty}\hat{\theta} = \theta. \tag{7.55}$$

Certain estimators such as Bayesian estimators are biased, but consistent. As the sample size increases, the parameter will converge to its true value. In the case of the Bayesian estimator the bias introduced by the prior becomes small as the sample size expands.

7.3.6 Maximum Likelihood

The basic concept behind maximum likelihood estimation is to choose that set of parameters that maximizes the likelihood of drawing a particular sample. For example, suppose that we know that a sample of random variables has a variance of one, but an unknown mean. Let the sample be $X = \{5, 6, 7, 8, 10\}$. The probability of each of these points based on the unknown mean (μ) can be written as

$$\begin{aligned} f(5|\mu) &= \frac{1}{\sqrt{2\pi}}\exp\left[-\frac{(5-\mu)^2}{2}\right] \\ f(6|\mu) &= \frac{1}{\sqrt{2\pi}}\exp\left[-\frac{(6-\mu)^2}{2}\right] \\ &\vdots \\ f(10|\mu) &= \frac{1}{\sqrt{2\pi}}\exp\left[-\frac{(10-\mu)^2}{2}\right]. \end{aligned} \tag{7.56}$$

Assuming that the sample is independent so that the joint distribution function can be written as the product of the marginal distribution functions, the probability of drawing the entire sample based on a given mean can then be written as

$$L(X|\mu) = \frac{1}{(2\pi)^{-5/2}} \exp\left[-\frac{(5-\mu)^2}{2} - \frac{(6-\mu)^2}{2} - \cdots \frac{(10-\mu)^2}{2}\right]. \quad (7.57)$$

The function $L(X|\mu)$ is typically referred to as the likelihood function. The value of μ that maximizes the likelihood function of the sample can then be defined by

$$\max_{\mu} L(X|\mu). \quad (7.58)$$

Under the current scenario, we find it easier, however, to maximize the natural logarithm of the likelihood function

$$\max_{\mu} \ln(L(X|\mu)) \Rightarrow \frac{\partial}{\partial\mu}\left[K - \frac{(5-\mu)^2}{2} - \frac{(6-\mu)^2}{2} - \cdots \frac{(10-\mu)^2}{2}\right]$$
$$= -(5-\mu) - (6-\mu) - \cdots (10-\mu) = 0$$
$$\hat{\mu}_{MLE} = \frac{5+6+7+8+9+10}{6} = 7.5 \quad (7.59)$$

where $K = -5/2\ln(2\pi)$. Note that the constant does not affect the estimate (i.e., the derivative $\partial K/\partial\mu = 0$).

7.4 Sufficient Statistics

There are a number of ways to classify statistical distribution. One of the most popular involves the number of parameters used to specify the distribution. For example, consider the set of distribution functions with two parameters such as the normal distribution, the gamma distribution, and the beta distribution. Each of these distributions is completely specified by two parameters. Intuitively, all their moments are functions of the two identifying parameters. The sufficient statistic is the empirical counterpart of this concept. Specifically, two empirical moments of the distribution contain all the relevent information regarding the distribution. Put slightly differently, two empirical moments (or functions of those moments) are sufficient to describe the distribution.

7.4.1 Data Reduction

The typical mode of operation in statistics is to use information from a sample $X_1, \cdots X_N$ to make inferences about an unknown parameter θ. The researcher summarizes the information in the sample (or the sample values) with a statistic. Thus, any statistic $T(X)$ summarizes the data, or reduces the information in the sample to a single number. We use only the information in the statistic instead of the entire sample. Put in a slightly more mathematical formulation, the statistic partitions the sample space into two sets defining the sample space for the statistic

$$T = \{t : t = T(x), x \in X\}. \tag{7.60}$$

Thus, a given value of a sample statistic $T(x)$ implies that the sample comes from a space of sets A_t such that $t \in T$, $A_t = \{x : T(x) = t\}$. The second possibility (that is ruled out by observing a sample statistic of $T(x)$) is $A_t^C = \{x : T(x) \neq t\}$. Thus, instead of presenting the entire sample, we could report the value of the sample statistic.

7.4.2 Sufficiency Principle

Intuitively, a sufficient statistic for a parameter is a statistic that captures all the information about a given parameter contained in the sample. *Sufficiency Principle*: If $T(X)$ is a sufficient statistic for θ, then any inference about θ should depend on the sample X only through the value of $T(X)$. That is, if x and y are two sample points such that $T(x) = T(y)$, then the inference about θ should be the same whether $X = x$ or $X = y$.

Definition 7.9 (Cassela and Berger). A statistic $T(X)$ is a sufficient statistic for θ if the conditional distribution of the sample X given $T(X)$ does not depend on θ [7, p. 272].

Definition 7.10 (Hogg, McKean, and Craig). Let $X_1, X_2, \cdots X_n$ denote a random sample of size n from a distribution that has a pdf (*probability density function*) or pmf (*probability mass function*) $f(x|\theta)$, $\theta \in \Theta$. Let $Y_1 = u_1(X_1, X_2, \cdots X_n)$ be a statistic whose pdf or pmf is $f_{y_1}(y_1|\theta)$. Then Y_1 is a sufficient statistic for θ if and only if

$$\frac{f(x_1|\theta) f(x_2|\theta) \cdots f(x_n|\theta)}{f_{Y_1}[u_1(x_1, x_2, \cdots x_n)|\theta]} = H(x_1, x_2, \cdots x_n) \tag{7.61}$$

where $H(x_1, x_2, \cdots x_n)$ does not depend on $\theta \in \Theta$ [18, p. 375].

Theorem 7.11 (Cassela and Berger). *If $p(x|\theta)$ is the joint pdf (probability density function) or pmf (probability mass function) of X and $q(t|\theta)$ is the pdf or pmf of $T(X)$, then $T(X)$ is a sufficient statistic for θ if, for every x in the sample space, the ratio of*

$$\frac{p(x|\theta)}{q(T(x)|\theta)} \tag{7.62}$$

is a constant as a function of θ [7, p. 274].

Example 7.12. Normal sufficient statistic: Let $X_1, \cdots X_n$ be independently and identically distributed $\text{N}\left(\mu, \sigma^2\right)$ where the variance is known. The sample mean $T(X) = \bar{X} = 1/n \sum_{i=1}^{n} X_i$ is the sufficient statistic for μ. Starting with the joint distribution function

$$\begin{aligned} f(x|\mu) &= \prod_{i=1}^{n} \frac{1}{\sqrt{2\pi\sigma^2}} \exp\left[-\frac{(x_i-\mu)^2}{2\sigma^2}\right] \\ &= \frac{1}{\left(2\pi\sigma^2\right)^{n/2}} \exp\left[\sum_{i=1}^{n} \frac{(x_i-\mu)^2}{2\sigma^2}\right]. \end{aligned} \tag{7.63}$$

Next, we add and subtract $\bar{x}$, yielding

$$\begin{aligned} f(x|\mu) &= \frac{1}{\left(2\pi\sigma^2\right)^{n/2}} \exp\left[-\sum_{i=1}^{n} \frac{(x_i-\bar{x}+\bar{x}-\mu)^2}{2\sigma^2}\right] \\ &= \frac{1}{\left(2\pi\sigma^2\right)^{n/2}} \exp\left[-\frac{\sum_{i=1}^{n}(x_i-\bar{x})^2 + n(\bar{x}-\mu)^2}{2\sigma^2}\right] \end{aligned} \tag{7.64}$$

where the last equality derives from

$$\sum_{i=1}^{n}(x_i-\bar{x})(\bar{x}-\mu) = (\bar{x}-\mu)\sum_{i=1}^{n}(x_i-\bar{x}) = 0. \tag{7.65}$$

The distribution of the sample mean is

$$q(T(X)|\theta) = \frac{1}{\left(2\pi\frac{\sigma^2}{n}\right)^{1/2}} \exp\left[-\frac{n(\bar{x}-\mu)^2}{2\sigma^2}\right]. \tag{7.66}$$

The ratio of the information in the sample to the information in the statistic becomes

$$\begin{aligned} \frac{f(x|\theta)}{q(T(x)|\theta)} &= \frac{\dfrac{1}{\left(2\pi\sigma^2\right)^{n/2}} \exp\left[-\sum_{i=1}^{n} \dfrac{(x_i-\bar{x}+\bar{x}-\mu)^2}{2\sigma^2}\right]}{\dfrac{1}{\left(2\pi\dfrac{\sigma^2}{n}\right)^{1/2}} \exp\left[-\dfrac{n(\bar{x}-\mu)^2}{2\sigma^2}\right]} \\ &= \frac{1}{n^{1/2}\left(2\pi\sigma^2\right)^{n-1/2}} \exp\left[-\frac{\sigma_{i=1}^{n}(x_i-\mu)^2}{2\sigma^2}\right] \end{aligned} \tag{7.67}$$

which does not depend on μ.

Theorem 7.13 (Cassela and Berger, Factorization Theorem). *Let $f(x|\theta)$ denote the joint pdf (probability density function) or pmf (probability mass function) of a sample X. A statistic $T(X)$ is a sufficient statistic for θ if and only if there exist functions $g(t|\theta)$ and $h(x)$ such that, for all sample points x and all parameter points θ,*

$$f(x|\theta) = g(T(x)|\theta)\, h(x) \tag{7.68}$$

[7, p. 276].

Definition 7.14 (Cassela and Berger). A sufficient statistic $T(X)$ is called a minimal sufficient statistic if, for any other sufficient statistic $T'(X)$, $T(X)$ is a function of $T'(X)$ [7, p. 280].

Basically, the mimimal sufficient statistics for the normal are the sum of the sample observations and the sum of the sample observations squared. All the parameters of the normal can be derived from these two sample moments. Similarly, the method of moments estimator for the gamma distribution presented in Section 7.2.1 uses the first two sample moments to estimate the parameters of the distribution.

7.5 Concentrated Likelihood Functions

In our development of the concept of maximum likelihood in Section 7.3.6 we assumed that we knew the variance of the normal distribution, but the mean was unknown. Undoubtedly, this framework is fictional. Even if we know that the distribution is normal, it would be a rare event to know the variance. Next, consider a scenario where we concentrate the variance out of the likelihood function. Essentially, we solve for the maximum likelihood estimate of the variance and substitute that estimate into the original normal specification to derive estimates of the sample mean. The more general form of the normal likelihood function can be written as

$$L\left(X|\mu,\sigma^2\right) = \prod_{i=1}^{n} \frac{1}{\sqrt{2\pi\sigma^2}} \exp\left[-\frac{(X_i-\mu)^2}{2\sigma^2}\right]. \tag{7.69}$$

Ignoring the constants, the natural logarithm of the likelihood function can be written as

$$\ln(L) = -\frac{n}{2}\ln\left(\sigma^2\right) - \frac{1}{2\sigma^2}\sum_{i=1}^{n}(X_i-\mu)^2. \tag{7.70}$$

This expression can be solved for the optimal choice of σ^2 by differentiating with respect to σ^2.

$$\begin{aligned}\frac{\partial \ln(L)}{\partial \sigma^2} &= -\frac{n}{2\sigma^2} + \frac{1}{2\left(\sigma^2\right)^2}\sum_{i=1}^{n}(X_i-\mu)^2 = 0 \\ &\Rightarrow -n\sigma^2 + \sum_{i=1}^{n}(X_i-\mu)^2 = 0 \\ &\Rightarrow \hat{\sigma}^2_{MLE} = \frac{1}{n}\sum_{i=1}^{n}(X_i-\mu)^2.\end{aligned} \tag{7.71}$$

Substituting this result into the original logarithmic likelihood yields

$$\begin{aligned}\ln(L) &= -\frac{n}{2}\ln\left(\frac{1}{n}\sum_{i=1}^{n}(X_i-\mu)^2\right) - \frac{1}{2\frac{1}{n}\sum_{j=1}^{n}(X_j-\mu)^2}\sum_{i=1}^{n}(X_i-\mu)^2 \\ &= -\frac{n}{2}\ln\left(\frac{1}{n}\sum_{i=1}^{n}(X_i-\mu)^2\right) - \frac{n}{2}.\end{aligned} \tag{7.72}$$

Intuitively, the maximum likelihood estimate of μ is that value that minimizes the mean squared error of the estimator. Thus, the least square estimate of the mean of a normal distribution is the same as the maximum likelihood estimator under the assumption that the sample is independently and identically distributed.

7.6 Normal Equations

If we extend the above discussion to multiple regression, we can derive the normal equations. Specifically, if

$$y_i = \alpha_0 + \alpha_1 x_i + \epsilon_i \tag{7.73}$$

where ϵ_i is distributed independently and identically normal, the concentrated likelihood function above can be rewritten as

$$\begin{aligned}\ln(L) &= -\frac{n}{2}\ln\left(\frac{1}{n}\sum_{i=1}^{n}\left[y_i-\alpha_0-\alpha_1 x_i\right]^2\right) \\ &= -\frac{n}{2}\ln\left(\frac{1}{n}\sum_{i=1}^{n}\left[y_i^2-2\alpha_0 y_i-2\alpha_1 x_i y_i+\alpha_0^2+2\alpha_0\alpha_1 x_i+\alpha_1^2 x_i^2\right]\right).\end{aligned} \tag{7.74}$$

Taking the derivative with respect to α_0 yields

$$-\frac{n}{2}\frac{n}{\sum_{j=1}^{n}\left[(y_i-\alpha_0-\alpha_1 x_i)^2\right]}\sum_{i=1}^{n}\left[-2y_i+2\alpha_0+2\alpha_1 x_i\right]=0$$

$$\Rightarrow -\frac{1}{n}\sum_{i=1}^{n}y_i+\alpha_0+\alpha_1\frac{1}{n}\sum_{i=1}^{n}x_i=0 \tag{7.75}$$

$$\Rightarrow \alpha_0=\frac{1}{n}\sum_{i=1}^{n}y_i-\alpha_1\frac{1}{n}\sum_{i=1}^{n}x_i.$$

Taking the derivative with respect to α_1 yields

$$-\frac{n}{2}\frac{n}{\sum_{j=1}^{n}\left[y_i-\alpha_0-\alpha_1 x_i\right]^2}\sum_{i=1}^{n}\left[-2x_iy_i+2\alpha_0x_i+2\alpha_1x_i^2\right]=0$$

$$\Rightarrow -\frac{1}{n}\sum_{i=1}^{n}x_iy_i+\frac{1}{n}\alpha_0\sum_{i=1}^{n}x_i+\frac{1}{n}\sum_{i=1}^{n}\alpha_1x_i^2. \tag{7.76}$$

Substituting for α_0 yields

$$\begin{aligned}&-\frac{1}{n}\sum_{i=1}^{n}x_iy_i+\left(\frac{1}{n}\sum_{i=1}^{n}y_i\right)\left(\frac{1}{n}\sum_{i=1}^{n}x_i\right)\\&+\alpha_1\left(\frac{1}{n}\sum_{i=1}^{n}x_i\right)\left(\frac{1}{n}\sum_{i=1}^{n}x_i\right)+\alpha_1\frac{1}{n}\sum_{i=1}^{n}x_i^2=0.\end{aligned} \tag{7.77}$$

Hence,

$$\alpha_1=\frac{\frac{1}{n}\left(\sum_{i=1}^{n}x_iy_i-\left[\sum_{i=1}^{n}x_i\right]\left[\sum_{i=1}^{n}y_i\right]\right)}{\frac{1}{n}\left(\sum_{i=1}^{n}x_i^2-\left[\sum_{i=1}^{n}x_i^2\right]^2\right)}. \tag{7.78}$$

Hence, the estimated coefficients for the linear model can be computed from the normal equations.

7.7 Properties of Maximum Likelihood Estimators

To complete our discussion of point estimators, we want to state some of the relevant properties of the general maximum likelihood estimator. First, the maximum likelihood provides a convenient estimator of the variance of the estimated parameters based on the Cramer–Rao Lower Bound.

Theorem 7.15 (Cramer–Rao Lower Bound). *Let $L(X_1, X_2, \cdots X_n|\theta)$ be the likelihood function and let $\hat{\theta}(X_1, X_2, \cdots X_n)$ be an unbiased estimator of θ. Then, under general conditions, we have*

$$\mathrm{V}\left(\hat{\theta}\right) \geq -\frac{1}{\mathrm{E}\left[\frac{\partial^2 \ln(L)}{\partial\theta^2}\right]}. \tag{7.79}$$

Intuitively, following the Lindeberg–Levy theorem, if the maximum likelihood estimator is consistent then the distribution of the estimates will converge to normality

Theorem 7.16 (Asymptotic Normality). *Let the likelihood function be $L(X_1, X_2, \cdots X_n|\theta)$. Then, under general conditions, the maximum likelihood estimator of θ is asymptotically distributed as*

$$\hat{\theta} \overset{A}{\sim} \mathrm{N}\left(\theta, -\left[\frac{\partial^2 \ln(L)}{\partial\theta^2}\right]^{-1}\right). \tag{7.80}$$

Using the second-order Taylor series expansion of the log of the likelihood function,

$$\begin{aligned}\ln(L(\theta)) \approx \ln(L(\theta_0)) &+ \left.\frac{\partial \ln(L(\theta))}{\partial\theta}\right|_{\theta=\theta_0}(\theta-\theta_0) \\ &+\frac{1}{2}\left.\frac{\partial^2 \ln(L(\theta))}{\partial\theta^2}\right|_{\theta=\theta_0}(\theta-\theta_0)^2.\end{aligned} \tag{7.81}$$

Letting θ_0 be the estimated value, as the estimated value approaches the true value (i.e., assume that θ_0 maximizes the log-likelihood function),

$$\left.\frac{\partial \ln(L(\theta))}{\partial\theta}\right|_{\theta=\theta_0} \to 0. \tag{7.82}$$

To meet the maximization conditions

$$\left.\frac{\partial^2 \ln(L(\theta))}{\partial\theta^2}\right|_{\theta=\theta_0} \ll 0. \tag{7.83}$$

Taking a little freedom with these results and imposing the fact that the maximum likelihood estimator is consistent,

$$2\frac{\ln(L(\theta)) - \ln\left(L\left(\hat{\theta}\right)\right)}{\left.\frac{\partial^2 \ln(L(\theta))}{\partial\theta^2}\right|_{\theta=\hat{\theta}}} \approx \left(\theta-\hat{\theta}\right)^2 \tag{7.84}$$

Taking the expectation of Equation 7.84, and then inserting the results into the characteristic function

$$f(\lambda) = \exp\left(i\lambda\left(\theta - \hat{\theta}\right) - \frac{\lambda^2\left(\theta - \hat{\theta}\right)^2}{2}\right) \tag{7.85}$$

yields a result consistent with the Cramer–Rao lower bound.

7.8 Chapter Summary

- A basic concept is that randomly drawing from a population allows the researcher to replicate the distribution of the full population in the sample.
- Many familiar estimators are point estimators. These estimators estimate the value of a specific sample parameter or statistic. Chapter 8 extends our discussion to interval estimators which allow the economist to estimate a range of parameter or statistic values.
- There are a variety of measures of the quality of an estimator. In this chapter we are primarily interested in measures of closeness (i.e., how close the estimator is to the true population value).
  - One measure of closeness is the mean squared error of the estimator.
  - An estimator is inadmissable if another estimator yields a smaller or equal mean squared error for all possible parameter values.
  - There may be more than one admissable estimator.
  - Measures of closeness allow for a variety of strategies for choosing among estimators including the minimax strategy – minimizing the maximum mean squared error.
- In econometrics we are often interested in the Best Linear Unbiased Estimator (BLUE).
- An estimator is unbiased if the expected value of the estimator is equal to its true value. Alternatively, estimators may be consistent, implying that the value of the estimator converges to the true value as the sample size grows.
- Sufficient statistics are the collection of sample statistics that are not dependent on a parameter of the distribution and contain all the information in the sample regarding a particular distribution. For example, the expected first and second moment of the sample are sufficient statistics for the normal distribution.

7.9 Review Questions

7-1R. Describe the relationship between $\mathrm{P}(|T-\theta|>\epsilon) \leq \mathrm{P}(|S-\theta|>\epsilon)$ and $\mathrm{E}(T-\theta)^2 \leq \mathrm{E}(S-\theta)^2$ using the convergence results in Chapter 6.

7-2R. A fellow student states that all unbiased estimators are consistent, but not all consistent estimators are unbiased. Is this statement true or false? Why?

7.10 Numerical Exercises

7-1E. Using the distribution function

$$f(x) = \frac{3}{4}\left(1-(1-x)^2\right), \; x \in (0,2) \tag{7.86}$$

generate a sample of 20 random variables.

a. Derive the cumulative distribution function.
b. Derive the inverse function of the cumulative distribution function.
c. Draw 20 $U[0,1]$ draws and map those draws back into x space using the inverse cumulative density function.
d. Derive the mean and variance of your new sample. Compare those values with the theoretical value of the distribution.

7-2E. Extend the estimation of the Bernoulli coefficient (θ) in Section 7.3.2 to three observations. Compare the MSE for two and three sample points graphically.

7-3E. Using a negative exponential distribution

$$f(x|\,\lambda) = \lambda \exp(-\lambda x) \tag{7.87}$$

compute the maximum likelihood estimator of λ using each column in Table 7.5.

7-4E. Compute the variance for the estimator of λ in Exercise 7-3E.

7-5E. Compute the normal equations for the regression

$$y_t = \alpha_0 + \alpha_1 x_t + \epsilon_t \tag{7.88}$$

where y_t is the interest rate on agricultural loans to Florida farmers and x_t is the interest rate on Baa Corporate bonds in Appendix D.

TABLE 7.5
Exercise 7-3E Data

Obs.	1	2	3
1	10.118	1.579	0.005
2	3.859	0.332	0.283
3	1.291	0.129	0.523
4	0.238	0.525	0.093
5	3.854	0.225	0.177
6	0.040	2.855	0.329
7	0.236	0.308	0.560
8	1.555	2.226	0.094
9	5.013	0.665	0.084
10	1.205	1.919	0.041
11	0.984	0.088	0.604
12	2.686	0.058	1.167
13	7.477	0.097	0.413
14	14.879	0.644	0.077
15	1.290	0.203	0.218
16	3.907	2.618	0.514
17	2.246	0.059	0.325
18	5.173	0.052	0.270
19	2.052	1.871	0.134
20	8.649	0.783	0.072
21	6.544	0.603	0.186
22	6.297	0.189	0.099
23	4.640	0.260	0.389
24	10.924	0.677	0.088
25	10.377	2.259	0.187

8

Interval Estimation

CONTENTS

As we discussed when we talked about continuous distribution functions, the probability of a specific number under a continuous distribution is zero. Thus, if we conceptualize any estimator, either a nonparametric estimate of the mean or a parametric estimate of a function, the probability that the true value is equal to the estimated value is obviously zero. Thus, we usually talk about estimated values in terms of confidence intervals. As in the case when we discussed the probability of a continuous variable, we define some range of outcomes. However, this time we usually work the other way around, defining a certain confidence level and then stating the values that contain this confidence interval.

8.1 Confidence Intervals

Amemiya [1, p. 160] notes a difference between confidence and probability. Most troubling is our classic definition of probability as "a probabilistic statement involving parameters." This is troublesome due to our inability, without some additional Bayesian structure, to state anything concrete about probabilities.

Example 8.1. Let X_i be distributed as a Bernoulli distribution, $i = 1, 2, \cdots N$. Then

$$T = \bar{X} \overset{A}{\sim} \mathrm{N}\left(\theta, \frac{\theta(1-\theta)}{N}\right). \tag{8.1}$$

TABLE 8.1
Confidence Levels

k	$\gamma/2$	γ
1.0000	0.1587	0.3173
1.5000	0.0668	0.1336
1.6449	0.0500	0.1000
1.7500	0.0401	0.0801
1.9600	0.0250	0.0500
2.0000	0.0228	0.0455
2.3263	0.0100	0.0200

Breaking this down a little more – we will construct the estimate of the Bernoulli parameter as

$$T = \bar{X} = \frac{1}{N}\sum_{i=1}^{N} X_i \tag{8.2}$$

where $T = \hat{\theta}$. If the X_i are independent, then

$$\mathrm{V}(T) = \frac{1}{N}\mathrm{V}(X_i) = \frac{1}{N}\theta(1-\theta). \tag{8.3}$$

Therefore, we can construct a random variable Z that is the difference between the true value of the parameter θ and the value of the observed estimate.

$$Z = \frac{T-\theta}{\sqrt{\dfrac{\theta(1-\theta)}{N}}} \overset{A}{\sim} \mathrm{N}(0,1). \tag{8.4}$$

Why? By the Central Limit Theory. Given this distribution, we can ask questions about the probability. Specifically, we know that if Z is distributed $\mathrm{N}(0,1)$, then we can define

$$\gamma_k = P(|Z| < k). \tag{8.5}$$

Essentially, we can either choose a k based on a target probability or we can define a probability based on our choice of k. Using the normal probability, the one tailed probabilities for the normal distribution are presented in Table 8.1. Taking a fairly standard example, suppose that I want to choose a k such that $\gamma/2 = 0.025$, or that we want to determine the values of k such that the probability is 0.05 that the true value of γ will lie outside the range. The value of k for this choice is 1.96. This example is comparable to the standard introductory example of a 0.95 confidence level.

The values of γ_k can be derived from the standard normal table as

$$\mathrm{P}\left[\frac{|T-\theta|}{\sqrt{\dfrac{\theta(1-\theta)}{n}}} < k\right] = \gamma_k. \tag{8.6}$$

Assuming that the sample value of T is t, the confidence interval (C[.]) is defined by

$$C\left[\frac{|T-\theta|}{\sqrt{\frac{\theta(1-\theta)}{N}}}<k\right]=\gamma_k. \tag{8.7}$$

Building on the first term,

$$\begin{aligned}
P\left[\frac{|t-\theta|}{\sqrt{\frac{\theta(1-\theta)}{N}}}<k\right] &= P\left[|t-\theta|<k\sqrt{\frac{\theta(1-\theta)}{N}}\right] \\
&= P\left[(t-\theta)^2<k^2\frac{\theta(1-\theta)}{N}\right] \\
&= P\left[t^2-2t\theta+\theta^2-\frac{k^2}{N}\theta+\frac{k^2}{N}\theta^2<0\right] \\
&= P\left[\theta^2\left(1+\frac{k^2}{N}\right)+\theta\left(2t+\frac{k^2}{N}\right)+t^2<0\right]<\gamma_k.
\end{aligned} \tag{8.8}$$

Using this probability, it is possible to define two numbers $h_1(t)$ and $h_2(t)$ for which this inequality holds. Mathematically, applying the quadratic equation,

$$P\left[h_1(t)<p<h_2(t)\right]\leq\gamma_k \text{ where}$$

$$h_1(t), h_2(t)=\frac{2t+\frac{k^2}{N}\pm\sqrt{\left(2t+\frac{k^2}{N}\right)^2-4\left(1+\frac{k^2}{N}\right)t^2}}{2\left(1+\frac{k^2}{N}\right)}. \tag{8.9}$$

In order to more fully develop the concept of the confidence interval, consider the sample estimates for four draws of two different Bernoulli distributions presented in Table 8.2. The population distribution for the first four columns is for $\theta = 0.40$ while the population distribution for the second four columns holds $\theta = 0.80$. Further, the samples are nested in that the sample of 100 for draw 1 includes the sample of 50 for draw 1. Essentially, each column represents an empirical limiting process.

Starting with draw 1 such that $\theta = 0.40$, the sample value for t is 0.3800. In this discussion, we are interested in constructing an interval that contains the true value of the parameter with some degree of confidence. The question is, what are our alternatives? First, we could use the overly simplistic version

TABLE 8.2
Sample Statistics for T for 4 Draws

Statistic	$\theta = 0.40$				$\theta = 0.60$			
	1	2	3	4	1	2	3	4
Sample Size 50								
t	0.3800	0.3800	0.4400	0.5200	0.6200	0.5400	0.5400	0.6000
S_t	0.4903	0.4903	0.5014	0.5047	0.4903	0.5035	0.5035	0.4949
$\sqrt{t(1-t)/N}$	0.0686	0.0686	0.0702	0.0707	0.0686	0.0705	0.0705	0.0693
Sample Size 100								
t	0.3400	0.4300	0.3900	0.4900	0.5700	0.6000	0.5900	0.5600
S_t	0.4761	0.4976	0.4902	0.5024	0.4976	0.4924	0.4943	0.4989
$\sqrt{t(1-t)/N}$	0.0474	0.0495	0.0488	0.0500	0.0495	0.0490	0.0492	0.0496
Sample Size 150								
t	0.3867	0.4400	0.3800	0.4467	0.5867	0.5933	0.5667	0.5867
S_t	0.4886	0.4980	0.4870	0.4988	0.4941	0.4929	0.4972	0.4941
$\sqrt{t(1-t)/N}$	0.0398	0.0405	0.0396	0.0406	0.0402	0.0401	0.0405	0.0402
Sample Size 200								
t	0.3900	0.4500	0.3900	0.4300	0.6000	0.5800	0.5750	0.6050
S_t	0.4890	0.4987	0.4890	0.4963	0.4911	0.4948	0.4956	0.4901
$\sqrt{t(1-t)/N}$	0.0345	0.0352	0.0345	0.0350	0.0346	0.0349	0.0350	0.0346

of the standard normal to conclude that

$$\theta \in (t - 1.96 \times S_t, t + 1.96 \times S_t) \tag{8.10}$$

where 1.96 corresponds to a "two-sided" confidence region under normality. Obviously the range in Equation 8.10 is much too broad (i.e., the range includes values outside legitimate values of θ). Why is this the case? The confidence interval implicitly assumes that t is normally distributed. Next, if we use the estimate of the variance associated with the Bernoulli distribution in Equation 8.4, we have

$$\theta \in \left(t - 1.96 \times \sqrt{\frac{t(1-t)}{N}}, t + 1.96 \times \sqrt{\frac{t(1-t)}{N}} \right) \Rightarrow \theta \in (0.2455, 0.5145) \tag{8.11}$$

for $N = 50$ of draw 1. This interval includes the true value of θ, but we would not know that in an application. Next, consider what happens to the confidence interval as we increase the number of draws to $N = 100$.

$$\theta \in (0.2472, 0.4328). \tag{8.12}$$

Notice that the confidence region is somewhat smaller and still contains the true value of θ.

Next, we consider the confidence interval computed from the results in Equation 8.9 for the same distributions presented in Table 8.3. In this case we assume $k = 0.95$ as in Equations 8.11 and 8.12. However, the linear term in Table 8.3 is computed as

$$\frac{2t + \frac{k^2}{N}}{2\left(1 + \frac{k^2}{N}\right)} \tag{8.13}$$

while the square root involves the

$$\frac{\sqrt{\left(2t + \frac{k^2}{N}\right)^2 - 4\left(1 + \frac{k^2}{N}\right)t^2}}{2\left(1 + \frac{k^2}{N}\right)} \tag{8.14}$$

term. The lower and upper bounds are then computed by adding and subtracting Equation 8.14 from Equation 8.13. In the case of $N = 50$, the confidence interval is $(0.3175, 0.4468)$ while for $N = 100$, the confidence interval is $(0.2966, 0.3863)$.

It is clear that the values of the confidence intervals are somewhat different. In practice, the first approach, based on the limiting distribution of the maximum likelihood formulation, is probably more typical.

Next, consider the confidence interval for the mean of a normally distributed random variable where the variance is known.

TABLE 8.3
Empirical Confidence Intervals for Samples

Statistic	$\theta = 0.40$				$\theta = 0.60$			
	1	2	3	4	1	2	3	4
			Sample Size 50					
Linear	0.3821	0.3821	0.4411	0.5196	0.6179	0.5393	0.5393	0.5982
Square Root	0.1317	0.1317	0.1346	0.1355	0.1317	0.1351	0.1351	0.1329
Lower Bound	0.3175	0.3175	0.3750	0.4531	0.5532	0.4729	0.4729	0.5330
Upper Bound	0.4468	0.4468	0.5072	0.5862	0.6825	0.6057	0.6057	0.6635
			Sample Size 100					
Linear	0.3414	0.4306	0.3910	0.4901	0.5694	0.5991	0.5892	0.5595
Square Root	0.0905	0.0945	0.0931	0.0954	0.0945	0.0935	0.0939	0.0947
Lower Bound	0.2966	0.3838	0.3448	0.4428	0.5225	0.5528	0.5427	0.5125
Upper Bound	0.3863	0.4775	0.4371	0.5374	0.6162	0.6454	0.6357	0.6064
			Sample Size 150					
Linear	0.3873	0.4404	0.3807	0.4470	0.5861	0.5928	0.5663	0.5861
Square Root	0.0758	0.0772	0.0755	0.0774	0.0766	0.0764	0.0771	0.0766
Lower Bound	0.3497	0.4020	0.3432	0.4085	0.5481	0.5548	0.5279	0.5481
Upper Bound	0.4250	0.4787	0.4183	0.4854	0.6242	0.6308	0.6046	0.6242
			Sample Size 200					
Linear	0.3905	0.4502	0.3905	0.4303	0.5996	0.5796	0.5747	0.6045
Square Root	0.0657	0.0670	0.0657	0.0667	0.0660	0.0665	0.0666	0.0658
Lower Bound	0.3578	0.4169	0.3578	0.3971	0.5667	0.5466	0.5415	0.5718
Upper Bound	0.4232	0.4836	0.4232	0.4635	0.6324	0.6127	0.6078	0.6373

Example 8.2. Let $X_i \sim \mathrm{N}\left(\mu, \sigma^2\right)$, $i = 1, 2, \cdots n$ where μ is unknown and σ^2 is known. We have

$$T = \bar{X} \sim \mathrm{N}\left(\mu, \frac{\sigma^2}{N}\right). \tag{8.15}$$

Define

$$\mathrm{P}\left[\frac{|T-\mu|}{\sqrt{\frac{\sigma^2}{N}}} < k\right] = \gamma_k. \tag{8.16}$$

Example 8.2 is neat and tidy, but unrealistic. If we do not know the mean of the distribution, then it is unlikely that we will know the variance. Hence, we need to modify the confidence interval in Example 8.2 by introducing the **Student's *t*-distribution.**

Example 8.3. Suppose that $X_i \sim \mathrm{N}\left(\mu, \sigma^2\right)$, $i = 1, 2, \cdots n$ with both μ and σ^2 unknown. Let

$$T = \bar{X} \tag{8.17}$$

be an estimator of μ and

$$S^2 = \frac{1}{n}\sum_{i=1}^{n}\left(X_i - \bar{X}\right)^2 \tag{8.18}$$

be the estimator of σ^2. Then the probability distribution is

$$t_{n-1} = S^{-1}\left(T-1\right)\sqrt{n-1}. \tag{8.19}$$

This distribution is known as the *Student's t-distribution with n−1 degrees of freedom.*

Critical to our understanding of the Student's t-distribution is the amount of information in the sample. To develop this, consider a simple two observation sample

$$\begin{aligned} S^2 &\rightarrow \left(X_1 - \bar{X}\right)^2 + \left(X_2 - \bar{X}\right)^2 \\ &= \left(X_1 - \frac{X_1 + X_2}{2}\right)^2 + \left(X_2 - \frac{X_1 + X_2}{2}\right)^2 \\ &= \left(\frac{X_1}{2} - \frac{X_2}{2}\right)^2 + \left(\frac{X_2}{2} - \frac{X_1}{2}\right)^2 \\ &= \frac{1}{2}\left(X_1 - X_2\right)^2. \end{aligned} \tag{8.20}$$

Thus, two observations on X_i only give us one observation on the variance after we account for the mean – two observations only give us one degree of freedom on the sample variance. Theorem 8.4 develops the concept in a slightly more rigorous fashion.

Theorem 8.4. *Let $X_1, X_2, \cdots X_n$ be a random sample from a $\mathrm{N}(\mu, \sigma^2)$ distribution, and let*

$$\bar{X} = \frac{1}{n}\sum_{i=1}^{n} X_i \text{ and } S^2 = \frac{1}{n-1}\sum_{i=1}^{n}\left(X_i - \bar{X}\right)^2. \tag{8.21}$$

Then

a) $\bar{X}$ and S^2 are independent random variables.

b) $\bar{X} \sim \mathrm{N}(\mu, \sigma^2/n)$.

c) $(n-1)S^2/\sigma^2$ has a chi-squared distribution with $n-1$ degrees of freedom.

The proof of independence is based on the fact that S^2 is a function of the deviations from the mean which, by definition, must be independent of the mean. More interesting is the discussion of the chi-squared statistic. The chi-squared distribution is defined as

$$f(x) = \frac{1}{\Gamma\left(\frac{p}{2}\right)2^{p/2}} x^{\frac{p}{2}-1} e^{-x/2}. \tag{8.22}$$

In general, the gamma distribution is defined through the gamma function

$$\Gamma(\alpha) = \int_0^{\infty} t^{\alpha-1} e^{-t} dt. \tag{8.23}$$

Dividing both sides of the expression by $\Gamma(\alpha)$ yields

$$1 = \frac{1}{\Gamma(\alpha)}\int_0^{\infty} t^{\alpha-1} e^{-t} dt \Rightarrow f(t) = \frac{t^{\alpha-1}e^{-t}}{\Gamma(\alpha)}. \tag{8.24}$$

Substituting $X = \beta t$ gives the traditional two parameter form of the distribution function

$$f(x|\alpha,\beta) = \frac{1}{\Gamma(\alpha)\beta^{\alpha}} x^{\alpha-1} e^{-x/\beta}. \tag{8.25}$$

The expected value of the gamma distribution is $\alpha\beta$ and the variance is $\alpha\beta^2$.

Lemma 8.5 (Facts about chi-squared random variables). *We use the notation χ_p^2 to denote a chi-squared random variable with p degrees of freedom.*

- *If Z is a $\mathrm{N}(0,1)$ random variable, then $Z^2 \sim \chi_1^2$, that is, the square of a standard normal random variable is a chi-squared random variable.*

- *If $X_1, X_2, \cdots X_n$ are independent, and $X_i^2 \sim \chi_{p_i}^2$, then $X_1^2 + X_2^2 + \cdots X_n^2 \sim \chi_{p_1+p_2+\cdots p_n}^2$, that is, independent chi-squared variables add to a chi-squared variable, and the degrees of freedom also add.*

The first part of Lemma 8.5 follows from the transformation of random variables for $Y = X^2$, which yields

$$f_Y(y) = \frac{1}{2\sqrt{y}}\left(f_X(\sqrt{y}) + f_X(-\sqrt{y})\right). \tag{8.26}$$

Returning to the proof at hand, we want to show that $(n-1)S^2/\sigma^2$ has a chi-squared distribution with $n-1$ degrees of freedom. To demonstrate this, note that

$$(n-1)S_n^2 = (n-2)S_{n-1}^2 + \left(\frac{n-1}{n}\right)\left(X_n - \bar{X}_{n-1}\right)^2$$

$$S_n^2 = \frac{1}{n-1}\sum_{i=1}^{n}\left(X_i - \frac{1}{n}\sum_{j=1}^{n}X_j\right)^2$$

$$(n-1)S_n^2 = \left(X_n - \frac{1}{n}\sum_{j=1}^{n}X_j - \frac{1}{n}X_n\right)^2 + \sum_{i=1}^{n}\left(X_i - \frac{1}{n}\sum_{j=1}^{n-1}X_j - \frac{1}{n}X_n\right)^2$$

$$(n-1)S_n^2 = \left(\frac{(n-1)}{n}X_n - \frac{(n-1)}{n}\bar{X}_{n-1}\right)^2 + \sum_{i=1}^{n-1}\left[\left(X_i - \frac{1}{n}\sum_{j=1}^{n-1}X_j\right) - \frac{1}{n}X_n\right]^2. \tag{8.27}$$

If $n = 2$, we get

$$S_2^2 = \frac{1}{2}\left(X_2 - X_1\right)^2. \tag{8.28}$$

Given $(X_2 - X_1)/\sqrt{2}$ is distributed $\mathrm{N}(0,1)$, $S_2^2 \sim \chi_1^2$ and by extension for $n = k$, $(k-1)S_k^2 \sim \chi_{k-1}^2$.

Given these results for the chi-squared, the distribution of the Student's t then follows.

$$\frac{\bar{X}-\mu}{\frac{S}{\sqrt{n}}} = \frac{\frac{(\bar{X}-\mu)}{\sigma/\sqrt{n}}}{\sqrt{\frac{S^2}{\sigma^2}}}. \tag{8.29}$$

Note that this creates a standard normal random variable in the numerator and a random variable distributed

$$\sqrt{\frac{\chi_{n-1}^2}{n-1}} \tag{8.30}$$

in the denominator. The complete distribution is found by multiplying the standard normal times the chi-squared distribution times the Jacobian of the transformation yields

$$f_T(t) = \frac{\Gamma\left(\frac{p+1}{2}\right)}{\Gamma\left(\frac{p}{2}\right)\sqrt{p\pi}} \frac{1}{\left(1+t^2/p\right)^{\frac{p+1}{2}}}. \tag{8.31}$$

8.2 Bayesian Estimation

Implicitly in our previous discussions about estimation, we adopted a classical viewpoint. We had some process generating random observations. This random process was a function of fixed, but unknown parameters. We then designed procedures to estimate these unknown parameters based on observed data. Specifically, if we assumed that a random process such as students admitted to the University of Florida generated heights, then this height process can be characterized by a normal distribution. We can estimate the parameters of this distribution using maximum likelihood. The likelihood of a particular sample can be expressed as

$$L\left(X_1, X_2, \cdots X_n | \mu, \sigma^2\right) = \frac{1}{(2\pi)^{n/2}\sigma^n} \exp\left[-\frac{1}{2\sigma^2}\sum_{i=1}^{n}\left(X_i-\mu\right)^2\right]. \tag{8.32}$$

Our estimates of μ and σ^2 are then based on the value of each parameter that maximizes the likelihood of drawing that sample.

Turning this process around slightly, Bayesian analysis assumes that we can make some kind of probability statement about parameters before we start. The sample is then used to update our prior distribution. First, assume that our prior beliefs about the distribution function can be expressed as a probability density function $\pi(\theta)$ where θ is the parameter we are interested in estimating. Based on a sample (the likelihood function), we can update our knowledge of the distribution using Bayes rule.

$$\pi(\theta|X) = \frac{L(X|\theta)\,\pi(\theta)}{\displaystyle\int_{-\infty}^{\infty} L(X|\theta)\,\pi(\theta)\,d\theta}. \tag{8.33}$$

To develop this concept, assume that we want to estimate the probability of a Bernoulli event (p) such as a coin toss. The standard probability is then

$$\mathrm{P}[x] = p^x (1-p)^{(1-x)}. \tag{8.34}$$

However, instead of estimating this probability using the sample mean, we use a Bayesian approach. Our prior is that p in the Bernoulli distribution is distributed $B(\alpha, \beta)$.

The beta distribution is defined similarly to the gamma distribution.

$$f(p\,|\alpha,\beta) = \frac{1}{B(\alpha,\beta)} p^{\alpha-1}(1-p)^{\beta-1}. \tag{8.35}$$

$B(\alpha,\beta)$ is defined as

$$B(\alpha,\beta) = \int_0^1 x^{\alpha-1}(1-x)^{\beta-1}\,dx = \frac{\Gamma(\alpha)\Gamma(\beta)}{\Gamma(\alpha+\beta)}. \tag{8.36}$$

Thus, the beta distribution is defined as

$$f(p\,|\alpha,\beta) = \frac{\Gamma(\alpha+\beta)}{\Gamma(\alpha)\Gamma(\beta)} p^{\alpha-1}(1-p)^{\beta-1}. \tag{8.37}$$

Assume that we are interested in forming the posterior distribution after a single draw.

$$\begin{aligned}\pi(p\,|X) &= \frac{p^X(1-p)^{1-X}\dfrac{\Gamma(\alpha+\beta)}{\Gamma(\alpha)\Gamma(\beta)}p^{\alpha-1}(1-p)^{\beta-1}}{\displaystyle\int_0^1 p^X(1-p)^{1-X}\frac{\Gamma(\alpha+\beta)}{\Gamma(\alpha)\Gamma(\beta)}p^{\alpha-1}(1-p)^{\beta-1}\,dp}\\ &= \frac{p^{X+\alpha-1}(1-p)^{\beta-X}}{\displaystyle\int_0^1 p^{X+\alpha-1}(1-p)^{\beta-X}\,dp}.\end{aligned} \tag{8.38}$$

Following the original specification of the beta function,

$$\begin{gathered}\int_0^1 p^{X+\alpha-1}(1-p)^{\beta-X}\,dp = \int_0^1 p^{\alpha^*}(1-p)^{\beta^*-1}\,dp\\ \text{where } \alpha^* = X+\alpha \text{ and } \beta^* = \beta - X + 1\end{gathered} \tag{8.39}$$

$$\Rightarrow \int_0^1 p^{X+\alpha-1}(1-p)^{\beta-X}\,dp = \frac{\Gamma(X+\alpha)\Gamma(\beta-X+1)}{\Gamma(\alpha+\beta+1)}.$$

The posterior distribution (the distribution of P after the value of the observation is known) is then

$$\pi(p\,|X) = \frac{\Gamma(\alpha+\beta+1)}{\Gamma(X+\alpha)\Gamma(\beta-X+1)} p^{X+\alpha-1}(1-p)^{\beta-X}. \tag{8.40}$$

The Bayesian estimate of p is then the value that minimizes a loss function. Several loss functions can be used, but we will focus on the quadratic loss function consistent with the mean squared error.

$$\begin{gathered}\min_p \mathrm{E}\left[(\hat{p}-p)^2\right] \Rightarrow \frac{\partial \mathrm{E}\left[(\hat{p}-p)^2\right]}{\partial \hat{p}} = 2\mathrm{E}[\hat{p}-p] = 0\\ \Rightarrow \hat{p} = \mathrm{E}[p].\end{gathered} \tag{8.41}$$

Taking the expectation of the posterior distribution yields

$$\begin{aligned} E[p] &= \int_0^1 \frac{\Gamma(\alpha+\beta+1)}{\Gamma(X+\alpha)\Gamma(\beta-X+1)} p^{X+\alpha}(1-p)^{\beta-X}\,dp \\ &= \frac{\Gamma(\alpha+\beta+1)}{\Gamma(X+\alpha)\Gamma(\beta-X+1)} \int_0^1 p^{X+\alpha}(1-p)^{\beta-X}\,dp. \end{aligned} \tag{8.42}$$

As before, we solve the integral by creating $\alpha^* = \alpha+X+1$ and $\beta^* = \beta-X+1$. The integral then becomes

$$\begin{aligned} \int_0^1 p^{\alpha^*-1}(1-p)^{\beta^*-1}\,dp &= \frac{\Gamma(\alpha^*)\Gamma(\beta^*)}{\Gamma(\alpha^*+\beta^*)} \\ &= \frac{\Gamma(\alpha+X+1)\Gamma(\beta-X+1)}{\Gamma(\alpha+\beta+2)}. \end{aligned} \tag{8.43}$$

Hence,

$$E[p] = \frac{\Gamma(\alpha+\beta+1)}{\Gamma(\alpha+\beta+2)} \frac{\Gamma(\alpha+X+1)}{\Gamma(\alpha+X)} \frac{\Gamma(\beta-X+1)}{\Gamma(\beta-X+1)} \tag{8.44}$$

which can be simplified using the fact

$$\Gamma(\alpha+1) = \alpha\Gamma(\gamma). \tag{8.45}$$

Therefore

$$\begin{aligned} \frac{\Gamma(\alpha+\beta+1)}{\Gamma(\alpha+\beta+2)} \frac{\Gamma(\alpha+X+1)}{\Gamma(\alpha+X)} &= \frac{\Gamma(\alpha+\beta+1)}{(\alpha+\beta+1)\Gamma(\alpha+\beta+1)} \\ &\quad \times \frac{(\alpha+X)\Gamma(\alpha+X)}{\Gamma(\alpha+X)} \\ &= \frac{(\alpha+X)}{(\alpha+\beta+1)}. \end{aligned} \tag{8.46}$$

To make this estimation process operational, assume that we have a prior distribution with parameters $\alpha = \beta = 1.4968$ that yields a beta distribution with a mean p of 0.5 and a variance of the estimate of 0.0625. Next assume that we flip a coin and it comes up heads ($X = 1$). The new estimate of p becomes 0.6252. If, on the other hand, the outcome is a tail ($X = 0$), the new estimate of p is 0.3747.

Extending the results to n Bernoulli trials yields

$$\pi(p|X) = \frac{\Gamma(\alpha+\beta+n)}{\Gamma(\alpha+Y)\Gamma(\beta-Y+n)} p^{Y+\alpha-1}(1-p)^{\beta-Y+n-1} \tag{8.47}$$

where Y is the sum of individual Xs or the number of heads in the sample. The estimated value of p then becomes

$$\hat{p} = \frac{Y+\alpha}{\alpha+\beta+n}. \tag{8.48}$$

If the first draw is $Y = 15$ and $n = 50$, then the estimated value of p is 0.3112. This value compares with the maximum likelihood estimate of 0.3000. Since the maximum likelihood estimator in this case is unbiased, the results imply that the Bayesian estimator is biased.

8.3 Bayesian Confidence Intervals

Apart from providing an alternative procedure for estimation, the Bayesian approach provides a direct procedure for the formulation of parameter confidence intervals. Returning to the simple case of a single coin toss, the probability density function of the estimator becomes

$$\pi\left(p|X\right) = \frac{\Gamma\left(\alpha+\beta+1\right)}{\Gamma\left(X+\alpha\right)\Gamma\left(\beta-X+1\right)} p^{X+\alpha-1}\left(1-p\right)^{\beta-X}. \tag{8.49}$$

As previously discussed, we know that given $\alpha = \beta = 1.4968$ and a head, the Bayesian estimator of p is 0.6252. However, using the posterior distribution function, we can also compute the probability that the value of p is less than 0.5 given a head.

$$\begin{aligned} \mathrm{P}\left[p < 0.5\right] &= \int_0^{0.5} \frac{\Gamma\left(\alpha+\beta+1\right)}{\Gamma\left(X+\alpha\right)\Gamma\left(\beta-X+1\right)} p^{X+\alpha-1}\left(1-p\right)^{\beta-X} dP \\ &= 0.2976. \end{aligned} \tag{8.50}$$

Hence, we have a very formal statement of confidence intervals.

8.4 Chapter Summary

- Interval estimation involves the estimation of a range of parameter values that contains the true population value. This range is typically referred to as the confidence interval.
- The Student's t-distribution is based on the fact that the variance coefficient used for a sample of normal random variables is estimated. If the variance parameter is known, then the confidence interval can be constructed using the normal distribution.
- The posterior distribution of the Bayesian estimator allows for a direct construction of the confidence interval based on the data.

TABLE 8.4
Data for Exercise 8-1E

Obs.	x_i	Obs	x_i	Obs.	x_i
1	1	11	1	21	1
2	0	12	0	22	1
3	1	13	1	23	1
4	1	14	1	24	1
5	0	15	0	25	1
6	1	16	1	26	1
7	1	17	1	27	0
8	1	18	1	28	1
9	1	19	1	29	1
10	1	20	1	30	1

8.5 Review Questions

8-1R. Demonstrate that

$$S^2 = \sum_{i=1}^{3} (x_i - \mu)^2 \tag{8.51}$$

can be written as

$$S^2 = (x_1 - x_2)^2 + (x_2 - x_3)^2 . \tag{8.52}$$

8-2R. Discuss the implications of 8-1R for the term degrees of freedom.

8-3R. Construct the posterior distribution for the parameter θ from the Bernoulli distribution based on a prior of U $[0, 1]$. Assume that T observations out of N are positive. (Hint, use the definition of the beta distribution.)

8.6 Numerical Exercises

8-1E. Compute the confidence interval for the θ parameter of the Bernoulli distribution given in Table 8.4 using the maximum likelihood estimate of the standard deviation of the parameter.

8-E2. Construct the confidence interval for the mean of each sample using the data in Table 8.5.

8-E3. Construct the posterior distribution from review question 8-3R using the first ten observations in Table 8.4.

TABLE 8.5
Normal Random Variables for Exercise 8-E2

	Sample		
Obs.	1	2	3
1	−2.231	−4.259	1.614
2	−1.290	−7.846	1.867
3	−0.317	−0.131	3.001
4	−1.509	−5.188	0.174
5	−1.324	−5.387	3.009
6	−0.396	4.795	−0.148
7	−2.048	−0.224	−0.110
8	−3.089	4.389	2.030
9	−0.717	−0.380	2.549
10	−2.311	−1.008	0.413
11	3.686	−0.464	−1.384
12	−1.985	1.577	2.313
13	2.153	−8.507	−3.697
14	−1.205	−6.396	3.075
15	−3.798	8.004	−1.167
16	−1.063	1.526	−0.897
17	−0.593	−2.890	0.589
18	0.213	2.600	2.357
19	0.175	8.166	−0.005
20	−1.804	−1.880	3.101
21	−1.566	0.266	−1.223
22	−1.953	1.814	1.936
23	1.045	4.248	−2.907
24	2.677	2.316	0.622
25	−6.429	3.454	2.658

9

Testing Hypotheses

CONTENTS

In general, there are two kinds of hypotheses: one type concerns the form of the probability distribution (i.e., is the random variable normally distributed) and the second concerns parameters of a distribution function (i.e., what is the mean of a distribution?).

The second kind of distribution is the traditional stuff of econometrics. We may be interested in testing whether the effect of income on consumption is greater than one, or whether the effect of price on the level consumed is equal to zero. The second kind of hypothesis is termed a simple hypothesis. Under this scenario, we test the value of a parameter against a single alternative. The first kind of hypothesis (whether the effect of income on consumption is greater than one) is termed a composite hypothesis. Implicit in this test is several alternative values.

Hypothesis testing involves the comparison between two competing hypotheses, or conjectures. The null hypothesis, denoted H_0, is sometimes referred to as the maintained hypothesis. The competing hypothesis to be accepted if the null hypothesis is rejected is called the alternative hypothesis.

The general notion of the hypothesis test is that we collect a sample of data $X_1, \cdots X_n$. This sample is a multivariate random variable, E_n (refers to this as an element of a Euclidean space). If the multivariate random variable is contained in space R, we reject the null hypothesis. Alternatively, if the random variable is in the complement of the space R, we fail to reject the null hypothesis. Mathematically,

$$\begin{aligned} &\text{if } X \in R \text{ then reject } H_0; \\ &\text{if } X \notin R \text{ then fail to reject } H_0. \end{aligned} \tag{9.1}$$

The set R is called the region of rejection or the critical region of the test.

In order to determine whether the sample is in a critical region, we construct a test statistic $T(X)$. Note that, like any other statistic, $T(X)$ is a random variable. The hypothesis test given this statistic can then be written as

$$\begin{aligned} &T(X) \in R \Rightarrow \text{reject } H_0; \\ &T(X) \in \bar{R} \Rightarrow \text{fail to reject } H_0. \end{aligned} \tag{9.2}$$

A statistic used to test hypotheses is called a test statistic.

Definition 9.1. A hypothesis is called simple if it specifies the values of all the parameters of a probability distribution. Otherwise, it is called composite.

As an example, consider constructing a standard t test for the hypothesis $H_0 : \mu = 0$ against the hypothesis $H_1 : \mu = 2$. To do this we will compute the t statistic for a sample (say 20 observations from the potential population). Given this scenario, we define a critical value of 1.79 (that is, the Student's t value for 19 degrees of freedom at a 0.95 level of confidence). This test is depicted graphically in Figure 9.1. If the sample value of the t statistic is greater than $t^* = 1.79$, we reject H_0. Technically, $t \in R$, so we reject H_0.

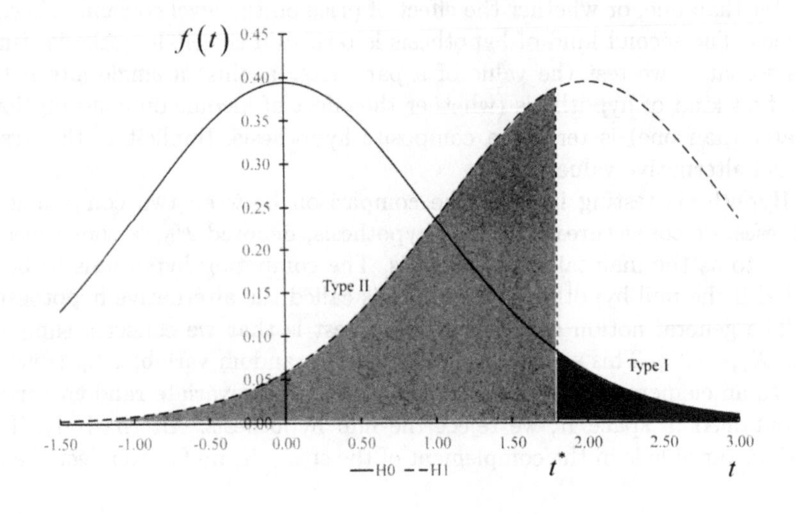

FIGURE 9.1
Type I and Type II Error.

9.1 Type I and Type II Errors

Whenever we develop a statistical test, there are two potential errors – the error of rejecting a hypothesis (typically the null hypothesis or the hypothesis of no effect) when it is true versus the error of failing to reject a hypothesis when it is false. These errors represent a tradeoff – we can make the first error essentially zero by increasing the amount of information (i.e., the level of t^* in Figure 9.1). However, increasing this critical value implies an increase in the second error – the probability of failing to reject the hypothesis when it is indeed false.

Definition 9.2. A Type I error is the error of rejecting H_0 when it is true. A Type II error is the error of accepting H_0 when it is false (that is, when H_1 is true).

We denote the probability of Type I error as α and the probability of Type II error as β. Mathematically,

$$\begin{aligned} \alpha &= \mathrm{P}\left[X \in R | H_0\right] \\ \beta &= \mathrm{P}\left[X \in \bar{R} | H_1\right]. \end{aligned} \tag{9.3}$$

The probability of a Type I error is also called the size of a test.

Assume that we want to compare two critical regions R_1 and R_2. Assume that we choose either confidence region R_1 or R_2 randomly with probabilities δ and $1-\delta$, respectively. This is called a randomized test. If the probabilities of the two types of errors for R_1 and R_2 are (α_1, β_1) and (α_2, β_2), respectively, the probability of each type of error becomes

$$\begin{aligned} \alpha &= \delta\alpha_1 + (1-\delta)\,\alpha_2 \\ \beta &= \delta\beta_1 + (1-\delta)\,\beta_2. \end{aligned} \tag{9.4}$$

The values (α, β) are called the characteristics of the test.

Definition 9.3. Let (α_1, β_1) and (α_2, β_2) be the characteristics of two tests. The first test is better (or more powerful) than the second test if $\alpha_1 \leq \alpha_2$, and $\beta_1 \leq \beta_2$ with a strict inequality holding for at least one point.

If we cannot determine that one test is better by the definition, we could consider the relative cost of each type of error. Classical statisticians typically do not consider the relative cost of the two errors because of the subjective nature of this comparison. Bayesian statisticians compare the relative cost of the two errors using a loss function.

As a starting point, we define the characteristics of a test in much the same way we defined the goodness of an estimator in Chapter 8.

Definition 9.4. A test is inadmissable if there exists another test which is better in the sense of Definition 9.3, otherwise it is called admissible.

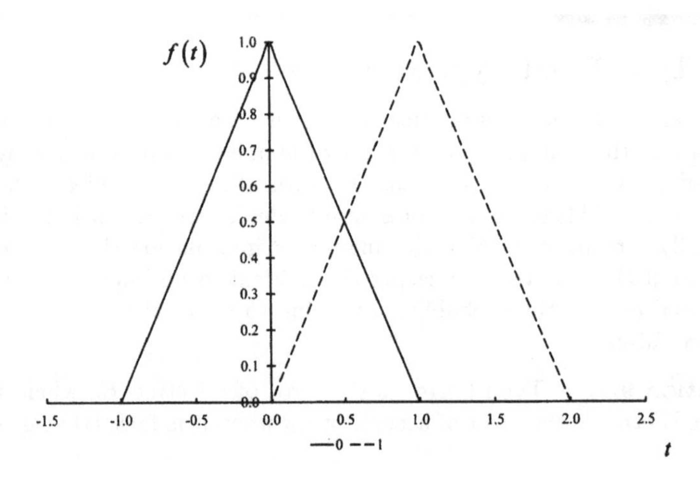

FIGURE 9.2
Hypothesis Test for Triangular Distribution.

Definition 9.5. R is the most powerful test of size α if $\alpha(R) = \alpha$ and for any test R_1 of size α, $\beta(R) \leq \beta(R_1)$.

Definition 9.6. R is the most powerful test of level α if for any test R_1 of level α (that is, such that $\alpha(R_1) \leq \alpha$), $\beta(R) \leq \beta(R_1)$.

Example 9.7. Let X have the density

$$\begin{aligned} f(x) &= 1 - \theta + x \text{ for } \theta - 1 \leq x \leq \theta \\ &= 1 + \theta - x \text{ for } \theta \leq x \leq \theta + 1. \end{aligned} \tag{9.5}$$

This funny looking beast is a triangular probability density function, as depicted in Figure 9.2. Assume that we want to test $H_0 : \theta = 0$ against $H_1 : \theta = 1$ on the basis of a single observation of X.

Type I and Type II errors are then defined by the choice of t, the cut off region

$$\begin{aligned} \alpha &= \frac{1}{2}(1-t)^2 \\ \beta &= \frac{1}{2}t^2. \end{aligned} \tag{9.6}$$

Specifically, assume that we define a sample statistic such as the mean of the sample or a single value from the distribution (x). Given this statistic, we fail to reject the null hypothesis ($H_0 : \theta = 0$) if the value is less than t. Alternatively, we reject the null hypothesis in favor of the alternative hypothesis

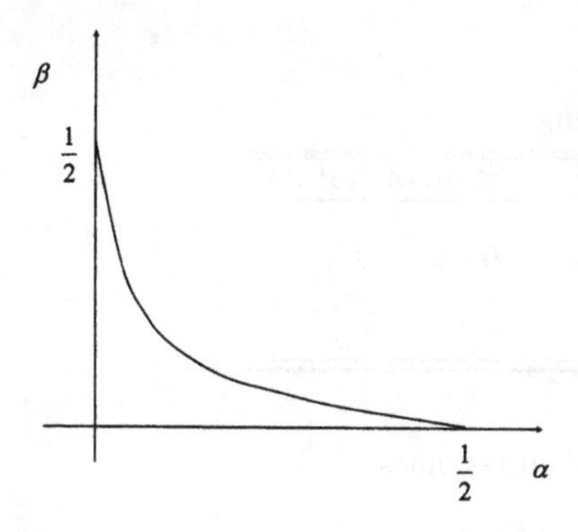

FIGURE 9.3
Tradeoff of the Power of the Test.

($H_1 : \theta = 1$) if the statistic is greater than t. In either case, we can derive the probability of the Type I error as the area of the triangle formed starting at $t = 1$ to the point t. Similarly, we can derive the Type II error starting from the origin ($t = 0$) to the point t.

Further, we can derive β in terms of α, yielding

$$\beta = \frac{1}{2}\left(1 - \sqrt{2\alpha}\right)^2. \tag{9.7}$$

Specifically, Figure 9.3 depicts the relationship between the Type I and Type II error for the hypothesis derived in Equation 9.7.

Theorem 9.8. *The set of admissible characteristics plotted on the α, β plane is a continuous, monotonically decreasing, convex function which starts at a point with $[0, 1]$ on the β axis and ends at a point within the $[0, 1]$ on the α axis.*

Note that the choice of any t yields an admissible test. However, any randomized test is inadmissible.

9.2 Neyman–Pearson Lemma

How does the Bayesian statistician choose between tests? The Bayesian chooses between the test H_0 and H_1 based on the posterior probability of the hypotheses: $\mathrm{P}(H_0|X)$ and $\mathrm{P}(H_1|X)$. Table 9.1 presents the loss matrix for hypothesis testing.

The Bayesian decision is then based on this loss function.

$$\text{Reject } H_0 \text{ if } \gamma_1 \mathrm{P}(H_0|X) < \gamma_2 \mathrm{P}(H_1|X). \tag{9.8}$$

TABLE 9.1
Loss Matrix in Hypothesis Testing

	State of Nature	
Decision	H_0	H_1
H_0	0	γ_1
H_1	γ_2	0

The critical region for the test then becomes

$$R_0 = \{x \,|\gamma_1 \text{P}(H_0|x) < \gamma_2 \text{P}(H_1|x)\}. \tag{9.9}$$

Alternatively, the Bayesian problem can be formulated as that of determining the critical region R in the domain X so as to

$$\begin{aligned}\min \phi(R) &= \gamma_1 \text{P}(H_0|X \in R)\, \text{P}(X \in R) \\ &+ \gamma_2 \text{P}(H_1|X \in \bar{R})\, \text{P}(X \in \bar{R}).\end{aligned} \tag{9.10}$$

We can write this expression as

$$\begin{aligned}\phi(R) &= \gamma_1 \text{P}(H_0)\, \text{P}(R\,|H_0) + \gamma_2 \text{P}(H_1)\, \text{P}(R\,|H_1) \\ &= \eta_0 \alpha(R) + \eta_1 \beta(R) \\ \eta_0 &= \gamma_1 \text{P}(H_0) \\ \eta_1 &= \gamma_2 \text{P}(H_1).\end{aligned} \tag{9.11}$$

Choosing between admissible test statistics in the (α, β) plane then becomes like the choice of a utility maximizing consumption point in utility theory. Specifically, the relative tradeoff between the two characteristics becomes $-\eta_0/\eta_1$.

The Bayesian optimal test R_0 can then be written as

$$R_0 = \left\{ x \left| \frac{L(x|H_1)}{L(x|H_0)} > \frac{\eta_0}{\eta_1} \right.\right\}. \tag{9.12}$$

Theorem 9.9 (Neyman–Pearson Lemma). *If testing $H_0 : \theta = \theta_0$ against $H_1 : \theta = \theta_1$, the best critical region is given by*

$$R = \left\{ x \left| \frac{L(x|\theta_1)}{L(x|\theta_0)} > c \right.\right\} \tag{9.13}$$

where L is the likelihood function and c (the critical value) is determined to satisfy

$$\text{P}(R\,|\theta_0) = \alpha \tag{9.14}$$

provided c exists.

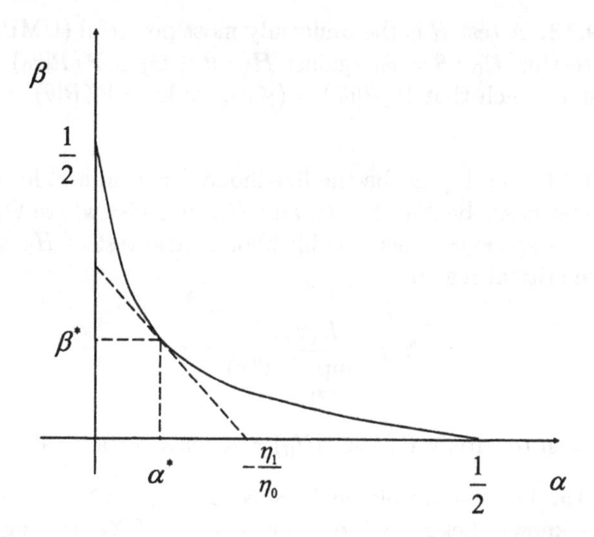

FIGURE 9.4
Optimal Choice of Type I and Type II Error.

Theorem 9.10. *The Bayes test is admissible.*

Thus, the choice of Type I and Type II error is depicted in Figure 9.4.

9.3 Simple Tests against a Composite

Mathematically, we now can express the tests as testing between $H_0 : \theta = \theta_0$ against $H_1 : \theta \in \Theta_1$, where Θ_1 is a subset of the parameter space. Given this specification, we must modify our definition of the power of the test because the β value (the probability of accepting the null hypothesis when it is false) is not unique. In this regard, it is useful to develop the power function.

Definition 9.11. If the distribution of the sample X depends on a vector of parameters θ, we define the power function of the test based on the critical region R by

$$Q(\theta) = \mathrm{P}(X \in R|\theta). \tag{9.15}$$

Definition 9.12. Let $Q_1(\theta)$ and $Q_2(\theta)$ be the power functions of two tests, respectively. Then we say that the first test is uniformly better (or uniformly most powerful) than the second in testing $H_0 : \theta = \theta_0$ against $H_1 : \theta \in \Theta_1$ if $Q_1(\theta_0) = Q_2(\theta_0)$ and $Q_1(\theta) \geq Q_2(\theta)$ for all $\theta \in \Theta_1$ and $Q_1(\theta) > Q_2(\theta)$ for at least one $\theta \in \Theta_1$.

Definition 9.13. A test R is the uniformly most powerful (UMP) test of size (level) α for testing $H_0 : \theta = \theta_0$ against $H_1 : \theta \in \Theta_1$ if $\mathrm{P}(R|\theta_0) = (\leq)\alpha$ and any other test R_1 such that $\mathrm{P}(R|\theta_0) = (\leq)\alpha$, we have $\mathrm{P}(R|\theta) \geq \mathrm{P}(R_1|\theta)$ for any $\theta \in \Theta_1$.

Definition 9.14. Let $L(x|\theta)$ be the likelihood function and let the null and alternative hypotheses be $H_0 : \theta = \theta_0$ and $H_1 : \theta \in \Theta_1$, where Θ_1 is a subset of the parameter space Θ. Then the likelihood ratio test of H_0 against H_1 is defined by the critical region

$$\Lambda = \frac{L(\theta_0|x)}{\sup\limits_{\theta_0 \cup \Theta_1} L(\theta|x)} < c \tag{9.16}$$

where c is chosen to satisfy $\mathrm{P}(\Lambda < c|H_0) = \alpha$ for a certain value of α.

Example 9.15. Let the sample be $X_i \sim \mathrm{N}(\mu, \sigma^2)$, $i = 1, 2, \cdots n$ where σ^2 is assumed to be known. Let x_i be the observed value of X_i. Testing $H_0 : \mu = \mu_0$ against $H_1 : \mu > \mu_0$, the likelihood ratio test is to reject H_0 if

$$\Lambda = \frac{\exp\left[-\frac{1}{2\sigma^2}\sum_{i=1}^{n}(x_i - \mu)^2\right]}{\sup\limits_{\mu \geq \mu_0} \exp\left[-\frac{1}{2\sigma^2}\sum_{i=1}^{n}(x_i - \mu)^2\right]} < c. \tag{9.17}$$

Assume that we had the sample $X = \{6, 7, 8, 9, 10\}$ from the preceding example and wanted to construct a likelihood ratio test for $\mu > 7.5$.

$$\begin{aligned} \Lambda &= \frac{\exp\left[-\frac{1}{2\sigma^2}\left([6-7.5]^2 + [7-7.5]^2 + \cdots [10-7.5]^2\right)\right]}{\exp\left[-\frac{1}{2\sigma^2}\left([6-8]^2 + [7-8]^2 + \cdots [10-8]^2\right)\right]} \\ &= \frac{\exp(-2.5000)}{\exp(-2.2222)} = 0.7574. \end{aligned} \tag{9.18}$$

where 8 is the maximum likelihood estimate of μ, assuming a standard deviation of 1.5 yields a likelihood ratio of 0.7574.

Theorem 9.16. *Let Λ be the likelihood ratio test statistic. Then $-2\ln(\Lambda)$ is asymptotically distributed as chi-squared with the degrees of freedom equal to the number of exact restrictions implied by H_0.*

Thus, the test statistic from Equation 9.18 is 0.5556, which is distributed χ_1^2. The probability of drawing a test statistic greater than 0.5556 is 0.46, so we fail to reject the hypothesis at any conventional level of significance.

9.4 Composite against a Composite

Testing a simple hypothesis against a composite is easy because the numerator has a single value – the value of the likelihood function evaluated at the single point θ_0. Adding a layer of complexity, consider a test that compares two possible ranges. For example, assume that we are interested in testing the hypothesis that the demand is inelastic. In this case we would test the hypothesis that $-1 \leq \eta \leq 0$ with the valid range of demand elasticities for normal goods $\eta \leq 0$. In this case we compare the likelihood function for the restricted range $(-1 \leq \eta \leq 0)$ with the general range $\eta \leq 0$.

Definition 9.17. A test R is the uniformly most powerful test of size (level) α if $\sup_{\theta \in \Theta} \mathrm{P}(R|\theta) = (\leq)\alpha$ and for any other test R_1 such that $\sup_{\theta \in \Theta} \mathrm{P}(R_1|\theta) = (\leq)\alpha$ we have $\mathrm{P}(R|\theta) \geq \mathrm{P}(R_1|\theta)$ for any $\theta \in \Theta$.

Definition 9.18. Let $L(x|\theta)$ be the likelihood function. Then the likelihood ratio test of H_0 against H_1 is defined by the critical region

$$\Lambda = \frac{\sup_{\theta_0} L(\theta|x)}{\sup_{\theta_0 \cup \Theta_1} L(\theta|x)} < c \tag{9.19}$$

where c is chosen to satisfy $\sup_{\Theta} \mathrm{P}(\Lambda < c|\theta)$ for a certain specified value of α.

Example 9.19. Let the sample $X_i \sim \mathrm{N}(\mu, \sigma^2)$ with unknown σ^2, $i = 1, 2, \cdots n$. We want to test $H_0 : \mu = \mu_0$ and $0 < \sigma^2 < \infty$ against $H_1 : \mu > \mu_0$ and $0 < \sigma^2 < \infty$.

$$L[\theta] = (2\pi)^{-n/2} \left(\sigma^2\right)^{-n/2} \exp\left[-\frac{1}{2\sigma^2} \sum_{i=1}^{n} (x_i - \mu)^2\right]. \tag{9.20}$$

Using the concentrated likelihood function at the null hypothesis,

$$\begin{aligned} \sup_{\Theta_0} L(\theta) &= (2\pi)^{-n/2} \left(\bar{\sigma}^2\right)^{-n/2} \exp\left[-\frac{n}{2}\right] \\ \bar{\sigma}^2 &= \frac{1}{n} \sum_{i=1}^{n} (x_i - \mu_0)^2 . \end{aligned} \tag{9.21}$$

The likelihood value can be compared with the maximum likelihood value of

$$\begin{aligned} \sup_{\Theta_0} L(\theta) &= (2\pi)^{-n/2} \left(\hat{\sigma}^2\right)^{-n/2} \exp\left[-\frac{n}{2}\right] \\ \hat{\sigma}^2 &= \frac{1}{n} \sum_{i=1}^{n} (x_i = \bar{x})^2 \\ \bar{x} &= \frac{1}{n} \sum_{i=1}^{n} x_i . \end{aligned} \tag{9.22}$$

The critical region then becomes

$$\left(\frac{\bar{\sigma}^2}{\hat{\sigma}^2}\right)^{-\frac{n}{2}} < c. \tag{9.23}$$

Turning back to the Bayesian model, the Bayesian would solve the problem testing $H_0 : \theta \leq \theta_0$ against $H_1 : \theta > \theta_0$. Let $L_2(\theta)$ be the loss incurred by choosing H_0 and $L_1(\theta)$ be the loss incurred by choosing H_1. The Bayesian rejects H_0 if

$$\int_{-\infty}^{\infty} L_1(\theta) f(\theta|x)\, d\theta < \int_{-\infty}^{\infty} L_2(\theta) f(\theta|x)\, d\theta \tag{9.24}$$

where $f(\theta|x)$ is the posterior distribution of θ.

Example 9.20 (Mean of a Binomial Distribution). Assume that we want to know whether a coin toss is biased based on a sample of ten tosses. Our null hypothesis is that the coin is fair ($H_0 : p = 1/2$) versus an alternative hypothesis that the coin toss is biased toward heads ($H_1 : p > 1/2$). Assume that you tossed the coin ten times and observed eight heads. What is the probability of drawing eight heads from ten tosses of a fair coin?

$$\mathrm{P}[n \geq 8] = \binom{10}{10} p^{10}(1-p)^0 + \binom{10}{9} p^9 (1-9)^1 + \binom{10}{8} p^8 (1-p)^2. \tag{9.25}$$

If $p = 1/2$, $\mathrm{P}[n \geq 8] = 0.054688$. Thus, we reject H_0 at a confidence level of 0.10 and fail to reject H_0 at a 0.05 confidence level.

Moving to the likelihood ratio test,

$$\Lambda = \frac{0.5^8 (1-0.5)^2}{0.8^8 (1-0.8)^2} = 0.1455 \Leftarrow \begin{cases} \bar{p} = 0.5 \\ \hat{p}_{MLE} = 0.8. \end{cases} \tag{9.26}$$

Given that

$$-2\ln(\Lambda) \sim \chi_1^2 \tag{9.27}$$

we reject the hypothesis of a fair coin toss at a 0.05 confidence level. $-2\ln(\Lambda) = 3.854$ and the critical region for a chi-squared distribution at one degree of freedom is 3.84.

Example 9.21. Suppose the heights of male Stanford students are distributed $N(\mu, \sigma^2)$ with a known variance of 0.16. Assume that we want to test whether the mean of this distribution is 5.8 against the hypothesis that the mean of the distribution is 6. What is the test statistic for a 5 percent level of confidence and a 10 percent level of confidence? Under the null hypothesis,

$$\bar{X} \sim \mathrm{N}\left(5.8, \frac{0.16}{10}\right). \tag{9.28}$$

The test statistic then becomes

$$Z = \frac{6 - 5.8}{0.1265} = 1.58 \sim \mathrm{N}(0, 1). \tag{9.29}$$

Given that $\mathrm{P}[Z \geq 1.58] = 0.0571$, we have the same decisions as above, namely, that we reject the hypothesis at a confidence level of 0.10 and fail to reject the hypothesis at a confidence level of 0.05.

Example 9.22 (Mean of Normal with Variance Unknown). Assume the same scenario as above, but that the variance is unknown. Given the estimated variance is 0.16, the test becomes

$$\frac{\left(\bar{X} - 6\right)}{\frac{0.16}{\sqrt{10}}} \sim t_9. \tag{9.30}$$

The computed statistic becomes $\mathrm{P}[t_9 > 1.58] = 0.074$.

Example 9.23 (Differences in Variances). In Section 8.1, we discussed the chi-squared distribution as a distribution of the sample variance. Following Theorem 8.5, let $X_1, X_2, \cdots X_n$ be a random sample from a $\mathrm{N}\left(\mu, \sigma^2\right)$ distribution, and let

$$\bar{X} = \frac{1}{n} \sum_{i=1}^{n} X_i \text{ and } S^2 = \frac{1}{n-1} \sum_{i=1}^{n} \left(X_i - \bar{X}\right)^2. \tag{9.31}$$

Then $\bar{X}$ and S^2 are independent random variables, $\bar{X} \sim \mathrm{N}\left(\mu, \frac{\sigma^2}{n}\right)$, and $\frac{(n-1) S^2}{\sigma^2}$ has a chi-squared distribution with $n - 1$ degrees of freedom.

Given the distribution of the sample variance, we may want to compare two sample variances,

$$\frac{n_X S_X^2}{\sigma_X^2} \sim \chi^2_{n_X - 1} \text{ and } \frac{n_Y S_Y^2}{\sigma_Y^2} \sim \chi^2_{n_Y - 1}. \tag{9.32}$$

Dividing the first by the second and correcting for degrees of freedom yields

$$\frac{(n_Y - 1) n_X S_X^2}{(n_X - 1) n_Y S_Y^2} \sim \mathrm{F}(n_X - 1, n_Y - 1). \tag{9.33}$$

This statistic is used to determine the statistical significance of regressions. Specifically, let S_X^2 be the standard error of a restricted model and S_Y^2 be the standard error of an unrestricted model. The ratio of the standard errors becomes a test of the restrictions. However, the test actually tests for differences in estimated variances.

9.5 Testing Hypotheses about Vectors

Extending the test results beyond the test of a single parameter, we now want to test $H_0 : \theta = \theta_0$ against $H_1 : \theta \neq \theta_0$ where θ is a $k \times 1$ vector of parameters. We begin by assuming that

$$\hat{\theta} \sim \mathrm{N}(\theta, \Sigma) \tag{9.34}$$

where Σ is a known variance matrix.

First, assuming that $k = 2$, we have

$$\begin{pmatrix} \hat{\theta}_1 \\ \hat{\theta}_2 \end{pmatrix} \sim \mathrm{N}\left[\begin{pmatrix} \theta_1 \\ \theta_2 \end{pmatrix}, \begin{pmatrix} \sigma_{11} & \sigma_{12} \\ \sigma_{12} & \sigma_{22} \end{pmatrix}\right]. \tag{9.35}$$

A simple test of the null hypothesis, assuming that the parameters are uncorrelated, would then be

$$R : \frac{\left(\hat{\theta}_1 - \theta_1\right)^2}{\sigma_{11}} + \frac{\left(\hat{\theta}_2 - \theta_2\right)^2}{\sigma_{22}} > c. \tag{9.36}$$

Building on this concept, assume that we can design a matrix A such that $A\Sigma A' = I$. This theorem relies on the eigenvalues of the matrix (see Chapter 10). Specifically, the eigenvalues of the matrix (λ) are defined by the solution of the equation $\det(\Sigma - I\lambda) = 0$. These values are real if the Σ matrix is symmetric, and positive if the Σ matrix is positive definite. In addition, if the matrix is positive definite, there are k distinct eigenvalues. Associated with each eigenvalue is an eigenvector u, defined by $u(\Sigma - I\lambda) = 0$. Carrying the eigenvector multiplication through implies $A\Sigma - A\Lambda = 0$ where A is a matrix of eigenvectors and Λ is a diagonal matrix of eigenvalues. By construction, the eigenvectors are orthogonal so that $AA' = I$. Thus, $A\Sigma A' = \Lambda$. The above decomposition is guaranteed by the diagonal nature of Λ.

This transformation implies

$$\begin{aligned} R &: \left(A\hat{\theta} - A\theta\right)'\left(A\hat{\theta} - A\theta\right) > c \\ &\Rightarrow \left[A\left(\hat{\theta} - \theta\right)\right]'\left[A\left(\hat{\theta} - \theta\right)\right] = \left(\hat{\theta} - \theta\right)' A'A\left(\hat{\theta} - \theta\right) \\ &= \left(\hat{\theta} - \theta\right)'\Sigma^{-1}\left(\hat{\theta} - \theta\right) > c. \end{aligned} \tag{9.37}$$

Note that the likelihood ratio test for this scenario becomes

$$\Lambda = \frac{\exp\left[-\frac{1}{2}\left(\hat{\theta} - \theta\right)'\Sigma^{-1}\left(\hat{\theta} - \theta\right)\right]}{\max\limits_{\hat{\theta}_{MLE}} \exp\left[-\frac{1}{2}\left(\hat{\theta}_{MLE} - \theta\right)'\Sigma^{-1}\left(\hat{\theta}_{MLE} - \theta\right)\right]}. \tag{9.38}$$

Given that $\hat{\theta}_{MLE}$ goes to θ, the numerator of the likelihood ratio test becomes one and

$$\left(\hat{\theta}-\theta_0\right)'\Sigma^{-1}\left(\hat{\theta}-\theta_0\right)\sim\chi^2_k. \tag{9.39}$$

A primary problem in the construction of these statistics is the assumption that we know the variance matrix. If we assume that we know the variance matrix to a scalar ($\Sigma=\sigma^2 Q$ where Q is known and σ^2 is unknown), the test becomes

$$\frac{\left(\hat{\theta}-\theta_0\right)'Q^{-1}\left(\hat{\theta}-\theta_0\right)}{\sigma^2}>c. \tag{9.40}$$

Using the traditional chi-squared result,

$$\frac{W}{\sigma^2}\sim\chi^2_M \tag{9.41}$$

dividing Equation 9.40 by Equation 9.41 yields

$$\frac{\left(\hat{\theta}-\theta_0\right)'Q^{-1}\left(\hat{\theta}-\theta_0\right)}{\frac{W}{M}}\sim\mathrm{F}\left(K,M\right). \tag{9.42}$$

9.6 Delta Method

The hypotheses presented above are all linear – $H_0:\beta=2$ or in vector space $H_0: 2\beta_1+\beta_2=0$. For the test in vector space we have

$$\begin{bmatrix}2\hat{\beta}_1\\ \hat{\beta}_2\end{bmatrix}'\Sigma^{-1}\begin{bmatrix}2\hat{\beta}_1\\ \hat{\beta}\end{bmatrix}\sim\chi^2_1. \tag{9.43}$$

A little more complex scenario involves the testing of nonlinear constraints or hypotheses.

Nonlinearity in estimation may arise from a variety of sources. One example involves the complexity of estimation, as discussed in Chapter 12. One frequent problem involves the estimation of the standard deviation instead of the variance in maximum likelihood. Specifically, using the normal density function to estimate the Cobb–Douglas production function with normal errors yields

$$L\propto-\frac{N}{2}\ln\left[\sigma^2\right]-\frac{1}{2\sigma^2}\sum_{i=1}^{N}\left(y_i-\alpha_0 x_{1i}^{\alpha_1}x_{2i}^{1-\alpha_1}\right)^2. \tag{9.44}$$

This problem is usually solved with iterative techniques as presented in Appendix E. These techniques attempt to improve on an initial guess by computing a step (i.e., changes in the parameters σ^2, α_0, and α_1) based on the derivatives of the likelihood function. Sometimes the step takes the parameters into an infeasible region. For example, the step may cause σ^2 to become negative. As a result, we often estimate the standard deviation rather than the variance (i.e., σ instead of σ^2). No matter the estimate of σ (i.e., negative or positive), the likelihood function presented in Equation 9.44 is always valid (i.e., $\hat{\sigma}^2 > 0$). The problem is that we are usually interested in the distribution of the variance and not the standard deviation.

The delta method is based on the first order Taylor series approximation

$$g(\beta) = g\left(\hat{\beta}\right) + \left.\frac{\partial g(\beta)}{\partial \beta}\right|_{\beta=\hat{\beta}} \left(\beta - \hat{\beta}\right). \tag{9.45}$$

Resolving Equation 9.45, we conjecture that

$$\lim_{\hat{\beta} \to \beta} \left[g(\beta) - g\left(\hat{\beta}\right)\right] = \lim_{\beta \to \hat{\beta}} \left.\frac{\partial g(\beta)}{\partial \beta}\right|_{\beta=\hat{\beta}} \left(\beta - \hat{\beta}\right) = 0. \tag{9.46}$$

Using the limit in Equation 9.46, we conclude that

$$\mathrm{V}(g(\beta)) = \left[\left.\frac{\partial g(\beta)}{\partial \beta}\right|_{\beta=\hat{\beta}}\right] \Sigma_\beta \left[\left.\frac{\partial g(\beta)}{\partial \beta}\right|_{\beta=\hat{\beta}}\right]' \tag{9.47}$$

where Σ_β is the variance (or variance matrix) for the β parameter(s).

In our simple case, assume that we estimate s (the standard deviation). The variance of the variance is then computed given that $\sigma^2 = s^2$. The variance of σ^2 is then $2^2\hat{s}^2\Sigma_s$.

9.7 Chapter Summary

- The basic concept in this chapter is to define regions for statistics such that we can fail to reject or reject hypotheses. This is the stuff of previous statistics classes – do we reject the hypothesis that the mean is zero based on a sample?
- Constructing these regions involves balancing two potential errors:
  - Type I error – the possibility of rejecting the null hypothesis when it is correct.
  - Type II error – the possibility of failing to reject the null hypothesis when it is incorrect.
- From an economic perspective there is no free lunch. Decreasing Type I error implies increasing Type II error.

- A simple hypothesis test involves testing against a single valued alternative hypothesis – $H_0 : \mu = 2$.

- A complex hypothesis involves testing against a range of alternatives $H_0 : \mu \in [0, 1]$.

9.8 Review Questions

9-1R. Given the sample $s = \{6.0, 7.0, 7.5, 8.0, 8.5, 10.0\}$, derive the likelihood test for $H_0 : \mu = 7.0$ versus $H_1 : \mu = 8.0$ assuming normality with a variance of $\sigma^2 = 2$.

9-2R. Using the sample from review question 9-1R, derive the likelihood test for $H_0 : \mu = 7.0$ versus $H_1 : \mu \neq 7.0$ assuming normality with a known variance of $\sigma^2 = 2$.

9.9 Numerical Exercises

9-1E. Consider the distribution functions

$$\begin{aligned} f(x) &= \frac{3}{4}\left(1 - (x+1)^2\right) \quad x \in [-2, 0] \\ g(x) &= \frac{1}{18}\left(\frac{9}{2} - \frac{(x-2)^2}{2}\right) \quad x \in [-1, 5] \end{aligned} \tag{9.48}$$

where the mean of $f(x)$ is -1 and $g(x)$ is 2. What value of T defined as a draw from the sample gives a Type I error of 0.10? What is the associated Type II error?

9-2E. Using the data in Table 8.4, test $H_0 : \theta = 0.50$ versus $H_1 : \theta = 0.75$.

9-3E. Using the data in Table 8.5, test whether the mean of sample 1 equals the mean of sample 3.

9-4E. Using the sample from review question 9-1R, compute the same test with an unkown variance.

Part III

Econometric Applications

10

Elements of Matrix Analysis

CONTENTS

Many of the traditional econometric applications involve the estimation of linear equations or systems of equations to describe the behavior of individuals or groups of individuals. For example, we can specify that the quantity of an input demanded by a firm is a linear function of the firm's output price, the price of the input, and the price of other inputs

$$x_t^D = \alpha_0 + \alpha_1 p_t + \alpha_2 w_{1t} + \alpha_2 w_{2t} + \alpha_3 w_{3t} + \epsilon_t \tag{10.1}$$

where x_t^D is the quantity of the input demanded at time t, p_t is the price of the firm's output, w_{1t} is the price of the input, w_{2t} and w_{3t} are the prices of other inputs used by the firm, and ϵ_t is a random error. Under a variety of assumptions such as those discussed in Chapter 6 or by assuming that $\epsilon_t \sim \mathrm{N}\left(0, \sigma^2\right)$, Equation 10.1 can be estimated using matrix methods. For example, assume that we have a simple linear specification

$$y_i = \alpha_0 + \alpha_1 x_i + \epsilon_i. \tag{10.2}$$

Section 7.6 depicts the derivation of two sets of normal equations.

$$\begin{aligned} -\frac{1}{N}\sum_{i=1}^{N} y_i + \alpha_0 + \alpha_1 \frac{1}{N}\sum_{i=1}^{N} x_i &= 0 \\ -\frac{1}{N}\sum_{i=1}^{N} x_i y_i + \alpha_0 \frac{1}{N}\sum_{i=1}^{N} x_i + \alpha_1 \frac{1}{N}\sum_{i=1}^{N} x_i^2 &= 0. \end{aligned} \tag{10.3}$$

Remembering our discussion regarding sufficient statistics in Section 7.4, we defined $T_y = 1/N \sum_{i=1}^{N} y_i$, $T_x = 1/N \sum_{i=1}^{N} x_i$, $T_{xy} = 1/N \sum_{i=1}^{N} x_i y_i$, and $T_{xx} = 1/N \sum_{i=1}^{N} x_i^2$. Given these sufficient statistics, we can rewrite the normal equations in Equation 10.3 as a linear system of equations.

$$\begin{aligned} T_y &= \alpha_0 + \alpha_1 T_x \\ T_{xy} &= \alpha_0 T_x + \alpha_1 T_{xx}. \end{aligned} \tag{10.4}$$

The system of normal equations in Equation 10.4 can be further simplified into a matrix form

$$\begin{bmatrix} T_y \\ T_{xy} \end{bmatrix} = \begin{bmatrix} 1 & T_x \\ T_x & T_{xx} \end{bmatrix} \begin{bmatrix} \alpha_0 \\ \alpha_1 \end{bmatrix}. \tag{10.5}$$

Further, we can solve for the set of parameters in Equation 10.5 using some fairly standard (and linear) operations.

This linkage between linear models and linear estimators has led to a historical reliance of econometrics on a set of linear estimators including ordinary least squares and generalized method of moments. It has also rendered the Gauss–Markov proof of the ordinary least squares (the proof that orindary least squares is the **Best Linear Unbiased Estimator** (BLUE)) an essential element of the econometrician's toolbox. This chapter reviews the basic set of matrix operations; Chapter 11 provides two related proofs of the BLUE property of ordinary least squares.

10.1 Review of Elementary Matrix Algebra

It is somewhat arbitrary and completely unnecessary for our purposes to draw a sharp demarkation between linear and matrix algebra. To introduce matrix algebra, consider the general class of linear problems similar to Equation 10.4:

$$\begin{aligned} y_1 &= \alpha_{10} + \alpha_{11} x_1 + \alpha_{12} x_2 + \alpha_{13} x_3 \\ y_2 &= \alpha_{20} + \alpha_{21} x_1 + \alpha_{22} x_2 + \alpha_{23} x_3 \\ y_3 &= \alpha_{30} + \alpha_{31} x_1 + \alpha_{32} x_2 + \alpha_{33} x_3 \\ y_4 &= \alpha_{40} + \alpha_{41} x_1 + \alpha_{42} x_2 + \alpha_{43} x_4. \end{aligned} \tag{10.6}$$

In this section, we develop the basic mechanics of writing systems of equations such as those depicted in Equation 10.6 as matrix expressions and develop solutions to these expressions.

10.1.1 Basic Definitions

As a first step, we define the concepts that allow us to write Equation 10.6 as a matrix equation.

Matrices and Vectors

A matrix A of size $m \times n$ is an $m \times n$ rectangular array of scalars:

$$A = \begin{bmatrix} a_{11} & a_{12} & \cdots & a_{1n} \\ a_{21} & a_{22} & \cdots & a_{2n} \\ \vdots & \vdots & \ddots & \vdots \\ a_{m1} & a_{m2} & \cdots & a_{mn} \end{bmatrix}. \tag{10.7}$$

As an example, we can write the coefficients from Equation 10.6 as a matrix:

$$A = \begin{bmatrix} \alpha_{10} & \alpha_{11} & \alpha_{12} & \alpha_{13} \\ \alpha_{20} & \alpha_{21} & \alpha_{22} & \alpha_{23} \\ \alpha_{30} & \alpha_{31} & \alpha_{32} & \alpha_{33} \\ \alpha_{40} & \alpha_{41} & \alpha_{42} & \alpha_{43} \end{bmatrix}. \tag{10.8}$$

It is sometimes useful to partition matrices into vectors.

$$A = \begin{bmatrix} a_{\cdot 1} & a_{\cdot 2} & \cdots & a_{\cdot n} \end{bmatrix} \Rightarrow a_{\cdot 1} = \begin{bmatrix} a_{11} \\ a_{21} \\ \vdots \\ a_{m1} \end{bmatrix} \cdots a_{\cdot n} = \begin{bmatrix} a_{1n} \\ a_{2n} \\ \vdots \\ a_{mn} \end{bmatrix} \tag{10.9}$$

$$A = \begin{bmatrix} a_{1\cdot} \\ a_{2\cdot} \\ \vdots \\ a_{m\cdot} \end{bmatrix} \Rightarrow \begin{array}{c} a_{1\cdot} = \begin{bmatrix} a_{11} & a_{12} & \cdots & a_{1n} \end{bmatrix} \\ a_{2\cdot} = \begin{bmatrix} a_{21} & a_{22} & \cdots & a_{2n} \end{bmatrix} \\ \vdots \\ a_{m\cdot} = \begin{bmatrix} a_{m1} & a_{m2} & \cdots & a_{mn} \end{bmatrix} \end{array} \tag{10.10}$$

Operations of Matrices

The sum of two identically dimensioned matrices can be expressed as

$$A + B = [a_{ij} + b_{ij}]. \tag{10.11}$$

In order to multiply a matrix by a scalar, multiply each element of the matrix by the scalar. In order to discuss matrix multiplication, we first discuss vector multiplication. Two vectors x and y can be multiplied together to form z ($z = x \cdot y$) only if they are conformable. If x is of order $1 \times n$ and y is of order $n \times 1$, then the vectors are conformable and the multiplication becomes

$$z = xy = \sum_{i=1}^{n} x_i y_i. \tag{10.12}$$

Extending this discussion to matrices, two matrices A and B can be multiplied if they are conformable. If A is of order $k \times n$ and B is of order $n \times 1$ then the matrices are conformable. Using the partitioned matrix above, we have

$$C = AB = \begin{bmatrix} a_{1\cdot} \\ a_{2\cdot} \\ \vdots \\ a_{k\cdot} \end{bmatrix} \begin{bmatrix} b_{\cdot 1} & b_{\cdot 2} & \cdots & b_{\cdot l} \end{bmatrix}$$

$$= \begin{bmatrix} a_{1\cdot}b_{\cdot 1} & a_{1\cdot}b_{\cdot 2} & \cdots & a_{1\cdot}b_{\cdot l} \\ a_{2\cdot}b_{\cdot 1} & a_{2\cdot}b_{\cdot 2} & \cdots & a_{2\cdot}b_{\cdot l} \\ \vdots & \vdots & \ddots & \vdots \\ a_{k\cdot}b_{\cdot 1} & a_{k\cdot}b_{\cdot 2} & \cdots & a_{k\cdot}b_{\cdot l} \end{bmatrix}. \tag{10.13}$$

These mechanics allow us to rewrite the equations presented in Equation 10.6 in true matrix form.

$$y = \begin{bmatrix} y_1 \\ y_2 \\ y_3 \\ y_4 \end{bmatrix} = \begin{bmatrix} \alpha_{10} & \alpha_{11} & \alpha_{12} & \alpha_{13} \\ \alpha_{20} & \alpha_{21} & \alpha_{22} & \alpha_{23} \\ \alpha_{30} & \alpha_{31} & \alpha_{32} & \alpha_{33} \\ \alpha_{40} & \alpha_{41} & \alpha_{42} & \alpha_{43} \end{bmatrix} \begin{bmatrix} 1 \\ x_1 \\ x_2 \\ x_3 \end{bmatrix} = Ax. \tag{10.14}$$

Theorem 10.1 presents some general matrix results that are useful. Basically, we can treat some matrix operations much the same way we treat scalar (i.e., single number) operations. The difference is that we always have to be careful that the matrices are conformable.

Theorem 10.1. *Let α and β be scalars and A, B, and C be matrices. Then when the operations involved are defined, the following properties hold:*

a) $A + B = B + A$

b) $(A + B) + C = A + (B + C)$

c) $\alpha (A + B) = \alpha A + \alpha B$

d) $(\alpha + \beta) = \alpha A + \beta B$

e) $A - A = A + (-A) = [0]$

f) $A (B + C) = AC + BC$

g) $(A + B) C = AC + BC$

h) $(AB) C = A (BC)$

The transpose of an $m \times n$ matrix is a $n \times m$ matrix with the rows and columns interchanged. The transpose of A is denoted A'.

Theorem 10.2. *Let α and β be scalars and A and B be matrices. Then when defined, the following hold:*

a) $(\alpha A)' = \alpha A'$

b) $(A')' = A$

c) $(\alpha A + \beta B)' = \alpha A' + \beta B'$

d) $(AB)' = B'A'$

Traces of Matrices

The trace is a function defined as the sum of the diagonal elements of a square matrix.

$$\text{tr}(A) = \sum_{i=1}^{m} a_{ii}. \tag{10.15}$$

Theorem 10.3. *Let α be scalar and A and B be matrices. Then when the appropriate operations are defined, we have:*

a) $\text{tr}(A') = \text{tr}(A)$

b) $\text{tr}(\alpha A) = \alpha\text{tr}(A)$

c) $\text{tr}(A + B) = \text{tr}(A) + \text{tr}(B)$

d) $\text{tr}(AB) = \text{tr}(B'A')$

e) $\text{tr}(A'A) = 0$ *if and only if* $A = [0]$

Traces can be very useful in statistical applications. For example, the natural logarithm of the normal distribution function can be written as

$$\begin{aligned} \Lambda_n(\mu, \Omega) &= -\frac{1}{2}mn\ln(2\pi) - \frac{1}{2}n\ln(|\Omega|) - \frac{1}{2}\text{tr}(\Omega^{-1}Z) \\ Z &= \sum i = 1^n (y_i - \mu)(y_i - \mu)'. \end{aligned} \tag{10.16}$$

Determinants of Matrices

The determinant is another function of square matrices. In its most technical form, the determinant is defined as

$$\begin{aligned} |A| &= \sum (-1)^{f(i_1, i_2, \cdots i_m)} a_{1i_1} a_{2i_2} \cdots a_{mi_m} \\ &= \sum (-1)^{f(i_1, i_2, \cdots i_m)} a_{i_1 1} a_{i_2 2} \cdots a_{i_m m} \end{aligned} \tag{10.17}$$

where the summation is taken over all permutations $(i_1, i_2, \cdots i_m)$ of the set of integers $(1, 2, \cdots m)$, and the function $f(i_1, i_2, \cdots i_m)$ equals the number of transpositions necessary to change $(i_1, i_2, \cdots i_m)$ to $(1, 2, \cdots m)$.

In the simple case of a 2×2, we have two possibilities, $(1,2)$ and $(2,1)$. The second requires one transposition. Under the basic definition of the determinant,

$$|A| = (-1)^0 a_{11}a_{22} + (-1)^1 a_{12}a_{21}. \tag{10.18}$$

In the slightly more complicated case of a 3×3, we have six possibilities, $(1,2,3)$, $(2,1,3)$, $(2,3,1)$, $(3,2,1)$, $(3,1,2)$, $(1,3,2)$. Each one of these differs from the previous one by one transposition. Thus, the number of transpositions is 0, 1, 2, 3, 4, 5. The determinant is then defined as

$$\begin{aligned} |A| &= (-1)^0 a_{11}a_{22}a_{33} + (-1)^1 a_{12}a_{21}a_{33} + (-1)^2 a_{12}a_{23}a_{31} \\ &\quad + (-1)^3 a_{13}a_{22}a_{31} (-1)^4 a_{13}a_{21}a_{32} + (-1)^5 a_{11}a_{23}a_{32} \\ &= a_{11}a_{22}a_{33} - a_{12}a_{21}a_{33} + a_{12}a_{23}a_{31} - a_{13}a_{22}a_{31} \\ &\quad + a_{13}a_{21}a_{32} - a_{11}a_{23}a_{32}. \end{aligned} \tag{10.19}$$

A more straightforward definition involves the expansion down a column or across the row. In order to do this, I want to introduce the concept of principal minors. The principal minor of an element in a matrix is the matrix with the row and column of the element removed. The determinant of the principal minor times negative one raised to the row number plus the column number is called the cofactor of the element. The determinant is then the sum of the cofactors times the elements down a particular column or across the row.

$$|A| = \sum_{j=1}^{m} a_{ij}A_{ij} = \sum a_{ij}\left[(-1)^{i+j} m_{ij}\right]. \tag{10.20}$$

In the 3×3 case

$$\begin{aligned} |A| &= a_{11}(-1)^{1+1}\begin{vmatrix} a_{22} & a_{23} \\ a_{32} & a_{33} \end{vmatrix} + a_{21}(-1)^{2+1}\begin{vmatrix} a_{12} & a_{13} \\ a_{32} & a_{33} \end{vmatrix} \\ &\quad + a_{31}(-1)^{3+1}\begin{vmatrix} a_{12} & a_{13} \\ a_{22} & a_{23} \end{vmatrix}. \end{aligned} \tag{10.21}$$

Expanding this expression yields

$$\begin{aligned} |A| &= a_{11}a_{22}a_{33} - a_{11}a_{23}a_{32} - a_{12}a_{21}a_{33} + a_{13}a_{21}a_{32} \\ &\quad + a_{12}a_{23}a_{31} - a_{13}a_{22}a_{31}. \end{aligned} \tag{10.22}$$

Theorem 10.4. *If α is a scalar and A is an $m \times m$ matrix, then the following properties hold:*

a) $|A'| = |A|$

b) $|\alpha A| = \alpha^m |A|$

c) If A is a diagonal matrix $|A| = a_{11}a_{22} \cdots a_{mm}$

d) If all the elements of a row (or column) of A are zero then $|A| = 0$

e) If two rows (or columns) of A are proportional then $|A| = 0$

f) The interchange of two rows (or columns) of A changes the sign of $|A|$

g) If all the elements of a row (or a column) of A are multiplied by α *then the determinant is multiplied by* α

h) The determinant of A is unchanged when a multiple of one row (or column) is added to another row (or column)

The Inverse

Any $m \times m$ matrix A such that $|A| \neq 0$ is said to be a nonsingular matrix and possesses an inverse denoted A^{-1}.

$$AA^{-1} = A^{-1}A = I_m. \tag{10.23}$$

Theorem 10.5. *If* α *is a nonzero scalar, and A and B are nonsingular* $m \times m$ *matrices, then:*

a) $(\alpha A)^{-1} = \alpha^{-1}A^{-1}$

b) $(A')^{-1} = (A^{-1})'$

c) $(A^{-1})^{-1} = A$

d) $|A^{-1}| = |A|^{-1}$

e) If $A = \text{diag}(a_{11}, \cdots a_{mm})$ *then* $A^{-1} = \text{diag}(a_{11}^{-1}, \cdots a_{mm}^{-1})$

f) If $A = A'$ *then* $A^{-1} = (A^{-1})'$

g) $(AB)^{-1} = B^{-1}A^{-1}$

The most general definition of an inverse involves the adjoint matrix (denoted $A_{\#}$). The adjoint matrix of A is the transpose of the matrix of cofactors of A. By construction of the adjoint, we know that

$$AA_{\#} = A_{\#}A = \text{diag}(|A|, |A|, \cdots |A|) = |A| I_m. \tag{10.24}$$

In order to see this identity, note that

$$\begin{aligned} a_{i\cdot}b_{\cdot i} &= |A| \text{ where } B = A_{\#} \\ a_{j\cdot}b_{\cdot i} &= 0 \text{ where } B = A_{\#} \; i \neq j. \end{aligned} \tag{10.25}$$

Focusing on the first point,

$$[AA_{\#}]_{11} = \begin{bmatrix} a_{11} & a_{12} & a_{13} \end{bmatrix} \begin{bmatrix} (-1)^{1+1} \begin{vmatrix} a_{22} & a_{23} \\ a_{32} & a_{33} \end{vmatrix} \\ (-1)^{1+2} \begin{vmatrix} a_{21} & a_{23} \\ a_{31} & a_{33} \end{vmatrix} \\ (-1)^{1+3} \begin{vmatrix} a_{21} & a_{22} \\ a_{31} & a_{32} \end{vmatrix} \end{bmatrix} \tag{10.26}$$

$$= (-1)^{1+1} a_{11} \begin{vmatrix} a_{22} & a_{23} \\ a_{32} & a_{33} \end{vmatrix} + (-1)^{1+2} a_{12} \begin{vmatrix} a_{21} & a_{23} \\ a_{31} & a_{33} \end{vmatrix}$$

$$+ (-1)^{1+3} a_{13} \begin{vmatrix} a_{21} & a_{22} \\ a_{31} & a_{32} \end{vmatrix} = |A|.$$

Given this expression, we see that

$$A^{-1} = |A|^{-1} A_{\#}. \tag{10.27}$$

A more applied view of the inverse involves row operations. For example, suppose we are interested in finding the inverse of a matrix

$$A = \begin{bmatrix} 1 & 9 & 5 \\ 3 & 7 & 8 \\ 2 & 3 & 5 \end{bmatrix}. \tag{10.28}$$

As a first step, we form an augmented solution matrix with the matrix we want on the left-hand side and an identity on the right-hand side, as depicted in Equation 10.29.

$$\left(\begin{array}{ccc|ccc} 1 & 9 & 5 & 1 & 0 & 0 \\ 3 & 7 & 8 & 0 & 1 & 0 \\ 2 & 3 & 5 & 0 & 0 & 1 \end{array} \right). \tag{10.29}$$

Next, we want to derive a sequence of elementry matrix operations to transform the left-hand matrix into an identity. These matrix operations will leave the inverse matrix in the right-hand side. From the matrix representation in Equation 10.29, the first series of operations is to subtract 3 times the first row from the second row and 2 times the first row from the third row. The elemental row operation to accomplish this transformation is

$$\begin{pmatrix} 1 & 0 & 0 \\ -3 & 1 & 0 \\ -2 & 0 & 1 \end{pmatrix}. \tag{10.30}$$

Multiplying Equation 10.29 by Equation 10.30 yields

$$\left(\begin{array}{ccc|ccc} 1 & 9 & 5 & 1 & 0 & 0 \\ 0 & -20 & -7 & -3 & 1 & 0 \\ 0 & -15 & -5 & -2 & 0 & 1 \end{array} \right) \tag{10.31}$$

which implies the next elemental transformation matrix

$$\begin{pmatrix} 1 & 9/20 & 0 \\ 0 & -1/20 & 0 \\ 0 & -15/20 & 1 \end{pmatrix}. \tag{10.32}$$

Multiplying Equation 10.31 by Equation 10.32 yields

$$\left(\begin{array}{ccc|ccc} 1 & 0 & -37/5 & -7/20 & 9/20 & 0 \\ 0 & 1 & -7/5 & 3/20 & -1/20 & 0 \\ 0 & 0 & 1/4 & 1/4 & -3/4 & 1 \end{array}\right) \tag{10.33}$$

which implies the final elementry operation matrix

$$\begin{pmatrix} 1 & 0 & -37/5 \\ 0 & 1 & -7/5 \\ 0 & 0 & 4 \end{pmatrix}. \tag{10.34}$$

The final result is then

$$\left(\begin{array}{ccc|ccc} 1 & 0 & 0 & -11/5 & 6 & -37/5 \\ 0 & 1 & 0 & -1/5 & 1 & -7/5 \\ 0 & 0 & 1 & 1 & -3 & 4 \end{array}\right) \tag{10.35}$$

so the inverse matrix becomes

$$A^{-1} = \begin{pmatrix} -11/5 & 6 & -37/5 \\ -1/5 & 1 & -7/5 \\ 1 & -3 & 4 \end{pmatrix}. \tag{10.36}$$

Checking this result,

$$\begin{aligned} &\begin{bmatrix} 1 & 9 & 5 \\ 3 & 7 & 8 \\ 2 & 3 & 5 \end{bmatrix} \begin{bmatrix} -11/5 & 6 & -37/5 \\ -1/5 & 1 & -7/5 \\ 1 & -3 & 4 \end{bmatrix} = \\ &\begin{bmatrix} -11/5 & 6 & -37/5 \\ -1/5 & 1 & -7/5 \\ 1 & -3 & 4 \end{bmatrix} \begin{bmatrix} 1 & 9 & 5 \\ 3 & 7 & 8 \\ 2 & 3 & 5 \end{bmatrix} = \begin{bmatrix} 1 & 0 & 0 \\ 0 & 1 & 0 \\ 0 & 0 & 1 \end{bmatrix}. \end{aligned} \tag{10.37}$$

Rank of a Matrix

The rank of a matrix is the number of linearly independent rows or columns. One way to determine the rank of any general matrix $m \times n$ is to delete rows or columns until the resulting $r \times r$ matrix has a nonzero determinant. What is the rank of the above matrix? If the above matrix had been

$$A = \begin{pmatrix} 1 & 9 & 5 \\ 3 & 7 & 8 \\ 4 & 16 & 13 \end{pmatrix} \tag{10.38}$$

note $|A| = 0$. Thus, to determine the rank, we delete the last row and column, leaving

$$A_1 = \begin{pmatrix} 1 & 9 \\ 3 & 7 \end{pmatrix} \Rightarrow |A_1| = 7 - 27 = -20 \tag{10.39}$$

The rank of the matrix is 2.

The rank of a matrix A remains unchanged by any of the following operations, called elementary transformations: (a) the interchange of two rows (or columns) of A, (b) the multiplication of a row (or column) of A by a nonzero scalar, and (c) the addition of a scalar multiple of a row (or column) of A to another row (or column) of A.

For example, we can derive the rank of the matrix in Equation 10.38 using a series of elemental matrices. As a starting point, consider the first elementary matrix to construct the inverse as discussed above.

$$\begin{pmatrix} 1 & 0 & 0 \\ -3 & 1 & 0 \\ -4 & 0 & 1 \end{pmatrix} \begin{pmatrix} 1 & 9 & 5 \\ 3 & 7 & 8 \\ 4 & 16 & 13 \end{pmatrix} = \begin{pmatrix} 1 & 9 & 5 \\ 0 & -20 & -7 \\ 0 & -20 & -7 \end{pmatrix}. \tag{10.40}$$

It is obvious in Equation 10.40 that the third row is equal to the second row so that

$$\begin{pmatrix} 1 & 0 & 0 \\ 0 & 1 & 0 \\ 0 & -1 & 1 \end{pmatrix} \begin{pmatrix} 1 & 9 & 5 \\ 0 & -20 & -7 \\ 0 & -20 & -7 \end{pmatrix} = \begin{pmatrix} 1 & 9 & 5 \\ 0 & -20 & -7 \\ 0 & 0 & 0 \end{pmatrix}. \tag{10.41}$$

Hence, the rank of the original matrix is 2 (there are two leading nonzero elements in the reduced matrix).

Orthogonal Matrices

An $m \times 1$ vector p is said to be a normalized vector or a unit vector if $p'p = 1$. The $m \times 1$ vectors $p_1, p_2, \cdots p_n$ where n is less than or equal to m are said to be orthogonal if $p_i' p_j = 0$ for all i not equal to j. If a group of n orthogonal vectors is also normalized, the vectors are said to be orthonormal. An $m \times m$ matrix consisting of orthonormal vectors is said to be orthogonal. It then follows

$$P'P = I. \tag{10.42}$$

It is possible to show that the determinant of an orthogonal matrix is either 1 or -1.

Quadratic Forms

In general, a quadratic form of a matrix can be written as

$$x'Ay = \sum_{i=1}^{m} \sum_{j=1}^{m} x_i y_j a_{ij}. \tag{10.43}$$

We are most often interested in the quadratic form $x'Ax$.

Every matrix A can be classified into one of five categories:

a) If $x'Ax > 0$ for all $x \neq 0$, then A is positive definite.

b) If $x'Ax \geq 0$ for all $x \neq 0$ and $x'Ax = 0$ for some $x \neq 0$, then A is positive semidefinite.

c) If $x'Ax < 0$ for all $x \neq 0$, then A is negative definite.

d) If $x'Ax \leq 0$ for all $x \neq 0$ and $x'Ax = 0$ for some $x \neq 0$, then A is negative semidefinite.

e) If $x'Ax > 0$ for some x and $x'Ax < 0$ for some x, then A is indefinite.

Positive and negative definiteness have a wide array of applications in econometrics. By its very definition, variance matrices are positive definite.

$$\mathrm{E}\left(x - \bar{x}\right)'\left(x - \bar{x}\right) \Rightarrow \frac{Z'Z}{N} = \Sigma_{xx} \tag{10.44}$$

where x is a vector of random variables, $\bar{x}$ is the vector of means for those random variables, and Σ_{xx} is the variance matrix. Typically, we write the sample as

$$Z = \begin{bmatrix} x_{11} - \bar{x}_1 & x_{12} - \bar{x}_2 & \cdots & x_{1r} - \bar{x}_4 \\ x_{21} - \bar{x}_1 & x_{22} - \bar{x}_2 & \cdots & x_{2r} - \bar{x}_r \\ \vdots & \vdots & \ddots & \vdots \\ x_{N1} - \bar{x}_1 & x_{N2} - \bar{x}_2 & \cdots & x_{Nr} - \bar{x}_r \end{bmatrix}. \tag{10.45}$$

Therefore $Z'Z/N = \Sigma_{xx}$ is an $r \times r$ matrix. Further, we know that the matrix is at least positive semidefinite because

$$x'\left(Z'Z\right)x \;\; \left(Zx\right)'\left(Zx\right) \geq 0. \tag{10.46}$$

Essentially, Equation 10.46 is the matrix equivalent to squaring a scalar number.

10.1.2 Vector Spaces

To develop the concept of a vector space, consider a simple linear system:

$$\begin{bmatrix} y_1 \\ y_2 \\ y_3 \end{bmatrix} = \begin{bmatrix} 5 & 3 & 2 \\ 4 & 1 & 6 \\ 9 & 4 & 8 \end{bmatrix} \begin{bmatrix} x_1 \\ x_2 \\ x_3 \end{bmatrix}. \tag{10.47}$$

Consider a slight reformulation of Equation 10.47:

$$\begin{bmatrix} y_1 \\ y_2 \\ y_3 \end{bmatrix} = \begin{bmatrix} 5 \\ 4 \\ 9 \end{bmatrix} x_1 + \begin{bmatrix} 3 \\ 1 \\ 4 \end{bmatrix} x_2 + \begin{bmatrix} 2 \\ 7 \\ 8 \end{bmatrix} x_3. \tag{10.48}$$

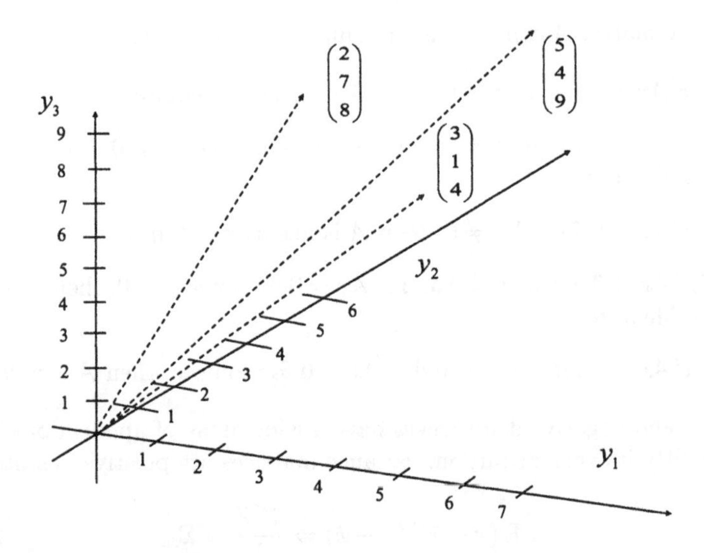

FIGURE 10.1
Vector Space.

This formulation is presented graphically in Figure 10.1. The three vectors in Equation 10.48 represent three points in a three-dimensional space. The question is what space does this set of three points span – can we explain any point in this three-dimensional space using a combination of these three points?

Definition 10.6. Let S be a collection of $m \times 1$ vectors satisfying the following: (1) $x_1 \in S$ and $x_2 \in S$, then $x_1 + x_2 \in S$ and (2) if $x \in S$ and α is a real scalar then $\alpha x \in S$. Then S is called a vector space in m-dimensional space. If S is a subset of T, which is another vector space in m-dimensional space, S is called a vector subspace of T.

Definition 10.7. Let $\{x_1, \cdots x_n\}$ be a set of $m \times 1$ vectors in the vector space S. If each vector in S can be expressed as a linear combination of the vectors $x_1, \cdots x_n$, then the set $\{x_1, \cdots x_n\}$ is said to span or generate the vector space S, and $\{x_1, \cdots x_n\}$ is called a spanning set of S.

Linear Independence and Dependence

At most the set of vectors presented in Equation 10.48 can span a three-dimensional space. However, it is possible that the set of vectors may only span a two-dimensional space. In fact, the vectors in Equation 10.48 only span a two-dimensional space – one of the dimensions is linearly dependent. If all the vectors are linearly independent, then the vectors would span a three-dimensional space. Taking it further, it is even possible that all three points lie on the same ray from the origin, so the set of vectors only spans a one-dimensional space. More formally, we state the conditions for linear independence in Definition 10.8.

Definition 10.8. The set of $m \times 1$ vectors $\{x_1, \cdots x_n\}$ is said to be linearly independent if the only solution to the equation

$$\sum_{i=1}^{n} \alpha_i x_i = 0 \tag{10.49}$$

is the zero vector $\alpha_1 = \cdots = \alpha_n = 0$.

Going back to the matrix in Equations 10.40 and 10.41,

$$\begin{pmatrix} 1 & 9 & 5 \\ 3 & 8 & 8 \\ 4 & 16 & 13 \end{pmatrix} \Rightarrow \begin{pmatrix} 1 & 0 & 37/20 \\ 0 & 1 & 7/20 \\ 0 & 0 & 0 \end{pmatrix}. \tag{10.50}$$

This reduction implies that

$$\frac{37}{20}\begin{pmatrix} 1 \\ 3 \\ 4 \end{pmatrix} + \frac{7}{20}\begin{pmatrix} 9 \\ 7 \\ 16 \end{pmatrix} = \begin{pmatrix} 5 \\ 8 \\ 13 \end{pmatrix} \tag{10.51}$$

or that the third column of the matrix is a linear combination of the first two.

Orthonormal Bases and Projections

Assume that a set of vectors $\{x_1, \cdots x_r\}$ forms a basis for some space S in R^m space such that $r \leq m$. For mathematical simplicity, we may want to forms an orthogonal basis for this space. One way to form such a basis is the Gram–Schmit orthonormalization. In this procedure, we want to generate a new set of vectors $\{y_1, \cdots y_r\}$ that are orthonormal. The Gram–Schmit process is

$$\begin{aligned} y_1 &= x_1 \\ y_2 &= x_2 - \frac{x_2' y_1}{y_1' y_1} y_1 \\ y_3 &= x_3 - \frac{x_3' y_1}{y_1' y_1} y_1 - \frac{x_3' y_2}{y_2' y_2} y_2 \end{aligned} \tag{10.52}$$

which produces a set of orthogonal vectors. Then the set of vectors z_i defined as

$$z_i = \frac{y_i}{\sqrt{y_i' y_i}} \tag{10.53}$$

spans a plane in three-dimensional space. Setting $y_1 = x_1$ (from Equation 10.48), y_2 is derived as

$$y_2 = \begin{pmatrix} 9 \\ 7 \\ 16 \end{pmatrix} - \frac{\begin{pmatrix} 9 & 7 & 16 \end{pmatrix}\begin{pmatrix} 1 \\ 3 \\ 4 \end{pmatrix}}{\begin{pmatrix} 1 & 3 & 4 \end{pmatrix}\begin{pmatrix} 1 \\ 3 \\ 4 \end{pmatrix}} \begin{pmatrix} 1 \\ 3 \\ 4 \end{pmatrix} = \begin{pmatrix} 70/13 \\ -50/13 \\ 20/13 \end{pmatrix}. \tag{10.54}$$

The vectors can then be normalized to one. However, to test for orthogonality,

$$\begin{pmatrix} 1 & 3 & 4 \end{pmatrix} \begin{pmatrix} 70/13 \\ -50/13 \\ 20/13 \end{pmatrix} = 0. \tag{10.55}$$

Theorem 10.9. *Every r-dimensional vector space, except the zero-dimensional space $\{0\}$, has an orthonormal basis.*

Theorem 10.10. *Let $\{z_1, \cdots z_r\}$ be an orthonomal basis for some vector space S, of R^m. Then each $x \in R^m$ can be expressed uniquely as*

$$x = u + v \tag{10.56}$$

where $u \in S$ and v is a vector that is orthogonal to every vector in S.

Definition 10.11. Let S be a vector subspace of R^m. The orthogonal complement of S, denoted $S^{\perp}$, is the collection of all vectors in R^m that are orthogonal to every vector in S: that is, $S^{\perp} = (x : x \in R^m \text{ and } x'y = 0, \forall y \in S\}$.

Theorem 10.12. *If S is a vector subspace of R^m, then its orthogonal complement $S^{\perp}$ is also a vector subspace of R^m.*

10.2 Projection Matrices

The orthogonal projection of an $m \times 1$ vector x onto a vector space S can be expressed in matrix form. Let $\{z_1, \cdots z_r\}$ be any othonormal basis for S while $\{z_1, \cdots z_m\}$ is an orthonormal basis for R^m. Any vector x can be written as

$$x = (\alpha_1 z_1 + \cdots \alpha_r z_r) + (\alpha_{r+1} z_{r+1} + \cdots \alpha_m z_m) = u + v. \tag{10.57}$$

Aggregating $\alpha = (\alpha_1' \quad \alpha_2')'$ where $\alpha_1 = (\alpha_1 \quad \cdots \quad \alpha)$ and $\alpha_2 = (\alpha_{r+1} \quad \cdots \quad \alpha_m)$ and assuming a similar decomposition of $Z = [Z_1 \quad Z_2]$, the vector x can be written as

$$\begin{aligned} X &= Z\alpha = Z_1\alpha_1 + Z_2\alpha_2 \\ u &= Z_1\alpha_1 \\ v &= Z_2\alpha_2. \end{aligned} \tag{10.58}$$

Given orthogonality, we know that $Z_1'Z_1 = I_r$ and $Z_1'Z_2 = [0]$, and so

$$Z_1 Z_1' x = Z_1 Z_1' [Z_1 \quad Z_2] \begin{bmatrix} \alpha_1 \\ \alpha_2 \end{bmatrix} = [Z_1 \quad 0] \begin{bmatrix} \alpha_1 \\ \alpha_2 \end{bmatrix} = Z_1 \alpha_1 = u. \tag{10.59}$$

Theorem 10.13. *Suppose the columns of the $m \times r$ matrix Z_1 form an orthonormal basis for the vector space S, which is a subspace of R^m. If $x \in R^m$, the orthogonal projection of x onto S is given by $Z_1 Z_1' x$.*

Projection matrices allow the division of the space into a spanned space and a set of orthogonal deviations from the spanning set. One such separation involves the Gram–Schmit system. In general, if we define the $m \times r$ matrix $X_1 = (x_1, \cdots x_r)$ and define the linear transformation of this matrix that produces an orthonormal basis as A,

$$Z_1 = X_1 A. \tag{10.60}$$

We are left with the result that

$$Z_1' Z_1 = A' X_1' X_1 A = I_r. \tag{10.61}$$

Given that the matrix A is nonsingular, the projection matrix that maps any vector x onto the spanning set then becomes

$$P_S = Z_1 Z_1' = X_1 A A' X_1' = X_1 \left(X'X\right)^{-1} X_1'. \tag{10.62}$$

Ordinary least squares is also a spanning decomposition. In the traditional linear model

$$\begin{aligned} y &= X\beta + \epsilon \\ \hat{y} &= X\hat{\beta} \end{aligned} \tag{10.63}$$

within this formulation β is chosen to minimize the error between y and estimated $\hat{y}$.

$$(y - X\beta)' \, (y - X\beta). \tag{10.64}$$

This problem implies minimizing the distance between the observed y and the predicted plane $X\beta$, which implies orthogonality. If X has full column rank, the projection space becomes $X\left(X'X\right)^{-1} X'$ and the projection then becomes

$$X\beta = X\left(X'X\right)^{-1} X'y. \tag{10.65}$$

Premultiplying each side by X' yields

$$\begin{aligned} X'X\beta &= X'X\left(X'X\right)^{-1} X'y \\ \beta &= \left(X'X\right)^{-1} X'X\left(X'X\right)^{-1} X'y \\ \beta &= \left(X'X\right)^{-1} X'y. \end{aligned} \tag{10.66}$$

Essentially, the projection matrix is defined as that spanning space where the unexplained factors are orthogonal to the space $\alpha_1 z_1 + \cdots \alpha_r z_4$. The spanning space defined by Equation 10.65 is identical to the definition of the spanning space in Equation 10.57. Hence, we could justify ordinary least squares as its name implies as that set of coefficients that minimizes the sum squared error, or as that space such that the residuals are orthogonal to the predicted values. These spanning spaces will also become important in the construction of instrumental variables in Chapter 11.

10.3 Idempotent Matrices

Idempotent matrices can be defined as any matrix such that $AA = A$. Note that the sum of square errors (SSE) under the projection can be expressed as

$$\begin{aligned} SSE &= (y - X\beta)'(y - X\beta) \\ &= \left(y - X(X'X)^{-1}X'y\right)'\left(y - X(X'X)^{-1}X'y\right) \\ &= \left(\left(I_n - X(X'X)^{-1}X'\right)y\right)'\left(I_n - X(X'X)^{-1}X'\right)y \\ & y'\left(I_n - X(X'X)^{-1}X'\right)\left(I_n - X(X'X)^{-1}X'\right)y. \end{aligned} \tag{10.67}$$

The matrix $I_n - X(X'X)^{-1}X'$ is an idempotent matrix.

$$\begin{aligned} &\left(I_n - X(X'X)^{-1}X'\right)\left(I_n - X(X'X)^{-1}X'\right) = \\ &I_n - X(X'X)^{-1}X' - X(X'X)^{-1}X' + X(X'X)^{-1}X'X(X'X)^{-1}X' \\ &= I_n - X(X'X)^{-1}X'. \end{aligned} \tag{10.68}$$

Thus, the SSE can be expressed as

$$\begin{aligned} SSE &= y'\left(I_n - X(X'X)^{-1}X'\right)y \\ &= y'y - y'X(X'X)^{-1}X'y = v'v \end{aligned} \tag{10.69}$$

which is the sum of the orthogonal errors from the regression.

10.4 Eigenvalues and Eigenvectors

Eigenvalues and eigenvectors (or more appropriately latent roots and characteristic vectors) are defined by the solution

$$Ax = \lambda x \tag{10.70}$$

for a nonzero x. Mathematically, we can solve for the eigenvalue by rearranging the terms

$$\begin{aligned} Ax - \lambda x &= 0 \\ (A - \lambda I)x &= 0. \end{aligned} \tag{10.71}$$

Solving for λ then involves solving the characteristic equation that is implied by

$$|A - \lambda I| = 0. \tag{10.72}$$

Again using the matrix in the previous example,

$$\left| \begin{bmatrix} 1 & 9 & 5 \\ 3 & 7 & 8 \\ 2 & 3 & 5 \end{bmatrix} - \lambda \begin{bmatrix} 1 & 0 & 0 \\ 0 & 1 & 0 \\ 0 & 0 & 1 \end{bmatrix} \right| = \left| \begin{matrix} 1-\lambda & 9 & 5 \\ 3 & 7-\lambda & 8 \\ 2 & 3 & 5-\lambda \end{matrix} \right| \tag{10.73}$$

$$= -5 + 14\lambda + 13\lambda^2 - \lambda^3 = 0.$$

In general, there are m roots to the characteristic equation. Some of these roots may be the same. In the above case, the roots are complex. Turning to another example,

$$A = \begin{bmatrix} 5 & -3 & 3 \\ 4 & -2 & 3 \\ 4 & -4 & 5 \end{bmatrix} \Rightarrow \lambda = \{1, 2, 5\}. \tag{10.74}$$

The eigenvectors are then determined by the linear dependence in the $A - \lambda I$ matrix. Taking the last example (with $\lambda = 1$),

$$[A - \lambda I] = \begin{bmatrix} 4 & -3 & 3 \\ 4 & -3 & 3 \\ 4 & -4 & 4 \end{bmatrix}. \tag{10.75}$$

The first and second rows are linear. The reduced system then implies that as long as $x_1 = x_2$ and $x_3 = 0$, the resulting matrix is zero.

Theorem 10.14. *For any symmetric matrix A there exists an orthogonal matrix H (that is, a square matrix satisfying $H'H = I$) such that*

$$H'AH = \Lambda \tag{10.76}$$

where Λ is a diagonal matrix. The diagonal elements of Λ are called the characteristic roots (or eigenvalues) of A. The ith column of H is called the characteristic vector (or eigenvector) of A corresponding to the characteristic root of A.

This proof follows directly from the definition of eigenvalues. Letting H be a matrix with eigenvalues in the columns, it is obvious that

$$AH = \Lambda H \tag{10.77}$$

by our original discussion of eigenvalues and eigenvectors. In addition, eigenvectors are orthogonal and can be normalized to one. H is an orthogonal matrix. Thus,

$$H'AH = H'\Lambda H = \Lambda H'H = \Lambda. \tag{10.78}$$

One useful application of eigenvalues and eigenvectors is the fact that the eigenvalues and eigenvectors of a real symmetric matrix are also real. Further, if all the eigenvalues are positive, the matrix is positive definite. Alternatively, if all the eigenvalues are negative, the matrix is negative definite. This is particularly useful for econometrics because most of our matrices are symmetric. For example, the sample variance matrix is symmetric and positive definite.

10.5 Kronecker Products

Two special matrix operations that you will encounter are the Kronecker product and vec (.) operators. The Kronecker product is a matrix of an element by element multiplication of the elements of the first matrix by the entire second matrix

$$A \otimes B = \begin{bmatrix} a_{11}B & a_{12}B & \cdots & a_{1n}B \\ a_{21}B & a_{22}B & \cdots & a_{2n}B \\ \vdots & \vdots & \ddots & \vdots \\ a_{m1}B & a_{m2}B & \cdots & a_{mn}B \end{bmatrix}. \quad (10.79)$$

The vec (.) operator then involves stacking the columns of a matrix on top of one another.

The Kronecker product and vec (.) operators appear somewhat abstract, but are useful in certain specifications of systems of equations. For example, suppose that we want to estimate a system of two equations,

$$\begin{aligned} y_1 &= \alpha_{10} + \alpha_{11}x_1 + \alpha_{12}x_2 \\ y_2 &= \alpha_{20} + \alpha_{21}x_1 + \alpha_{22}x_2. \end{aligned} \quad (10.80)$$

Assuming a small sample of three observations, we could express the system of equations in a sample as

$$\begin{bmatrix} y_{11} & y_{12} \\ y_{21} & y_{22} \\ y_{31} & y_{32} \end{bmatrix} = \begin{bmatrix} 1 & x_{11} & x_{12} \\ 1 & x_{21} & x_{22} \\ 1 & x_{31} & x_{32} \end{bmatrix} \begin{bmatrix} \alpha_{10} & \alpha_{20} \\ \alpha_{11} & \alpha_{21} \\ \alpha_{21} & \alpha_{22} \end{bmatrix}. \quad (10.81)$$

As will be developed more fully in Chapter 11, we can estimate both equations at once if we rearrange the system in Equation 10.81 as

$$\begin{bmatrix} y_{11} \\ y_{12} \\ y_{21} \\ y_{22} \\ y_{31} \\ y_{32} \end{bmatrix} = \begin{bmatrix} 1 & x_{11} & x_{12} & 0 & 0 & 0 \\ 0 & 0 & 0 & 1 & x_{11} & x_{12} \\ 1 & x_{21} & x_{22} & 0 & 0 & 0 \\ 0 & 0 & 0 & 1 & x_{21} & x_{22} \\ 1 & x_{31} & x_{32} & 0 & 0 & 0 \\ 0 & 0 & 0 & 1 & x_{31} & x_{32} \end{bmatrix} \begin{bmatrix} \alpha_{10} \\ \alpha_{11} \\ \alpha_{12} \\ \alpha_{20} \\ \alpha_{21} \\ \alpha_{22} \end{bmatrix}. \quad (10.82)$$

Taking each operation in turn, the first operation is usually written as vecr (y) or making a vector by rows

$$\text{vecr}\left(\begin{bmatrix} y_{11} & y_{12} \\ y_{21} & y_{22} \\ y_{31} & y_{32} \end{bmatrix}\right) \begin{bmatrix} \begin{bmatrix} y_{11} \\ y_{12} \end{bmatrix} \\ \begin{bmatrix} y_{21} \\ y_{22} \end{bmatrix} \\ \begin{bmatrix} y_{31} \\ y_{32} \end{bmatrix} \end{bmatrix} = \begin{bmatrix} y_{11} \\ y_{12} \\ y_{21} \\ y_{22} \\ y_{31} \\ y_{32} \end{bmatrix} \tag{10.83}$$

(this is equivalent to vec (y')). The next term involves the Kronecker product

$$I_{2\times 2} \otimes X = \begin{bmatrix} 1 & 0 \\ 0 & 1 \end{bmatrix} \otimes \begin{bmatrix} 1 & x_{11} & x_{12} \\ 1 & x_{21} & x_{22} \\ 1 & x_{31} & x_{32} \end{bmatrix}$$

$$= \begin{bmatrix} 1\times\begin{bmatrix}1 & x_{11} & x_{12}\end{bmatrix} & 0\times\begin{bmatrix}1 & x_{11} & x_{12}\end{bmatrix} \\ 0\times\begin{bmatrix}1 & x_{11} & x_{12}\end{bmatrix} & 1\times\begin{bmatrix}1 & x_{11} & x_{12}\end{bmatrix} \\ 1\times\begin{bmatrix}1 & x_{21} & x_{22}\end{bmatrix} & 0\times\begin{bmatrix}1 & x_{21} & x_{22}\end{bmatrix} \\ 0\times\begin{bmatrix}1 & x_{21} & x_{22}\end{bmatrix} & 1\times\begin{bmatrix}1 & x_{21} & x_{22}\end{bmatrix} \\ 1\times\begin{bmatrix}1 & x_{11} & x_{12}\end{bmatrix} & 0\times\begin{bmatrix}1 & x_{11} & x_{12}\end{bmatrix} \\ 0\times\begin{bmatrix}1 & x_{11} & x_{12}\end{bmatrix} & 1\times\begin{bmatrix}1 & x_{11} & x_{12}\end{bmatrix} \end{bmatrix} \tag{10.84}$$

Following the operations through gives the X matrix in Equation 10.82. Completing the formulation vec (α) is the standard vectorization

$$\text{vec}\left(\begin{bmatrix} \alpha_{10} & \alpha_{20} \\ \alpha_{11} & \alpha_{21} \\ \alpha_{21} & \alpha_{22} \end{bmatrix}\right) = \begin{bmatrix} \begin{bmatrix} \alpha_{10} \\ \alpha_{11} \\ \alpha_{12} \end{bmatrix} \\ \begin{bmatrix} \alpha_{20} \\ \alpha_{21} \\ \alpha_{22} \end{bmatrix} \end{bmatrix} = \begin{bmatrix} \alpha_{10} \\ \alpha_{11} \\ \alpha_{12} \\ \alpha_{20} \\ \alpha_{21} \\ \alpha_{22} \end{bmatrix}. \tag{10.85}$$

Thus, while we may not refer to it as simplified, Equation 10.81 can be written as

$$\text{vecr}\,(y) = [I_{2\times 2} \otimes X]\,\text{vec}\,(\alpha)\,. \tag{10.86}$$

At least it makes for simpler computer coding.

10.6 Chapter Summary

- The matrix operations developed in this chapter are used in ordinary least squares and the other linear econometric models presented in Chapter 11.
- Matrix algebra allows us to specify multivariate regression equations and solve these equations relatively efficiently.
- Apart from solution techniques, matrix algebra has implications for spanning spaces – regions that can be explained by a set of vectors.
  - Spanning spaces are related to the standard ordinary least squares estimator – the mechanics of the ordinary least squares estimator guarantees that the residuals are orthogonal to the estimated regression relationship.
  - The concept of an orthogonal projection will be used in instrumental variables techniques in Chapter 11.

10.7 Review Questions

10-1R. If $\text{tr}[A] = 13$, then what is $\text{tr}[A + k \times I_{m \times m}]$?

10.8 Numerical Exercises

10-1E. Starting with the matrix

$$A = \begin{bmatrix} 5 & 3 & 2 \\ 3 & 1 & 4 \\ 2 & 4 & 7 \end{bmatrix} \tag{10.87}$$

compute the determinant of matrix A.

10-2E. Compute $B = A + 2 \times I_{3\times3}$ where $I_{3\times3}$ is the identity matrix (i.e., a matrix of zeros with a diagonal of 1).

10-3E. Compute the inverse of $B = A + 2 \times I_{3\times3}$ using row operations.

10-4E. Compute the eigenvalues of B. Is B negative definite, positive definite, or indefinite?

10-5E. Demonstrate that $AA_{\#} = |A|\, I$.

10-6E. Compute the inverse of A using the cofactor matrix (i.e., the fact that $AA_{\#} = |A|\, I$). Hint – remember that A is symmetric.

10-7E. As a starting point for our discussion, consider the linear model

$$r_t = \alpha_0 + \alpha_1 R_t + \alpha_2 \Delta \left(D/A\right)_t + \epsilon_t \tag{10.88}$$

where r_t is the interest rate paid by Florida farmers, R_t is the Baa Corporate bond rate, and $\Delta \left(D/A\right)_t$ is the change in the debt to asset ratio for Florida agriculture. For our discussion, we assume

$$\begin{bmatrix} r_t \\ 1 \\ R_t \\ \Delta \left(D/a\right)_t \end{bmatrix} \sim N\left(\begin{bmatrix} 0.07315 \\ 1.00000 \\ 0.08534 \\ 0.00904 \end{bmatrix}, \begin{bmatrix} 0.00559 & 0.07315 & 0.00655 & 0.00045 \\ 0.07315 & 1.00000 & 0.08535 & 0.00905 \\ 0.00655 & 0.08535 & 0.00793 & 0.00063 \\ 0.00045 & 0.00905 & 0.00063 & 0.00323 \end{bmatrix} \right). \tag{10.89}$$

- Defining

$$\Sigma_{XX} = \begin{bmatrix} 1.00000 & 0.08535 & 0.00905 \\ 0.08535 & 0.00793 & 0.00063 \\ 0.00905 & 0.00063 & 0.00323 \end{bmatrix} \tag{10.90}$$

and

$$\Sigma_{XY} = \begin{bmatrix} 0.07315 \\ 0.00655 \\ 0.00045 \end{bmatrix} \tag{10.91}$$

compute $\beta = \Sigma_{XX}^{-1} \Sigma_{XY}$.

- Compute $\Sigma_{YY} - \Sigma_{YX} \Sigma_{XX}^{-1} \Sigma_{XY}$ where $\Sigma_{YY} = 0.00559$.

11

Regression Applications in Econometrics

CONTENTS

The purpose of regression analysis is to explore the relationship between two variables. In this course, the relationship that we will be interested in can be expressed as

$$y_i = \alpha + \beta x_i + \epsilon_i \tag{11.1}$$

where y_i is a random variable and x_i is a variable hypothesized to affect or drive y_i.

(a) The coefficients α and β are the intercept and slope parameters, respectively.

(b) These parameters are assumed to be fixed, but unknown.

(c) The residual ϵ_i is assumed to be an unobserved, random error.

(d) Under typical assumptions $\mathrm{E}[\epsilon_i] = 0$.

(e) Thus, the expected value of y_i given x_i then becomes

$$\mathrm{E}[y_i] = \alpha + \beta x_i. \tag{11.2}$$

The goal of regression analysis is to estimate α and β and to say something about the significance of the relationship. From a terminology standpoint, y is typically referred to as the dependent variable and x is referred to as the independent variable. Casella and Berger [7] prefer the terminology of y as the response variable and x as the predictor variable. This relationship is a linear regression in that the relationship is linear in the parameters α and β. Abstracting for a moment, the traditional Cobb–Douglas production function can be written as

$$y_i = \alpha x_i^{\beta}. \tag{11.3}$$

Taking the natural logarithm of both sides yields

$$\ln(y_i) = \ln(\alpha) + \beta \ln(x_i). \tag{11.4}$$

Noting that $\ln(\alpha) = \alpha^*$, this relationship is linear in the estimated parameters and thus can be estimated using a simple linear regression.

11.1 Simple Linear Regression

The setup for simple linear regression is that we have a sample of n pairs of variables $(x_1, y_1), \cdots (x_n, y_n)$. Further, we want to summarize this relationship by fitting a line through the data. Based on the sample data, we first describe the data as follows:

1. The sample means

$$\bar{x} = \frac{1}{n}\sum_{i=1}^{n} x_i,\ \bar{y} = \frac{1}{n}\sum_{i=1}^{n} y_i. \tag{11.5}$$

2. The sums of squares

$$\begin{aligned} S_{xx} &= \sum_{i=1}^{n}(x_i - \bar{x})^2,\ S_{yy} = \sum_{i=1}^{n}(y_i - \bar{y})^2 \\ S_{xy} &= \sum_{i=1}^{n}(x_i - \bar{x})(y_i - \bar{y}). \end{aligned} \tag{11.6}$$

3. The most common estimators given this formulation are then given by

$$\beta = \frac{S_{xy}}{S_{xx}},\ \alpha = \bar{y} - \beta\bar{x}. \tag{11.7}$$

11.1.1 Least Squares: A Mathematical Solution

Following our theme in the discussion of linear projections, this definition involves minimizing the **Residual Squared Error** (RSS) by the choice of α and β.

$$\min_{\alpha,\beta} RSS = \sum_{i=1}^{n} \left(y_i - (\alpha + \beta x_i)\right)^2. \tag{11.8}$$

Focusing on α first,

$$\sum_{i=1}^{n} \left(y_i - (\alpha - \beta x_i)\right)^2 = \sum_{i=1}^{n} \left((y_i - \beta x_i) - \alpha\right)^2$$

$$\begin{aligned} &\Rightarrow \frac{\partial RSS}{\partial \alpha} = 2\sum_{i=1}^{n} \left((y_i - \beta x_i) - \alpha\right) = 0 \\ &\Rightarrow \sum_{i=1}^{n} y_i - \beta \sum_{i=1}^{n} x_i = n\alpha \\ &\bar{y} - \hat{\beta}\bar{x} = \hat{\alpha}. \end{aligned} \tag{11.9}$$

Taking the first-order conditions with respect to β yields

$$\begin{aligned} \frac{\partial RSS}{\beta} &= \sum_{i=1}^{n} \left((y_i - \beta x_i) - (\bar{y} - \beta \bar{x})\right) x_i \\ &= \sum_{i=1}^{n} \left((y_i - \bar{y}) - \beta (x_i - \bar{x})\right) x_i \\ &- \sum_{i=1}^{n} (y_i - \bar{y}) x_i - \beta \sum_{i=1}^{n} (x_i - \bar{x}) x_i. \end{aligned} \tag{11.10}$$

Going from this result to the traditional estimator requires the statement that

$$\begin{aligned} S_{xy} = \sum_{i=1}^{n} (y_i - \bar{y})(x_i - \bar{x}) &= \sum_{i=1}^{n} \left((y_i - \bar{y}) x_i - (y_i - \bar{y}) \bar{x}\right) \\ &= \sum_{i=1}^{n} (y_i - \bar{y}) x_i \end{aligned} \tag{11.11}$$

since $n\bar{y} - \sum_{i=1}^{N} y_i = 0$ by definition of $\bar{y}$. The least squares estimator of β then becomes

$$\hat{\beta} = \frac{S_{xy}}{S_{xx}}. \tag{11.12}$$

TABLE 11.1
U.S. Consumer Total and Food Expenditures, 1984 through 2002

Year	Total Expenditure (E)	Food Expenditure	$\ln(E)$	Food Share of Expenditure
1984	21,975	3,290	10.00	14.97
1985	23,490	3,477	10.06	14.80
1986	23,866	3,448	10.08	14.45
1987	24,414	3,664	10.10	15.01
1988	25,892	3,748	10.16	14.48
1989	27,810	4,152	10.23	14.93
1990	28,381	4,296	10.25	15.14
1991	29,614	4,271	10.30	14.42
1992	29,846	4,273	10.30	14.32
1993	30,692	4,399	10.33	14.33
1994	31,731	4,411	10.37	13.90
1995	32,264	4,505	10.38	13.96
1996	33,797	4,698	10.43	13.90
1997	34,819	4,801	10.46	13.79
1998	35,535	4,810	10.48	13.54
1999	36,995	5,031	10.52	13.60
2000	38,045	5,158	10.55	13.56
2001	39,518	5,321	10.58	13.46
2002	40,677	5,375	10.61	13.21

Example 11.1 (Working's Law of Demand). Working's law of demand is an economic conjecture that the percent of the consumer's budget spent on food declines as income increases. One variant of this formulation presented in Theil, Chung, and Seale [50] is that

$$w_{food,t} = \frac{p_{food,t}q_{food,t}}{E = \sum_i p_i q_i} = \alpha + \beta \ln(E) \tag{11.13}$$

where $w_{food,t}$ represents the consumer's budget share for food in time period t, and E is the total level of expenditures on all consumption categories. In this representation $\beta < 0$, implying that $\alpha > 0$. Table 11.1 presents consumer income, food expenditures, the natural logarithm of consumer income, and food expenditures as a percent of total consumer expenditures for the United States for 1984 through 2002. The sample statistics for these data are

$$\begin{aligned} \bar{x} &= 10.3258 \ \bar{y} = 14.1984 \\ S_{xx} &= 0.03460 \ S_{xy} = -0.09969 \\ \hat{\beta} &= -2.8812 \ \hat{\alpha} = 43.9506. \end{aligned} \tag{11.14}$$

The results of this regression are depicted in Figure 11.1. Empirically the data appear to be fairly consistent with Working's law. Theil, Chung, and Seale find

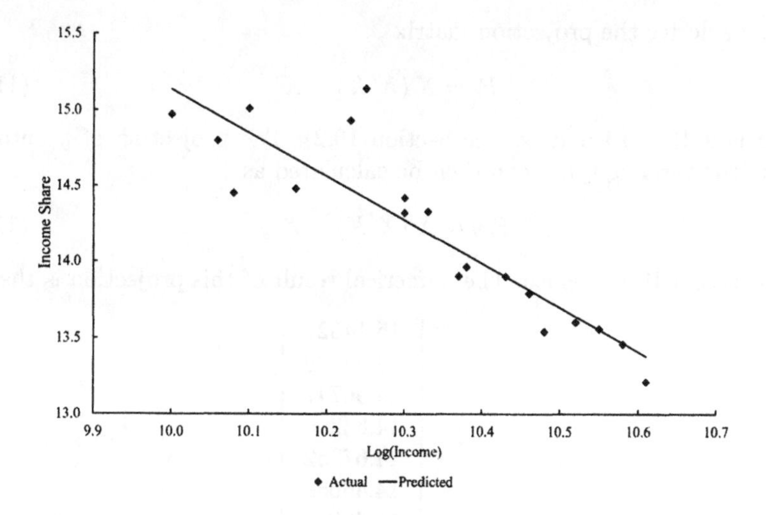

FIGURE 11.1
Working's Model of Food Expenditures.

that a large portion of the variation in consumption shares across countries can be explained by Working's law.

In order to bring the results from matrix algebra into the discussion, we are going to use the unity regressor form and rewrite the x and y matrices as

$$X = \begin{bmatrix} 1 & 9.9977 \\ 1 & 10.0643 \\ 1 & 10.0802 \\ 1 & 10.1029 \\ 1 & 10.1617 \\ 1 & 10.2332 \\ 1 & 10.2535 \\ 1 & 10.2960 \\ 1 & 10.3038 \\ 1 & 10.3318 \\ 1 & 10.3650 \\ 1 & 10.3817 \\ 1 & 10.4281 \\ 1 & 10.4579 \\ 1 & 10.4783 \\ 1 & 10.5185 \\ 1 & 10.5465 \\ 1 & 10.5845 \\ 1 & 10.6134 \end{bmatrix} \quad y = \begin{bmatrix} 14.9700 \\ 14.8000 \\ 14.4500 \\ 15.0100 \\ 14.4800 \\ 14.9300 \\ 15.1400 \\ 14.4200 \\ 14.3200 \\ 14.3300 \\ 13.9000 \\ 13.9600 \\ 13.9000 \\ 13.7900 \\ 13.5400 \\ 13.6000 \\ 13.5600 \\ 13.4600 \\ 13.2100 \end{bmatrix}. \tag{11.15}$$

First, we derive the projection matrix

$$P_c = X\left(X'X\right)^{-1}X' \tag{11.16}$$

which is a 19×19 matrix (see Section 10.2). The projection of y onto the dependent variable space can then be calculated as

$$P_c y = X\left(X'X\right)^{-1}X'y \tag{11.17}$$

in this case a 19×1 space. The numerical result of this projection is then

$$P_c y = \begin{bmatrix} 15.14521 \\ 14.95329 \\ 14.90747 \\ 14.84206 \\ 14.67262 \\ 14.46659 \\ 14.40809 \\ 14.28563 \\ 14.26315 \\ 14.18247 \\ 14.08680 \\ 14.03867 \\ 13.90497 \\ 13.81910 \\ 13.76031 \\ 13.64447 \\ 13.56379 \\ 13.45429 \\ 13.37101 \end{bmatrix}. \tag{11.18}$$

Comparing these results with the estimated values of y from the linear model yields

$$\begin{bmatrix} \alpha + 9.9977\beta = 15.14521 \\ \alpha + 10.0643\beta = 14.95329 \\ \vdots \\ \alpha + 10.6134\beta = 13.37101 \end{bmatrix}. \tag{11.19}$$

11.1.2 Best Linear Unbiased Estimator: A Statistical Solution

From Equation 11.2, the linear relationship between the xs and ys is

$$\mathrm{E}\left[y_i\right] = \alpha + \beta x_i \tag{11.20}$$

and we assume that

$$\mathrm{V}\left(y_i\right) = \sigma^2. \tag{11.21}$$

The implications of this variance assumption are significant. Note that we assume that each observation has the same variance regardless of the value of the independent variable. In traditional regression terms, this implies that the errors are homoscedastic.

One way to state these assumptions is

$$\begin{gathered} y_i = \alpha + \beta x_i + \epsilon_i \\ \mathrm{E}\left[\epsilon_i\right] = 0,\ \mathrm{V}\left(\epsilon_i\right) = \sigma^2. \end{gathered} \tag{11.22}$$

This specification is consistent with our assumptions, since the model is homoscedastic and linear in the parameters.

Based on this formulation, we can define the linear estimators of α and β as

$$\sum_{i=1}^{n} d_i y_i. \tag{11.23}$$

An unbiased estimator of β can further be defined as those linear estimators whose expected value is the true value of the parameter

$$\mathrm{E}\left[\sum_{i=1}^{n} d_i y_i\right] = \beta. \tag{11.24}$$

This implies that

$$\begin{aligned} \beta &= \mathrm{E}\left[\sum_{i=1}^{n} d_i y_i\right] \\ &= \sum_{i=1}^{n} d_i \mathrm{E}\left[y_i\right] \\ &= \sum_{i=1}^{n} d_i \left(\alpha + \beta x_i\right) \\ &= \alpha\left(\sum_{i=1}^{n} d_i\right) + \beta\left(\sum_{i=1}^{n} d_i x_i\right) \Rightarrow \left\{ \begin{array}{l} \sum_{i=1}^{n} d_i = 0 \\ \sum_{i=1}^{n} d_i x_i = 1. \end{array} \right. \end{aligned} \tag{11.25}$$

The linear estimator that satisfies these unbiasedness conditions and yields the smallest variance of the estimate is referred to as the best linear unbiased estimator (or BLUE). In this example, we need to show that

$$d_i = \frac{\left(x_i - \bar{x}\right)}{S_{xx}} \Rightarrow \hat{\beta} = \frac{\sum_{i=1}^{n} \left(x_i - \bar{x}\right) y_i}{S_{xx}} \tag{11.26}$$

minimizes the variance for all such linear models. Given that the y_is are uncorrelated, the variance of the linear model can be written as

$$V\left(\sum_{i=1}^{n} d_i y_i\right) = \sum_{i=1}^{n} d_i^2 V(y_i) = \sigma^2 \sum_{i=1}^{n} d_i^2. \tag{11.27}$$

The problem of minimizing the variance then becomes choosing the d_is to minimize this sum subject to the unbiasedness constraints

$$\begin{aligned} &\min_{d_i} \sigma^2 \sum_{i=1}^{n} d_i^2 \\ &\text{s.t.} \sum_{i=1}^{n} d_i x_i = 1 \\ &\sum_{i=1}^{n} d_i = 0. \end{aligned} \tag{11.28}$$

Transforming Equation 11.28 into a Lagrangian form,

$$L = \sigma^2 \sum_{i=1}^{n} d_i^2 + \lambda\left(1 - \sum_{i=1}^{n} d_i x_i\right) - \mu\left(\sum_{i=1}^{n} d_i\right)$$

$$\frac{\partial L}{\partial d_i} = 2\sigma^2 d_i - \lambda x_i - \mu = 0$$

$$\Rightarrow d_i = \frac{\lambda}{2\sigma^2} x_i + \frac{\mu}{2\sigma^2} \tag{11.29}$$

$$\frac{\partial L}{\partial \lambda} = 1 - \sum_{i=1}^{n} d_i x_i = 0$$

$$\frac{\partial L}{\partial \mu} = -\sum_{i=1}^{n} d_i = 0.$$

Using the results from the first n first-order conditions and the second constraint, we have

$$-\sum_{i=1}^{n}\left(\frac{\lambda}{2\sigma^2} x_i + \frac{\mu}{2\sigma^2}\right) = 0$$

$$\Rightarrow \frac{\lambda}{2\sigma^2} \sum_{i=1}^{n} x_i = -\frac{n\mu}{2\sigma^2} \tag{11.30}$$

$$\Rightarrow \mu = -\frac{\lambda \sum_{i=1}^{n} x_i}{n} = -\lambda \bar{x}.$$

Substituting this result into the first n first-order conditions yields

$$\begin{aligned} d_i &= \frac{\lambda}{2\sigma^2} x_i - \frac{\lambda}{2\sigma^2} \bar{x} \\ &= \frac{\lambda}{2\sigma^2} (x_i - \bar{x}) . \end{aligned} \tag{11.31}$$

Substituting these conditions into the first constraint, we get

$$\begin{aligned} & 1 - \sum_{i=1}^{n} \frac{\lambda}{2\sigma^2} (x_i - \bar{x}) x_i = 0 \\ & \Rightarrow \lambda = \frac{2\sigma^2}{\sum_{i=1}^{n} (x_i - \bar{x}) x_i} \\ & \Rightarrow d_i = \frac{(x_i - \bar{x})}{\sum_{i=1}^{n} (x_i - \bar{x}) x_i} = \frac{(x_i - \bar{x})}{S_{xx}} . \end{aligned} \tag{11.32}$$

This proves that the simple least squares estimator is BLUE on a fairly global scale. Note that we did not assume normality in this proof. The only assumptions were that the expected error term is equal to zero and that the variances were independently and identically distributed.

11.1.3 Conditional Normal Model

The conditional normal model assumes that the observed random variables are distributed

$$y_i \sim \mathrm{N}\left(\alpha + \beta x_i, \sigma^2\right) . \tag{11.33}$$

The expected value of y given x is $\alpha + \beta x$ and the conditional variance of y_i equals σ^2. The conditional normal can be expressed as

$$\mathrm{E}\left[y_i | x_i\right] = \alpha + \beta x_i . \tag{11.34}$$

Further, the ϵ_i are independently and identically distributed:

$$\begin{aligned} \epsilon_i &= y_i - \alpha - \beta x_i \\ \epsilon_i &\sim N\left(0, \sigma^2\right) \end{aligned} \tag{11.35}$$

(consistent with our BLUE proof).

Given this formulation, the likelihood function for the simple linear model can be written

$$L\left(\alpha, \beta, \sigma^2 | x\right) = \prod_{i=1}^{n} \frac{1}{\sqrt{2\pi}\sigma} \exp\left[-\frac{\left(y_i - \left(\alpha + \beta x_i\right)\right)^2}{2\sigma^2}\right] . \tag{11.36}$$

Taking the log of this likelihood function yields

$$\ln(L) = -\frac{n}{2}\ln(2\pi) - \frac{n}{2}\ln(\sigma^2) - \frac{1}{2\sigma^2}\sum_{i=1}^{n}(y_i - \alpha - \beta x_i)^2. \tag{11.37}$$

Thus, under normality the ordinary least squares estimator is also the maximum likelihood estimator.

11.1.4 Variance of the Ordinary Least Squares Estimator

The variance of β can be derived from the results presented in Section 11.1.2. Note from Equation 11.32,

$$\begin{aligned} \hat{\beta} &= \sum_{i=1}^{n} d_i y_i = \sum_{i=1}^{n} \frac{(x_i - \bar{x})}{S_{xx}}(\alpha + \beta x_i + \epsilon_i) \\ &= \sum_{i=1}^{n} d_i \alpha + \sum_{i=1}^{n} d_i \beta x_i + \sum_{i=1}^{n} d_i \epsilon_i. \end{aligned} \tag{11.38}$$

Under our standard assumptions about the error term, we have

$$\mathrm{E}\left(\sum_{i=1}^{n} d_i \epsilon_i\right) = \sum_{i=1}^{n} d_i \mathrm{E}(\epsilon_i) = 0. \tag{11.39}$$

In addition, by the unbiasedness constraint of the estimator, we have

$$\sum_{i=1}^{n} d_i \alpha = 0. \tag{11.40}$$

Leaving the unbiasedness result

$$\mathrm{E}\left(\hat{\beta}\right) = \beta \text{ if } \sum_{i=1}^{n} d_i = 1. \tag{11.41}$$

However, remember that the objective function of the minimization problem that we solved to get the results was the variance of parameter estimate

$$\mathrm{V}\left(\hat{\beta}\right) = \sigma^2 \sum_{i=1}^{n} d_i^2. \tag{11.42}$$

This assumes that the errors are independently distributed. Thus, substituting the final result for d_i into this expression yields

$$\mathrm{V}\left(\hat{\beta}\right) = \sigma^2 \sum_{i=1}^{n} \frac{(x_i - \bar{x})^2}{S_{xx}^2} = \sigma^2 \frac{S_{xx}}{S_{xx}^2} = \frac{\sigma^2}{S_{xx}}. \tag{11.43}$$

Noting that the numerator of this fraction is the true sample variance yields the Student's t-distribution for statistical tests of the linear model. Specifically, the slope coefficient is distributed t with $n - 2$ degrees of freedom.

11.2 Multivariate Regression

Given that the single cause model is restrictive, we next consider a multivariate regression. In general, the multivariate relationship can be written in matrix form as

$$y = \begin{pmatrix} 1 & x_1 & x_2 \end{pmatrix} \begin{pmatrix} \beta_0 \\ \beta_1 \\ \beta_2 \end{pmatrix} = \beta_0 + \beta_1 x_1 + \beta_2 x_2. \tag{11.44}$$

If we expand the system to three observations, this system becomes

$$\begin{aligned} \begin{pmatrix} y_1 \\ y_2 \\ y_2 \end{pmatrix} &= \begin{pmatrix} 1 & x_{11} & x_{12} \\ 1 & x_{21} & x_{22} \\ 1 & x_{31} & x_{32} \end{pmatrix} \begin{pmatrix} \beta_0 \\ \beta_1 \\ \beta_2 \end{pmatrix} \\ &= \begin{pmatrix} \beta_0 + \beta_1 x_{11} + \beta_2 x_{12} \\ \beta_0 + \beta_1 x_{21} + \beta_2 x_{22} \\ \beta_0 + \beta_1 x_{31} + \beta_2 x_{32} \end{pmatrix}. \end{aligned} \tag{11.45}$$

Given that the X matrix is of full rank, we can solve for the βs. In a statistical application, we have more rows than coefficients.

Expanding the exactly identified model in Equation 11.45, we get

$$\begin{pmatrix} y_1 \\ y_2 \\ y_3 \\ y_4 \end{pmatrix} = \begin{pmatrix} 1 & x_{11} & x_{12} \\ 1 & x_{21} & x_{22} \\ 1 & x_{31} & x_{32} \\ 1 & x_{41} & x_{42} \end{pmatrix} \begin{pmatrix} \beta_0 \\ \beta_1 \\ \beta_2 \end{pmatrix} + \begin{pmatrix} \epsilon_1 \\ \epsilon_2 \\ \epsilon_3 \\ \epsilon_4 \end{pmatrix}. \tag{11.46}$$

In matrix form, this can be expressed as

$$y = X\beta + \epsilon. \tag{11.47}$$

The sum of squared errors can then be written as

$$\begin{aligned} SSE &= (y - \hat{y})' (y - \hat{y}) = (y - X\beta)' (y - X\beta) \\ &= (y' - \beta' X') (y - X\beta). \end{aligned} \tag{11.48}$$

Using a little matrix calculus,

$$\begin{aligned} dSSE &= d\{(y - X\beta)'\} \ (y - X\beta) + (y - X\beta)' \, d\{(y - X\beta)\} \\ &= -(d\beta)' X' (y - X\beta) - (y - X\beta)' X d\beta \end{aligned} \tag{11.49}$$

(see Magnus and Nuedecker [29] for a full development of matrix calculus). Note that each term on the left-hand side is a scalar. Since the transpose of a scalar is itself, the left-hand side can be rewritten as

$$dSSE = -2(y - X\beta)' X d\beta$$

$$\begin{aligned} \Rightarrow \frac{dSSE}{d\beta} &= -2(y - X\beta)' X = 0 \\ y'X - \beta' X' X &= 0 \\ y'X &= \beta' X' X \\ X'y &= X' X \beta \\ (X'X)^{-1} X'y &= \beta. \end{aligned} \tag{11.50}$$

Thus, we have the standard result $\hat{\beta} = (X'X)^{-1} X'y$. Note that as in the two-parameter system we do not make any assumptions about the distribution of the error (ϵ).

11.2.1 Variance of Estimator

The variance of the parameter matrix can be written as

$$V\left(\hat{\beta}\right) = \mathrm{E}\left[\left(\hat{\beta} - \beta\right)\left(\hat{\beta} - \beta\right)'\right]. \tag{11.51}$$

Working backward,

$$\begin{aligned} y = X\beta + \epsilon \Rightarrow \hat{\beta} &= (X'X)^{-1} X'y \\ &= (X'X)^{-1} X' (X\beta + \epsilon) \\ &= (X'X)^{-1} X'X\beta + (X'X)^{-1} X'\epsilon \\ &= \beta + (X'X)^{-1} X'\epsilon. \end{aligned} \tag{11.52}$$

Substituting this back into the variance relationship in Equation 11.51 yields

$$\mathrm{V}\left(\hat{\beta}\right) = \mathrm{E}\left[(X'X)^{-1} X'\epsilon\epsilon' X (X'X)^{-1}\right]. \tag{11.53}$$

Note that $\epsilon\epsilon' = \sigma^2 I$ (i.e., assuming homoscedasticity); therefore

$$\begin{aligned} \mathrm{V}\left(\hat{\beta}\right) &= \mathrm{E}\left[(X'X)^{-1} X'\epsilon\epsilon' X (x'x)^{-1}\right] \\ &= (X'X)^{-1} X'\sigma_2 I X (X'X)^{-1} \\ &= \sigma^2 (X'X)^{-1} X'X (X'X)^{-1} \\ &= \sigma^2 (X'X)^{-1}. \end{aligned} \tag{11.54}$$

Again, notice that the construction of the variance matrix depends on the assumption that the errors are independently and identically distributed, but we do not assume a specific distribution of the ϵ (i.e., we do not assume normality of the errors).

11.2.2 Gauss–Markov Theorem

The fundamental theorem for most econometric applications is the Gauss–Markov theorem, which states that the ordinary least squares estimator is the best linear unbiased estimator. The theory developed in this section represents a generalization of the result presented in Section 11.1.2.

Theorem 11.2 (Gauss–Markov). *Let $\beta^* = C'y$ where C is a $T \times K$ constant matrix such that $C'X = I$. Then, $\hat{\beta}$ is better than β^* if $\beta^* \neq \hat{\beta}$.*

Proof. Starting with

$$\beta^* = \beta + C'u \tag{11.55}$$

the problem is how to choose C. Given the assumption $C'X = 1$, the choice of C guarantees that the estimator β^* is an unbiased estimator of β. The variance of β^* can then be written as

$$\begin{aligned} \mathrm{V}(\beta^*) &= \mathrm{E}[C'uu'C] \\ &= C'\mathrm{E}[uu']C \\ &= \sigma^2 C'C. \end{aligned} \tag{11.56}$$

To complete the proof, we want to add a special form of zero. Specifically, we want to add $\sigma^2 (X'X)^{-1} - \sigma^2 (X'X)^{-1} = 0$.

$$\mathrm{V}(\beta^*) = \sigma^2 (X'X)^{-1} - \sigma^2 C'C - \sigma^2 (X'X)^{-1}. \tag{11.57}$$

Focusing on the last terms, we note that by the orthogonality conditions for the C matrix,

$$\begin{aligned} Z'Z &= \left(C - X(X'X)^{-1}\right)' \left(C - X(X'X)^{-1}\right) \\ &= C'C - C'X(X'X)^{-1} - (X'X)^{-1}X'C + (X'X)^{-1}X'X(X'X)^{-1}. \end{aligned} \tag{11.58}$$

Substituting backwards,

$$\begin{aligned} \sigma^2 C'C - \sigma^2 (X'X)^{-1} &= \sigma^2 \left[C'C - (X'X)^{-1}\right] \\ &= \sigma^2 \left[C'C - C'X(X'X)^{-1} - (X'X)^{-1}X'C + (X'X)^{-1}X'X(X'X)^{-1}\right] \\ &= \sigma^2 \left[\left(C - X(X'X)^{-1}\right)' \left(C - X(X'X)^{-1}\right)\right]. \end{aligned} \tag{11.59}$$

Thus,

$$\mathrm{V}(\beta^*) = \sigma^2 (X'X)^{-1} + \sigma^2 \left[\left(C' - (X'X)^{-1} X' \right) \left(C - X (X'X)^{-1} \right) \right]. \tag{11.60}$$

The minimum variance estimator is then $C = X(X'X)^{-1}$, which is the ordinary least squares estimator.

□

Again, notice that the only assumption that we require is that the residuals are independently and identically distributed – **we do not need to assume normality to prove that ordinary least squares is BLUE.**

11.3 Linear Restrictions

Consider fitting the linear model

$$y = \beta_0 + \beta_1 x_1 + \beta_2 x_2 + \beta_3 x_3 + \beta_4 x_4 + \epsilon \tag{11.61}$$

to the data presented in Table 11.2. Solving for the least squares estimates,

$$\hat{\beta} = (X'X)^{-1}(X'y) = \begin{pmatrix} 4.7238 \\ 4.0727 \\ 3.9631 \\ 2.0185 \\ 0.9071 \end{pmatrix}. \tag{11.62}$$

Estimating the variance matrix,

$$\hat{s}^2 = \frac{y'y - (y'X)(X'X)^{-1}(X'y)}{30 - 5} = 1.2858$$

$$\mathrm{V}\left(\hat{\beta}\right) = \hat{s}^2 (X'X)^{-1}$$
$$= \begin{bmatrix} 0.5037 & -0.0111 & -0.0460 & 0.0252 & -0.0285 \\ -0.0111 & 0.0079 & -0.0068 & 0.0044 & -0.0033 \\ -0.0460 & -0.0068 & -0.0164 & -0.0104 & 0.0047 \\ 0.0252 & 0.0044 & -0.0104 & 0.0141 & -0.0070 \\ -0.0285 & -0.0033 & 0.0047 & -0.0070 & 0.0104 \end{bmatrix}. \tag{11.63}$$

Next, consider the hypothesis that $\beta_1 = \beta_2$ (which seems plausible given the results above). As a starting point, consider the least squares estimator

TABLE 11.2
Regression Data for Restricted Least Squares

Observation	y	x_1	x_2	x_3	x_4
1	75.72173	4.93638	9.76352	4.39735	2.27485
2	45.11874	6.95106	3.11080	−1.96920	3.59838
3	51.61298	4.69639	4.17138	3.84384	2.73787
4	92.53986	10.22038	8.93246	1.73695	5.36207
5	118.74310	12.05240	12.22066	6.40735	4.92600
6	80.78596	10.42798	5.58383	1.61742	9.30154
7	43.79312	2.94557	5.16446	1.21681	4.75092
8	47.84554	3.54233	5.58659	2.18433	3.65499
9	63.02817	4.56528	6.52987	4.40254	5.36942
10	88.83397	11.47854	8.82219	0.70927	2.94652
11	104.06740	11.87840	8.53466	5.21573	8.91658
12	57.40342	7.99115	7.42219	−3.62246	−2.19067
13	76.62745	7.14806	7.39096	5.19569	3.00548
14	109.96540	10.34953	9.82083	7.82591	7.09768
15	72.66822	7.74594	4.79418	5.39538	6.29685
16	68.22719	4.10721	8.51792	4.00252	3.88681
17	122.50920	12.77741	11.57631	6.85352	7.63219
18	70.71453	9.69691	6.54209	0.53160	0.79405
19	70.00971	6.46460	6.62652	4.31049	5.03634
20	75.82481	6.31186	8.49487	3.38461	5.53753
21	38.82780	3.04641	2.99413	2.69198	6.26460
22	79.15832	8.85780	7.29142	3.33994	2.86917
23	62.29580	5.82182	6.16096	4.18066	1.73678
24	80.63698	4.97058	9.83663	6.71842	3.47608
25	77.32687	5.90209	8.56241	5.42130	4.70082
26	23.34500	1.57363	2.82311	0.95729	0.69178
27	81.54044	9.25334	6.43342	5.02273	3.84773
28	67.16680	10.77622	5.21271	−0.87349	−1.17348
29	47.92786	6.96800	2.39798	−0.56746	6.08363
30	48.58950	7.06326	3.24990	−0.77682	3.09636

as a constrained minimization problem.

$$\begin{aligned} L(\beta) &= (y - X\beta)'(y - X\beta) + 2\lambda'(r - R\beta) \\ &= (y' - \beta'X')(y - X\beta) + 2\lambda'(r - R\beta) \\ \nabla_{\beta'} L(\beta) &= X'(y - X\beta) + X'(y - X\beta) - 2R'\lambda = 0 \end{aligned} \tag{11.64}$$

where $\nabla_{\beta'} L(\beta)$ is the gradient or a row vector of derivatives.[1] The second

[1] Technically, the gradient vector is defined as

$$\nabla_{\beta'} L(\beta) = \left[\begin{array}{cccc} \frac{\partial L(\beta)}{\partial \beta_1} & \frac{\partial L(\beta)}{\partial \beta_2} & \cdots & \frac{\partial L(\beta)}{\partial \beta_k} \end{array} \right].$$

term in the gradient vector depends on the vector derivative

$$\nabla_{\beta'}(X\beta) = \nabla_{\beta'}(\beta'X')' = (\nabla_{\beta'}(\beta'X'))' = (X')' = X. \tag{11.65}$$

Solving for the first-order condition for β,

$$\begin{aligned} X'(y - X\beta) + X'(y - X\beta) - 2R'\lambda = 0 \\ 2(X'y) - 2(X'X)\beta - 2R'\lambda = 0 \\ (X'X)\beta = (X'y) - R'\lambda \\ \beta = (X'X)^{-1}(X'y) - (X'X)^{-1}R'\lambda. \end{aligned} \tag{11.66}$$

Taking the gradient of the Lagrange formulation with respect to λ yields

$$\nabla_{\lambda'}L(\beta) = r - R\beta = 0. \tag{11.67}$$

Substituting the solution of β into the first-order condition with respect to the Lagrange multiplier,

$$\begin{aligned} r - R\left[(X'X)^{-1}(X'y) - (X'X)^{-1}R'\lambda\right] = 0 \\ R - R(X'X)^{-1}(X'y) + R(X'X)^{-1}R'\lambda = 0 \\ R(X'X)^{-1}R'\lambda = R(X'X)^{-1}(X'y) - r \\ \lambda = \left[R(X'X)^{-1}R'\right]\left(R(X'X)^{-1}(X'y) - r\right). \end{aligned} \tag{11.68}$$

Note that substituting $\beta = (X'X)^{-1}(X'y)$ into this expression yields

$$\lambda = \left[R(X'X)^{-1}R'\right]^{-1}(R\beta - r). \tag{11.69}$$

Substituting this result for λ back into the first-order conditions with respect to β yields

$$\begin{aligned} \beta_R = (X'X)^{-1}(X'y) - \\ (X'X)^{-1}R'\left[R'(X'X)^{-1}R'\right]^{-1}(R(X'X)(X'y) - r) \\ = \beta - (X'X)^{-1}R'[R(X'X)R']^{-1}(R\beta - r). \end{aligned} \tag{11.70}$$

Thus, the ordinary least squares estimates can be adjusted to impose the constraint $R\beta - r = 0$.

11.3.1 Variance of the Restricted Estimator

Start by deriving $\beta - \mathrm{E}[\beta]$ based on the previous results.

$$\begin{aligned}\beta = & (X'X)^{-1}(X'X\beta + X'\epsilon) - (X'X)^{-1} R' \left[R(X'X)^{-1} R'\right]^{-1} \\ & \times \left(R(X'X)^{-1} X'\beta + R(X'X)^{-1} X'\epsilon - r\right)\end{aligned}$$

$$\begin{aligned}\mathrm{E}[\beta] = & (X'X)^{-1}(X'X)\beta - (X'X)^{-1} R \left[R(X'X)^{-1} R'\right]^{-1} \\ & \times \left(R(X'X)^{-1} X'\beta - r\right)\end{aligned} \tag{11.71}$$

$$\begin{aligned}\beta - \mathrm{E}[\beta] = & (X'X)^{-1} X'\epsilon - (X'X)^{-1} R' \left[R(X'X)^{-1} R'\right]^{-1} R \\ & \times (X'X)^{-1} X'\epsilon.\end{aligned}$$

Computing $(\beta - \mathrm{E}[\beta])(\beta - \mathrm{E}[\beta])'$ based on this result,

$$\begin{aligned}(\beta - \mathrm{E}[\beta])(\beta - \mathrm{E}[\beta])' = & (X'X)^{-1} X'\epsilon\epsilon' X (X'X)^{-1} \\ & - (X'X)^{-1} X'\epsilon\epsilon' X (X'X)^{-1} R' [R(X'X)R']^{-1} R(X'X)^{-1} \\ & - (X'X)^{-1} R' \left[R(X'X)^{-1} R'\right]^{-1} R(X'X)^{-1} X'\epsilon\epsilon' X (X'X)^{-1} \\ & + (X'X)^{-1} R' \left[R(X'X)^{-1} R'\right]^{-1} R(X'X)^{-1} X'\epsilon\epsilon' \\ & \times X (X'X)^{-1} R' \left[R(X'X)^{-1} R'\right]^{-1} R(X'X)^{-1}.\end{aligned} \tag{11.72}$$

Taking the expectation of both sides and noting that $\mathrm{E}[\epsilon\epsilon'] = \sigma^2 I$,

$$\begin{aligned}V(\beta) = & \mathrm{E}\left[(\beta - \mathrm{E}[\beta])(\beta - \mathrm{E}[\beta])'\right] \\ = & \sigma^2 \left[(X'X)^{-1} - (X'X)^{-1} R' \left[R(X'X)^{-1} R'\right] R(X'X)^{-1}\right].\end{aligned} \tag{11.73}$$

Again, the traditional ordinary least squares estimate of the variance can be adjusted following the linear restriction $R\beta - r = 0$ to produce the variance matrix for the restricted least squares estimator.

11.3.2 Testing Linear Restrictions

In this section we derive the F-test of linear restrictions. Start with the derivation of the error under the restriction

$$\begin{aligned}\epsilon_R = & y - X\beta_R \\ = & y - X\beta - X(\beta_R - \beta).\end{aligned} \tag{11.74}$$

Compute the variance under the restriction

$$\begin{aligned}\epsilon_R'\epsilon_R &= (\epsilon - X(\beta_R - \beta))'(\epsilon - X(\beta_R - \beta)) \\ &= \epsilon'\epsilon - \epsilon'X(\beta_R - \beta) - (\beta_R - \beta)'X'\epsilon + (\beta_R - \beta)'X'X(\beta_R - \beta).\end{aligned} \tag{11.75}$$

Taking the expectation of both sides (with $\mathrm{E}[\epsilon] = 0$),

$$\begin{aligned}&\mathrm{E}(\epsilon_R'\epsilon_R) = \mathrm{E}(\epsilon'\epsilon) + (\beta_R - \beta)'X'X(\beta_R - \beta) \\ &\Rightarrow \mathrm{E}(\epsilon_R'\epsilon_R) - \mathrm{E}(\epsilon'\epsilon) = (\beta_R - \beta)'X'X(\beta_R - \beta).\end{aligned} \tag{11.76}$$

From our foregoing discussion,

$$\beta_R - \beta = (X'X)^{-1}R'\left[R(X'X)^{-1}R'\right]^{-1}(r - R\beta). \tag{11.77}$$

Substituting this result back into the previous equation yields

$$\begin{aligned}\mathrm{E}(\epsilon_R'\epsilon_R) - \mathrm{E}(\epsilon'\epsilon) &= (r - R\beta)'\left[R(X'X)^{-1}R'\right]^{-1} \\ R(X'X)^{-1}X'X(X'X)^{-1}R'&\left[R(X'X)^{-1}R'\right]^{-1}(r - R\beta) \\ &= (r - R\beta)'\left[R(X'X)^{-1}R'\right]^{-1}(r - R\beta).\end{aligned} \tag{11.78}$$

Therefore, the test for these linear restrictions becomes

$$\begin{aligned}F(q, n-k) &= \frac{(\epsilon_R'\epsilon_R - \epsilon'\epsilon)/q}{\epsilon'\epsilon/(n-k)} \\ &= \frac{(r - R\beta)'\left[R(X'X)^{-1}R'\right]^{-1}(r - R\beta)/q}{\epsilon'\epsilon/(n-k)}\end{aligned} \tag{11.79}$$

where there are q restrictions, n observations, and k independent variables.

11.4 Exceptions to Ordinary Least Squares

Several departures from the assumptions required for BLUE are common in econometrics. In this chapter we address two of the more significant departures – heteroscedasticity and endogeneity. Heteroscedasticity refers to the case where the errors are not identically distributed. This may happen due to a variety of factors such as risk in production. For example, we may want to estimate a production function for corn that is affected by weather events that are unequal across time. Alternatively, the risk of production may be partially a function of one of the input levels (i.e., the level of nitrogen interacting with weather may affect the residual by increasing the risk of production). Endogeneity refers to the scenario where one of the regressors is determined in part by the dependent variable. For example, the demand for an input is affected by the price, which is affected by the aggregate level of demand.

11.4.1 Heteroscedasticity

Using the derivation of the variance of the ordinary least squares estimator,

$$\begin{aligned}\hat{\beta} &= \beta + (X'X)^{-1}(X'\epsilon) \\ \Rightarrow \mathrm{V}\left(\hat{\beta}\right) &= (X'X)^{-1} X'\epsilon\epsilon' X (X'X)^{-1} \\ \mathrm{V}\left(\hat{\beta}\right) &= (X'X)^{-1} X'SX (X'X)^{-1} \ni: S = \mathrm{E}[\epsilon\epsilon']\end{aligned} \tag{11.80}$$

under the Gauss–Markov assumptions $S = \mathrm{E}[\epsilon\epsilon'] = \sigma^2 I_{T\times T}$.

However, if we assume that $S = \mathrm{E}[\epsilon\epsilon'] \neq \sigma^2 I_{T\times T}$, the ordinary least squares estimator is still unbiased, but is no longer efficient. In this case, we use the generalized least squares estimator

$$\tilde{\beta} = (X'AX)^{-1}(X'Ay). \tag{11.81}$$

The estimator under heteroscedasticity (generalized least squares) implies

$$\begin{aligned}\tilde{\beta} &= (X'AX)^{-1}(X'AX\beta + X'A\epsilon) \\ &= (X'AX)^{-1}(X'AX)\beta + (X'AX)^{-1} X'A\epsilon \\ &= \beta + (X'AX)^{-1} X'A\epsilon.\end{aligned} \tag{11.82}$$

The variance of the generalized least squares estimator then becomes

$$\begin{aligned}\mathrm{V}\left(\tilde{\beta} - \beta\right) &= (X'AX)^{-1} X'A\epsilon\epsilon'A'X (X'AX)^{-1} \\ &= (X'AX)^{-1} X'ASA'X (X'AX)^{-1}.\end{aligned} \tag{11.83}$$

Setting $A = S^{-1}$,

$$\begin{aligned}V\left(\tilde{\beta} - \beta\right) &= (X'AX)^{-1} X'AX (X'AX)^{-1} \ni: A' = A \\ &= (X'AX)^{-1}.\end{aligned} \tag{11.84}$$

The real problem is that the true A matrix is unknown and must be estimated. For example, consider Jorgenson's KLEM (Capital, Labor, Energy,

TABLE 11.3
Capital, Labor, Energy, and Materials Data for Agriculture

Year	Output	Capital	Labor	Energy	Materials
1960	153,474	35,474.5	100,351.0	5,196.6	105,589
1961	156,256	35,714.3	98,162.3	5,320.1	106,874
1962	155,497	36,121.5	96,396.5	5,508.9	107,933
1963	160,845	36,847.0	95,453.7	5,700.5	111,161
1964	159,818	37,528.9	94,510.9	5,655.4	109,132
1965	164,301	38,256.0	86,246.5	5,839.9	113,028
1966	163,392	39,281.7	84,037.2	6,102.9	115,790
1967	169,478	40,297.8	82,051.1	6,269.0	115,553
1968	171,661	41,925.8	78,578.8	6,491.2	115,996
1969	173,386	42,162.3	76,414.2	6,956.3	120,276
1970	174,451	42,943.0	75,021.7	7,155.6	119,990
1971	183,510	43,468.1	72,182.6	7,301.0	121,859
1972	186,534	43,516.7	73,255.9	8,234.6	135,623
1973	190,194	44,315.1	69,179.4	10,088.5	159,667
1974	183,488	45,337.8	64,548.4	9,430.8	149,627
1975	188,571	45,708.6	65,179.6	8,971.2	142,520
1976	192,838	46,163.3	71,072.6	9,272.9	143,341
1977	200,567	47,826.1	68,440.4	8,730.9	145,657
1978	203,682	48,754.6	67,991.2	9,480.2	151,936
1979	220,251	49,763.9	66,170.8	10,596.4	166,234
1980	208,824	50,125.1	67,244.9	10,186.8	157,684
1981	230,896	50,418.4	67,313.8	8,620.4	156,314
1982	234,524	50,717.8	66,520.9	9,312.6	166,379
1983	215,450	49,819.9	81,869.0	8,403.9	142,506
1984	231,061	48,978.0	73,590.2	9,945.0	145,318
1985	245,257	48,256.5	71,519.1	8,869.2	145,262
1986	245,492	47,715.7	70,930.2	7,439.1	138,931
1987	254,229	47,261.5	70,399.2	8,314.3	144,895
1988	243,603	46,627.4	76,095.8	8,120.2	145,481
1989	256,601	46,891.1	74,135.0	8,452.1	145,955
1990	268,591	47,213.8	72,955.2	8,621.7	150,062
1991	269,660	47,103.2	76,571.8	8,002.6	148,325
1992	285,108	47,192.3	74,753.0	8,462.2	151,953
1993	270,593	48,062.7	75,275.8	8,381.3	150,312
1994	299,636	48,510.2	77,155.3	8,603.5	156,693
1995	294,954	49,007.9	85,809.6	8,086.9	158,422
1996	305,745	49,929.8	82,818.2	8,852.2	164,145
1997	304,443	51,173.2	84,646.5	8,472.4	164,514
1998	307,862	52,505.3	83,157.0	8,728.9	159,117
1999	319,071	53,171.0	89,571.2	9,416.5	164,197
2000	348,475	53,515.9	87,635.4	9,622.8	177,534
2001	346,322	53,672.7	81,945.6	9,886.5	177,848
2002	346,868	53,844.9	68,581.6	9,374.6	170,610
2003	359,890	54,275.4	75,478.1	9,614.7	182,605
2004	382,321	55,219.1	67,169.8	10,207.2	199,310
2005	378,033	56,024.9	59,610.4	9,077.5	189,927

TABLE 11.4
Estimated Parameters for the Agricultural Production Function

Parameter	Estimate
α_0	-18.7157^{***}
	(2.3003)[a]
α_1	1.5999^{***}
	(0.3223)
α_2	0.3925^{**}
	(0.1364)
α_3	-0.9032^{***}
	(0.1741)
α_4	1.4807
	(0.2590)

Where *** and ** denotes statistical significance at the 0.01 and 0.05 level of respectively.
[a]Numbers in parenthesis denote standard errors.

and Materials) data for the agricultural sector presented in Table 11.3. Suppose that we want to estimate the Cobb–Douglas production function using the standard linearization.

$$\ln(y_t) = \alpha_0 + \alpha_1 \ln(x_{1t}) + \alpha_2 \ln(x_{2t}) + \alpha_3 \ln(x_{3t}) + \alpha_4 \ln(x_{4t}) + \epsilon_t. \quad (11.85)$$

The ordinary least squares estimates of Equation 11.85 are presented in Table 11.4. For a variety of reasons, there are reasons to suspect that the residuals may be correlated with at least one input. As depicted in Figure 11.2, in this case we suspect that the residuals are correlated with the level of energy used. In order to estimate the possibility of this relationship, we regress the estimated error squared from Equation 11.85 on the logarithm of energy used in agriculture production.

$$\hat{\epsilon}_t^2 = \beta_0 + \beta_1 \ln(x_{3t}) + \nu_t. \quad (11.86)$$

The estimated parameters in Equation 11.86 are significant at the 0.05 level of confidence. Hence, we can use this result to estimate the parameters of the A matrix in Equation 11.81.

$$\hat{A}_{tt} = \hat{\beta}_0 + \hat{\beta}_1 \ln(x_{3t}) \Rightarrow \tilde{\beta} = \left(X'\hat{A}X\right)^{-1}\left(X'\hat{A}y\right) \quad (11.87)$$

where $\tilde{\beta}$ is the estimated generalized least squares (EGLS) estimator.

One important point to remember is that generalied least squares is always at least as efficient as ordinary least squares (i.e., A could equal the identity matrix). However, estimated generalized least squares is not necessarily as efficient as ordinary least squares – there is error in the estimation of $\hat{A}$.

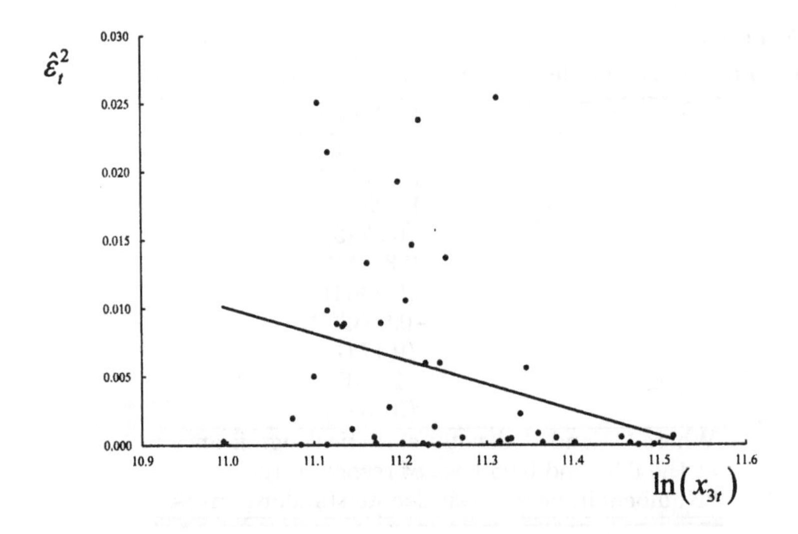

FIGURE 11.2
Estimated Residual Squared.

Seemingly Unrelated Regressions

One of the uses of generalized least squares is the estimation of simultaneous systems of equations without endogeneity. Derived input demand equations derived from cost minimization implies relationships between the parameters

$$\begin{aligned} x_1 &= \alpha_1 + A_{11}w_1 + A_{12}w_2 + \Gamma_{11}y + \epsilon_1 \\ x_2 &= \alpha_2 + A_{21}w_1 + A_{22}w_2 + \Gamma_{21}y + \epsilon_2 \end{aligned} \tag{11.88}$$

where x_1 and x_2 are input levels, w_1 and w_2 are the respective input prices, y is the level of output, and α_1, α_2, A_{11}, A_{12}, A_{21}, A_{22}, Γ_{11}, and Γ_{21} are estimated parameters.

Both relationships can be estimated simultaneously by forming the regression matrices as

$$\begin{bmatrix} x_{11} \\ x_{12} \\ \vdots \\ x_{1n} \\ x_{21} \\ x_{22} \\ \vdots \\ x_{2n} \end{bmatrix} = \begin{bmatrix} 1 & w_{11} & w_{22} & y_1 & 0 & 0 & 0 & 0 \\ 1 & w_{12} & w_{22} & y_2 & 0 & 0 & 0 & 0 \\ \vdots & \vdots & \vdots & \vdots & \vdots & \vdots & \vdots & \vdots \\ 1 & w_{1n} & w_{2n} & y_n & 0 & 0 & 0 & 0 \\ 0 & 0 & 0 & 0 & 1 & w_{11} & w_{21} & y_1 \\ 0 & 0 & 0 & 0 & 1 & w_{12} & w_{22} & y_2 \\ \vdots & \vdots & \vdots & \vdots & \vdots & \vdots & \vdots & \vdots \\ 0 & 0 & 0 & 0 & 1 & w_{1n} & w_{2n} & y_n \end{bmatrix} \begin{bmatrix} \alpha_1 \\ A_{11} \\ A_{12} \\ \Gamma_{11} \\ \alpha_2 \\ A_{21} \\ A_{22} \\ \Gamma_{21} \end{bmatrix} + \begin{bmatrix} \epsilon_{11} \\ \epsilon_{12} \\ \vdots \\ \epsilon_{1n} \\ \epsilon_{21} \\ \epsilon_{22} \\ \vdots \\ \epsilon_{2n} \end{bmatrix}. \tag{11.89}$$

It would be tempting to conclude that this formulation implies that the input demand system requires generalized least squares estimation. Specifically, using a two-step methodology, we can estimate the parameter vector using ordinary least squares. The ordinary least squares coefficients could then be used to estimate the variance for each equation. This variance could then be used to estimate the A matrix.

$$A = \begin{bmatrix} \frac{1}{s_1^2} & 0 & \cdots & 0 & 0 & 0 & \cdots & 0 \\ 0 & \frac{1}{s_1^2} & \cdots & 0 & 0 & 0 & \cdots & 0 \\ \vdots & \vdots & \ddots & \vdots & \vdots & \vdots & \ddots & \vdots \\ 0 & 0 & \cdots & \frac{1}{s_1^2} & 0 & 0 & \cdots & 0 \\ 0 & 0 & \cdots & 0 & \frac{1}{s_2^2} & 0 & \cdots & 0 \\ 0 & 0 & \cdots & 0 & 0 & \frac{1}{s_2^2} & \cdots & 0 \\ \vdots & \vdots & \ddots & \vdots & \vdots & \vdots & \ddots & \vdots \\ 0 & 0 & \cdots & 0 & 0 & 0 & \cdots & \frac{1}{s_2^2} \end{bmatrix}. \tag{11.90}$$

However, the separable nature of the estimation implies that there is no change in efficiency. To introduce changes in efficiency, we need to impose the restriction that $A_{12} = A_{21}$. Imposing this restriction on the matrix of independent variables implies

$$\begin{bmatrix} x_{11} \\ x_{12} \\ \vdots \\ x_{1n} \\ x_{21} \\ x_{22} \\ \vdots \\ x_{2n} \end{bmatrix} = \begin{bmatrix} 1 & w_{11} & w_{22} & y_1 & 0 & 0 & 0 \\ 1 & w_{12} & w_{22} & y_2 & 0 & 0 & 0 \\ \vdots & \vdots & \vdots & \vdots & \vdots & \vdots & \vdots \\ 1 & w_{1n} & w_{2n} & y_n & 0 & 0 & 0 \\ 0 & 0 & w_{11} & 0 & 1 & w_{21} & y_1 \\ 0 & 0 & w_{12} & 0 & 1 & w_{22} & y_2 \\ \vdots & \vdots & \vdots & \vdots & \vdots & \vdots & \vdots \\ 0 & 0 & w_{1n} & 0 & 1 & w_{2n} & y_n \end{bmatrix} \begin{bmatrix} \alpha_1 \\ A_{11} \\ A_{12} \\ \Gamma_{11} \\ \alpha_2 \\ A_{22} \\ \Gamma_{21} \end{bmatrix} + \begin{bmatrix} \epsilon_{11} \\ \epsilon_{12} \\ \vdots \\ \epsilon_{1n} \\ \epsilon_{21} \\ \epsilon_{22} \\ \vdots \\ \epsilon_{2n} \end{bmatrix}. \tag{11.91}$$

In the latter case, generalized least squares will yield efficiency gains.

The estimation Equation 11.91 requires a combination Kronecker product (see Equations 10.81 through 10.86), implying Equation 11.89 can be written as

$$\text{vecr}\,(y) = X \otimes I_{2\times 2}\text{vec}\beta + \text{vec}\,(\epsilon) \tag{11.92}$$

and then restricting $A_{12} = A_{21}$ using restricted least squares using Equation 11.70. Given these estimates, the researcher can then construct a sample estimate of the variance matrix to adjust for heteroscedasticity.

11.4.2 Two Stage Least Squares and Instrumental Variables

The foregoing example does not involve dependency between equations. For example, assume that the supply and demand curves for a given market can

TABLE 11.5
Ordinary Least Squares

Parameter	Ordinary Least Squares
α_0	1.2273
	(1.4657)
α_1	3.3280
	(0.1930)
α_2	−0.7181
	(0.1566)

be written as

$$\begin{aligned} q_s &= -3 + 4p_1 - p_2 + \epsilon_1 \\ q_d &= 10 - p_1 + 2y + \epsilon_2. \end{aligned} \tag{11.93}$$

Solving this two equation system yields

$$p_1 = \frac{13}{5} + \frac{1}{5}p_2 + \frac{2}{5}y - \epsilon_1 + \epsilon_2. \tag{11.94}$$

Ignoring the problem of simultaneity, the supply equation can be estimated as

$$q_s = \alpha_0 + \alpha_1 p_1 + \alpha_2 p_2 + \tilde{\epsilon}_1. \tag{11.95}$$

The results for this simple estimation are presented in Table 11.5. Obviously, these results are not close to the true values. Why? The basic problem is a simultaneous equation bias. Substituting the solution of p_1 into the estimated equation yields

$$q_s = \alpha_0 + \alpha_1 \left(\frac{13}{5} + \frac{1}{5}p_2 + \frac{2}{5}y - \epsilon_2 + \epsilon_2 \right) + \alpha_2 p_2 + \epsilon_1. \tag{11.96}$$

Substituting

$$\tilde{p}_1 = \frac{13}{5} + \frac{1}{5}p_2 + \frac{2}{5}y + \epsilon_2 \Rightarrow p_1 = \tilde{p}_1 - \epsilon_1 \tag{11.97}$$

we note that the x matrix is now correlated with the residual vector. Specifically

$$\mathrm{E}\left[p_1 \epsilon_1\right] = -\sigma_1^2 \neq 0. \tag{11.98}$$

Essentially the ordinary least squares results are biased.

Two Stage Least Squares

The first approach developed by Theil [55] was to estimate the reduced form of the price model and then use this estimated value in the regression. In this example,

$$\hat{p}_1 = \gamma_0 + \gamma_1 p_2 + \gamma_3 y + \nu. \tag{11.99}$$

TABLE 11.6
First-Stage Estimation

Parameters	Ordinary Least Squares
γ_0	2.65762
	(0.23262)
γ_1	0.15061
	(0.03597)
γ_2	0.40602
	(0.01863)

TABLE 11.7
Second-Stage Least Squares Estimator of the Demand Equation

Parameter	Two Stage Least Sqares
$\tilde{\beta}_0$	9.9121
	(0.9118)
$\tilde{\beta}_1$	−1.0096
	(0.2761)
$\tilde{\beta}_2$	2.0150
	(0.1113)

The parameter estimates for Equation 11.99 are presented in Table 11.6. Given the estimated parameters, we generate $\hat{p}_1$ and then estimate

$$q^S = \tilde{\alpha}_0 + \tilde{\alpha}_1 \hat{p}_1 + \tilde{\alpha}_2 p_2 + \epsilon_2. \tag{11.100}$$

In the same way, estimate the demand equation as

$$q^d = \tilde{\beta}_0 + \tilde{\beta}_1 \hat{p}_1 + \tilde{\beta}_2 y + \epsilon_2. \tag{11.101}$$

The results for the second stage estimates of the demand equation are presented in Table 11.7.

Generalized Instrumental Variables

The alternative would be to use variables as instruments to remove the correlation between endogenous variables. In this case, we assume that

$$y = X\beta + \epsilon. \tag{11.102}$$

Under the endogeneity assumption,

$$\frac{1}{N} X'\epsilon \nrightarrow 0. \tag{11.103}$$

But, we have a set of instruments (Z) which are correlated with the residuals and imperfectly correlated with X. Thegeneralized instrumental variable solution is

$$\beta_{IV} = \left(X'P_Z X\right)^{-1}\left(X'P_Z y\right) \tag{11.104}$$

where $P_Z = Z\left(Z'Z\right)^{-1}Z'$ (see the derivation of the projection matrix in ordinary least squares in Section 10.2).

In the current case, we use $Z = [1 \quad p_2 \quad y]$, yielding

$$\beta_{IV} = \begin{bmatrix} -3.2770 \\ 3.9531 \\ -0.7475 \end{bmatrix}. \tag{11.105}$$

The estimates in Equation 11.105 are very close to the original supply function in Equation 11.93.

11.4.3 Generalized Method of Moments Estimator

Finally we introduce the generalized method of moments estimator (GMM), which combines the generalized least squares estimator with a generalized instrumental variable approach. Our general approach follows that of Hall [17]. Starting with the basic linear model,

$$y_t = x_t'\theta_0 + u_t \tag{11.106}$$

where y_t is the dependent variable, x_t is the vector of independent variables, θ_0 is the parameter vector, and u_t is the residual. In addition to these variables, we will introduce the notion of a vector of instrumental variables denoted z_t. Reworking the original formulation slightly, we can express the residual as a function of the parameter vector.

$$u_t\left(\theta_0\right) = y_t - x_t'\theta_0. \tag{11.107}$$

Based on this expression, estimation follows from the population moment condition.

$$\mathrm{E}\left[z_t u_t\left(\theta_0\right)\right] = 0. \tag{11.108}$$

Or more specifically, we select the vector of parameters so that the residuals are orthogonal to the set of instruments.

Note the similarity between these conditions and the orthogonality conditions implied by the linear projection space.

$$P_c = X\left(X'X\right)^{-1}X'. \tag{11.109}$$

Further developing the orthogonality condition, note that if a single θ_0 solves the orthogonality conditions, or that θ_0 is unique, then

$$\mathrm{E}\left[z_t u_t\left(\theta\right)\right] = 0 \text{ if and only if } \theta = \theta_0. \tag{11.110}$$

Alternatively,

$$\mathrm{E}\left[z_t u_t\left(\theta\right)\right] \neq 0 \text{ if } \theta \neq \theta_0. \tag{11.111}$$

Going back to the original formulation,

$$\mathrm{E}\left[z_t u_t(\theta)\right] = \mathrm{E}\left[z_t\left(y_t - x_t'\theta\right)\right]. \tag{11.112}$$

Taking the first-order Taylor series expansion,

$$\begin{aligned} \mathrm{E}\left[z_t\left(y_t - x_t'\theta\right)\right] &= \mathrm{E}\left[z_t\left(y_t - x_t'\theta_0\right)\right] - \mathrm{E}\left[z_t x_t'\right]\left(\theta - \theta_0\right) \\ &\Leftarrow \frac{\partial}{\partial\theta}\left(y_t - x_t'\theta\right) = -x_t'. \end{aligned} \tag{11.113}$$

Given that $\mathrm{E}\left[z_t\left(y_t - x_t'\theta_0\right)\right] = \mathrm{E}\left[z_t u_t\left(\theta_0\right)\right] = 0$, this expression implies

$$\mathrm{E}\left[z_t\left(y_t - x_t'\theta\right)\right] = \mathrm{E}\left[z_t x_t'\right]\left(\theta_0 - \theta\right). \tag{11.114}$$

Given this background, the most general form of the minimand (objective function) of the GMM model ($Q_t(\theta)$) can be expressed as

$$Q_T(\theta) = \left\{\frac{1}{T}u(\theta)' Z\right\} W_T \left\{\frac{1}{T}Z'u(\theta)\right\} \tag{11.115}$$

where T is the number of observations, $u(\theta)$ is a column vector of residuals, Z is a matrix of instrumental variables, and W_T is a weighting matrix (akin to a variance matrix).

Given that W_T is a type of variance matrix, it is positive definite, guaranteeing that

$$z'W_T z > 0 \tag{11.116}$$

for any vector z. Building on the initial model,

$$\mathrm{E}\left[z_t u_t(\theta)\right] = \frac{1}{T}Z'u(\theta). \tag{11.117}$$

In the linear case,

$$\mathrm{E}\left[z_t u_t(\theta)\right] = \frac{1}{T}Z'\left(y - X\theta\right). \tag{11.118}$$

Given that W_T is positive definite, the optimality condition when the residuals are orthogonal to the variances based on the parameters is

$$\mathrm{E}\left[z_t u_t\left(\theta_0\right)\right] \Rightarrow \frac{1}{T}Z'u\left(\theta_0\right) = 0 \Rightarrow Q_T\left(\theta_0\right) = 0. \tag{11.119}$$

Working the minimization problem out for the linear case,

$$\begin{aligned} Q_T(\theta) &= \frac{1}{T^2}\left[\left(y - X\theta\right)' Z\right] W_T \left[Z'\left(y - X\theta\right)\right] \\ &= \frac{1}{T^2}\left[y'ZW_t - \theta'X'ZW_T\right]\left[Z'y - Z'X\theta\right] \\ &= \frac{1}{T^2}\left[y'ZW_T Z'y - \theta'X'ZW_T Z'y - y'ZW_T Z'X\theta + \theta'X'ZW_T Z'X\theta\right]. \end{aligned} \tag{11.120}$$

Note that since $Q_T(\theta)$ is a scalar, $\theta'X'ZW_T Z'y = y'ZW_T Z'X\theta$. Therefore,

$$Q_T(\theta) = \frac{1}{T^2}\left[y'ZW_T Z'y + \theta'X'ZW_T Z'X\theta - 2\theta'X'ZW_T Z'y\right]. \tag{11.121}$$

Solving the first-order conditions,

$$\begin{aligned} \nabla_\theta Q_T(\theta) &= \frac{1}{T^2}\left[2X'ZW_TZ'X\theta - 2X'ZW_TZ'y\right] = 0 \\ \Rightarrow \hat{\theta} &= \left(X'ZW_TZ'X\right)^{-1}\left(X'ZW_TZ'y\right). \end{aligned} \tag{11.122}$$

An alternative approach is to solve the implicit first-order conditions above. Starting with

$$\begin{aligned} \nabla_\theta Q_T(\theta) &= \frac{1}{T^2}\left[2X'ZW_TZ'X\theta - 2X'ZW_TZ'y\right] = 0 \\ \Rightarrow \frac{1}{2}\nabla_\theta Q_T(\theta) &= \left(\frac{1}{T}X'Z\right)W_T\left(\frac{1}{T}Z'X\theta\right) - \\ &\quad \left(\frac{1}{T}X'Z\right)W_T\left(\frac{1}{T}Z'y\right) = 0 \\ &= \left(\frac{1}{T}X'Z\right)W_T\left(\frac{1}{T}Z'X\theta - Z'y\right) = 0 \\ &= \left(\frac{1}{T}X'Z\right)W_T\left(\frac{1}{T}Z'\left\{y - X\theta\right\}\right) = 0 \\ \Rightarrow \frac{1}{2}\nabla_\theta Q_T(\theta) &= \left(\frac{1}{T}X'Z\right)W_T\left(\frac{1}{T}Z'\left\{y - X\theta\right\}\right) = 0 \\ &= \left(\frac{1}{T}X'Z\right)W_T\left(\frac{1}{T}Zu(\theta)\right) = 0. \end{aligned} \tag{11.123}$$

Substituting $u(\theta) = y - X\theta$ into Equation 11.123 yields the same relationship as presented in Equation 11.122.

The Limiting Distribution

By the Central Limit Theorem,

$$\frac{1}{\sqrt{T}}Z'u(\theta) = \frac{1}{\sqrt{T}}\sum_{t=1}^{T} z_t u_t(\theta) \xrightarrow{d} \mathrm{N}(0, S). \tag{11.124}$$

Therefore

$$\begin{aligned} \frac{1}{\sqrt{T}}\left(\hat{\theta} - \theta_0\right) &\xrightarrow{d} \mathrm{N}(0, MSM') \\ M &= \left(\mathrm{E}\left[x_t z_t'\right] W \mathrm{E}\left[z_t x_t'\right]\right)^{-1} \mathrm{E}\left[x_t z_t'\right] W \\ S &= \lim_{T\to\infty} \frac{1}{T}\sum_{s=1}^{T}\sum_{t=1}^{T} \mathrm{E}\left[u_t u_s z_t z_s'\right] = \mathrm{E}\left[u^2 z z'\right] \\ MSM' &= \left\{\mathrm{E}\left[z_t x_t'\right]\right\} S^{-1} \left\{\mathrm{E}\left[x_t z_t'\right]\right\}. \end{aligned} \tag{11.125}$$

Under the classical instrumental variable assumptions,

$$\begin{aligned}\hat{S}_T &= \frac{1}{T}\sum_{t=1}^{T}\hat{u}_t^2 z_t z_t' \\ \hat{S}_{CIV} &= \frac{\hat{\sigma}_T^2}{T}Z'Z.\end{aligned} \tag{11.126}$$

Example 11.3 (Differential Demand Model). Following Theil's model for the derived demands for inputs,

$$f_i d\ln[q_i] = \theta_i d\ln[O] + \sum_{j=1}^{n}\pi_{ij} d\ln[p_j] + \epsilon_i \tag{11.127}$$

where f_i is the factor share of input i ($f_i = p_i q_i / C$ such that p_i is the price of the input, q_i is the level of the input used, and C is the total cost of production), O is the level of output, and ϵ_i is the residual. The model is typically estimated as

$$\bar{f}_{it} D[q_{it}] = \theta_i D[O_t] + \sum_{j=1}^{n}\pi_{ij} D[p_{jt}] + \epsilon_t \tag{11.128}$$

such that $\bar{f}_{it} = \frac{1}{2}(f_{it} + f_{i,t-1})$ and $D[x_t] = \ln[x_t] - \ln[x_{t-1}]$.

Applying this to capital in agriculture from Jorgenson's [22] database, the output is an index of all outputs and the inputs are capital (p_{ct}), labor (p_{lt}), energy (p_{et}), and materials (p_{mt}). Thus,

$$\begin{aligned}X &= \begin{bmatrix} O_t & p_{ct} & p_{lt} & p_{et} & p_{mt}\end{bmatrix}_{t=1}^{T} \\ Z &= \begin{bmatrix} O_t & p_{ct} & p_{lt} & p_{et} & p_{mt} & O_t^2 & p_{ct}^2 & p_{lt}^2 & p_{et}^2 & p_{mt}^2\end{bmatrix}_{t=1}^{T}.\end{aligned} \tag{11.129}$$

Rewriting the demand model,

$$y = X\theta. \tag{11.130}$$

The objective function for the generalized method of moments estimator is

$$Q_T(\theta) = (y - X\theta)' Z W_T Z' (y - X\theta). \tag{11.131}$$

Initially we let $W_T = I$ and minimize $Q_T(\theta)$. This yields a first approximation to the estimates in the second column of Table 11.8. Updating W_T,

$$\hat{S}_{CIV} = \frac{\hat{\sigma}_T^2}{T}Z'Z \tag{11.132}$$

and resolving yields the second stage generalized method of moments estimates in the second column of Table 11.8.

TABLE 11.8
Generalized Methods of Moments Estimates of Differential Demand Equation

Parameter	First Stage GMM	Second Stage GMM	Ordinary Least Squares
Output	0.01588	0.01592	0.01591
	(0.00865)	(0.00825)	(0.00885)
Capital	−0.00661	−0.00675	−0.00675
	(0.00280)	(0.00261)	(0.00280)
Labor	0.00068	0.00058	0.00058
	(0.03429)	(0.00334)	(0.00359)
Energy	0.00578	0.00572	0.00572
	(0.00434)	(0.00402)	(0.00432)
Materials	0.02734	0.02813	0.02813
	(0.01215)	(0.01068)	(0.01146)

11.5 Chapter Summary

- Historically ordinary least squares has been the standard empirical method in econometrics.
- The classical assumptions for ordinary least squares are:
  - A general linear model $y = X\beta + \epsilon$.
    - Sometimes this model is generated by a first-order Taylor series expansion of an unknown function

$$y = f(x) = f(x_0) + X \begin{bmatrix} \beta_1 \approx \dfrac{\partial f(x)}{\partial x_{1i}} \\ \beta_2 \approx \dfrac{\partial f(x)}{\partial x_{2i}} \end{bmatrix} + \epsilon_i \tag{11.133}$$

      where the residual includes the approximation error. Note that this construction may lead to problems with heteroscedasticity.
    - Alternatively, it may be possible to transform the model in such a way as to yield a linear model. For example, taking the logarithm of the Cobb–Douglas production function ($y = \alpha_0 x_1^{\alpha_1} x_2^{\alpha_2}$) yields a linear model. Of course this also has implications for the residuals, as discussed in Chapter 12.
  - The independent variables have to be fixed (i.e., nonstochastic).
  - The residuals must be homoscedastic (i.e., $\epsilon\epsilon' = \sigma^2 I_{N\times N}$).
  - Given that the model obeys the assumptions, the estimates are best linear unbiased regardless of the distribution of the errors.

- If the residuals of a general linear model are normally distributed, the ordinary least squares estimates are also maximum likelihood.
- One frequently encountered exception to the conditions for best linear unbiased estimator involves differences in the variance ($\epsilon\epsilon' \neq \sigma^2 I_{N\times N}$).
  - This condition is referred to as heteroscedasticity. The problem is typically corrected by the design of a weighting matrix A such that $\epsilon A\epsilon' = \sigma^2 I_{N\times N}$.
  - This correction for heteroscedasticity opens the door to the estimation of simultaneous equations. Specifically, we can estimate two different equations at one time by realizing that the variance of the equation is different (i.e., seemingly unrelated regression).
  - It is important that generalized least squares is always at least as good as ordinary least squares (i.e., if A is known – in fact if regression is homoscedastic, then $A = I_{T\times T}$). However, estimated generalized least squares need not be as efficient as ordinary least squares because the estimate of A may contain error.
- One of the possible failures for the assumption that X is fixed involves possible correlation between the independent variables and the residual term (i.e., $\text{E}[X'\epsilon] \neq 0$).
  - These difficulties are usually referred to as endogeneity problems.
  - The two linear corrections for endogeneity are two-stage least squares and instrumental variables.

11.6 Review Questions

11-1R. True or false – we have to assume that the residuals are normally distributed for ordinary least squares to be best linear unbiased? Discuss.

11-2R. Why are the estimated ordinary least squares coefficients normally distributed in a small sample if we assume that the residuals are normally distributed?

11-3R. When are ordinary least squares coefficients normally distributed for large samples?

11-4R. Demonstrate Aitkin's theorem [49, p. 238] that $\beta = (X'AX)\, X'Ay$ yields a minimum variance estimator of β.

11.7 Numerical Exercises

10-1E. Regress

$$r_t = \alpha_0 + \alpha_1 R_t + \alpha_2 \Delta \left(D/A\right)_t + \epsilon_t \qquad (11.134)$$

using the data in Appendix A for Georgia. Under what conditions are the estimates best linear unbaised? How would you test for heteroscedasticity?

12

Survey of Nonlinear Econometric Applications

CONTENTS

The techniques developed in Chapter 11 estimate using linear (or iteratively linear in the case of two-stage least squares and the generalized method of moments) procedures. The linearity was particularly valuable before the widespread availability of computers and the development of more complex mathematical algorithms. However, the innovations in computer technology coupled with the development of statistical and econometric software have liberated our estimation efforts from these historical techniques. In this chapter we briefly develop three techniques that have no closed-form (or simple linear) solution: nonlinear least squares and maximum likelihood, applied Bayesian estimators, and least absolute deviation estimators.

12.1 Nonlinear Least Squares and Maximum Likelihood

Nonlinear least squares and maximum likelihood are related estimation techniques dependent on numerical optimization algorithms. To develop these routines, consider the Cobb–Douglas production function that is widely used in both theoretical and empirical economic literature:

$$y = \alpha_0 x_1^{\alpha_1} x_2^{\alpha_2} \tag{12.1}$$

where y is the level of output, and x_1 and x_2 are input levels. While this model is nonlinear, many applications transform the variable by taking the logarithm of both sides to yield

$$\ln(y) = \tilde{\alpha}_0 + \alpha_1 \ln(x_1) + \alpha_2 \ln(x_2) + \epsilon \tag{12.2}$$

where $\tilde{\alpha}_0 = \ln(\alpha_0)$. While the transformation allows us to estimate the production function using ordinary least squares, it introduces a significant assumption about the residuals. Specifically, if we assume that $\epsilon \sim N(0, \sigma^2)$ so that we can assume unbiasedness and use t-distributions and F-distributions to test hypotheses, the error in the original model becomes log-normal. Specifically,

$$\ln(y) = \tilde{\alpha}_0 + \alpha_1 \ln(x_1) + \alpha_2 \ln(x_2) + \epsilon \Rightarrow y = \alpha_0 x_1^{\alpha_1} x_2^{\alpha_2} e^{\epsilon} \tag{12.3}$$

which yields a variance of $\exp(\sigma^2 + 1/2\mu)$. In addition, the distribution is positively skewed, which is inconsistent with most assumptions about the error from the production function (i.e., the typical assumption is that most errors are to the left [negatively skewed] due to firm level inefficiencies).

The alternative is to specify the error as an additive term:

$$y = \alpha_0 x_1^{\alpha_1} x_2^{\alpha_2} + \epsilon. \tag{12.4}$$

The specification in Equation 12.4 cannot be estimated using a simple linear model. However, the model can be estimated using either nonlinear least squares,

$$\min_{\alpha_0, \alpha_1, \alpha_2} L(\alpha_0, \alpha_1, \alpha_2) = \sum_{i=1}^{N} \left(y_i - \alpha_0 x_{1i}^{\alpha_1} x_{2i}^{\alpha_2}\right)^2 \tag{12.5}$$

which must be solved using iterative nonlinear or maximum likelihood techniques.

Consider the corn production data presented in Table 12.1 (taken from the first 40 observations from Moss and Schmitz [34]). As a first step, we simplify the general form of the Cobb–Douglas in Equation 12.5 to

$$\min_{\alpha_0, \alpha_1, \alpha_2} L(\alpha_1) = \sum_{i=1}^{40} \left(y_i - 50 x_{1i}^{\alpha_1}\right)^2 \tag{12.6}$$

(i.e., we focus on the effect of nitrogen on production). Following the standard formulation, we take the first derivative of Equation 12.6, yielding

$$\frac{\partial L(\alpha_1)}{\partial \alpha_1} = 2 \sum_{i=1}^{40} \left(-\ln(x_{1i}) \times 50 x_{1i}^{\alpha_1}\right)\left(y_i - 50 x_{1i}^{\alpha_1}\right) \tag{12.7}$$

TABLE 12.1
Corn Production Data

Obs.	Nitrogen	Phosphorous	Potash	Yield	Obs.	Nitrogen	Phosphorous	Potash	Yield
1	348	75	124	134	21	203	46	150	129
2	127	60	90	140	22	125	69	90	160
3	202	104	120	110	23	175	80	80	120
4	88	24	90	61	24	167	51	48	168
5	150	69	120	138	25	214	138	180	150
6	153	53	126	102	26	195	104	135	149
7	139	35	90	160	27	196	92	180	142
8	150	60	120	115	28	60	46	60	132
9	160	40	50	165	29	88	138	180	120
10	180	37	120	140	30	196	23	5	155
11	160	30	60	135	31	176	92	180	138
12	182	77	128	160	32	168	46	60	96
13	154	60	80	151	33	185	115	180	135
14	196	92	90	143	34	186	92	120	137
15	172	58	120	154	35	161	92	200	112
16	227	69	60	155	36	201	92	120	121
17	147	69	150	46	37	167	69	90	133
18	166	72	160	115	38	157	69	90	101
19	105	58	21	115	39	194	138	120	149
20	72	69	90	105	40	225	115	150	146

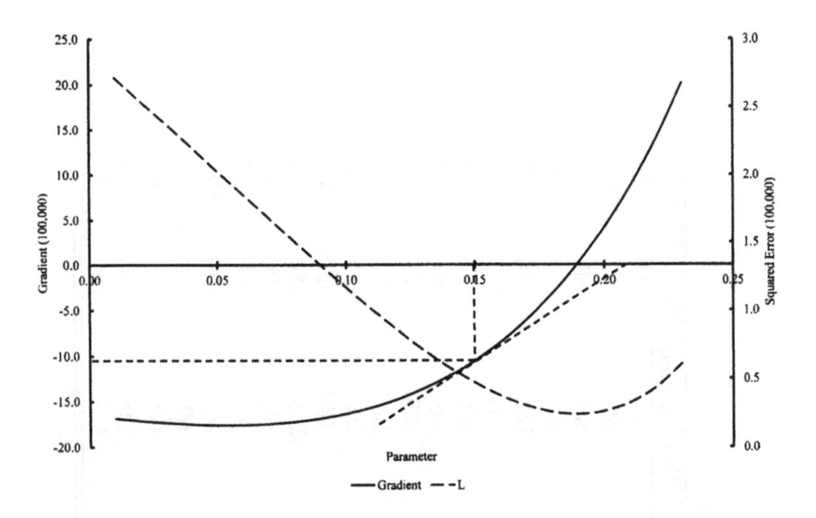

FIGURE 12.1
Minimum of the Nonlinear Least Squares Formulation.

and then solve this expression for the α_1 that yields $\partial L(\alpha_1)/\partial\alpha_1 = 0$. While this expression may be tractable, we typically solve for the α_1 using a numerical method known as Newton–Raphson.

To motivate this numerical procedure, notice that we are actually trying to find the zero of a function

$$g(\alpha_1) = -50\sum_{i=1}^{40}\left(\ln(x_{1i})\, x_{1i}^{\alpha_1}\right)\left(y_i - 50x_{1i}^{\alpha_1}\right). \tag{12.8}$$

Figure 12.1 presents the squared error and the derivative (gradient) of the least squares function for $\alpha_1 \in (0.01, 0.23)$. From the graphical depiction, it is clear that the minimum error squared occurs at around 0.185. The question is how to find the exact point in a systematic way.

Newton's method finds the zero of a function (in this case the zero of the gradient $g(\alpha_1)$ in Equation 12.8) using information in the derivative. For example, assume that we start at a value of $\alpha_1 = 0.15$, which yields a value of $g(\alpha_1)$ of -1,051,808. Graphically, draw a triangle based on the tangency of $g(\alpha_1)$ at that point and solve for the value of $\tilde{\alpha}_1$ such that $g(\tilde{\alpha}_1) = 0$. To develop this concept a little further, consider the first-order Taylor series expansion of $g(\alpha_1)$:

$$g(\alpha_1) \approx g(\alpha_1^0) + \left.\frac{\partial g(\alpha_1)}{\partial\alpha_1}\right|_{\alpha_1=\alpha_1^0}(\alpha_1 - \alpha_1^0). \tag{12.9}$$

Solving for $\tilde{\alpha}_1$ such that $g(\tilde{\alpha}_1) = 0$ implies

$$\tilde{\alpha}_1 = \alpha_1^0 - \frac{g(\alpha_1^0)}{\left.\dfrac{\partial g(\alpha_1)}{\partial\alpha_1}\right|_{\alpha_1=\alpha_1^0}}. \tag{12.10}$$

TABLE 12.2
Newton–Raphson Iterations for Simple Cobb–Douglas Form

α_1^0	$g(\alpha_1)$	$L(\alpha_1)$	$\partial g(\alpha_1)/\partial\alpha_1$	$\Delta\alpha_1$	$\tilde{\alpha}_1$
0.15000	−1,051,807.80	46,762.50	18,698,025.70	−0.056252	0.20625
0.20625	695,878.17	29,446.05	46,600,076.03	0.014933	0.19132
0.19132	71,859.09	23,887.35	37,253,566.76	0.001929	0.18939
0.18939	1,055.24	23,817.36	36,163,659.92	0.000029	0.18936
0.18936	0.24	23,817.35	36,147,366.06	0.000000	0.18936

Given that the derivative of $g(\alpha_1)$ is

$$\frac{\partial g(\alpha_1)}{\partial \alpha_1} = -100\sum_{i=1}^{40}\left[\left(\ln(x_{1i})^2 x_{1i}^{\alpha_1}\right)\left(y-50x_{1i}^{\alpha_1}\right) - 50\ln\left(\ln(x_{1i})\,x_{1i}^{\alpha_1}\right)^2\right] \tag{12.11}$$

the value of the derivative of $g(\alpha_1)$ at 0.15 is 18,698,026. Thus, the next estimate of the α_1 that minimizes the nonlinear least squares is

$$\tilde{\alpha}_1 = 0.15 - \frac{-1,051,808}{18,698,026} = 0.20625. \tag{12.12}$$

Evaluating $L(\alpha_1)$ at this point yields a smaller value (29,446.05 compared with 46,762.50). Table 12.2 presents the solution of the minimization problem following the Newton–Raphson algorithm. To clean up the proof a little, note that $g(\alpha_1) = \partial L(\alpha_1)/\partial\alpha_1$; the Newton–Raphson algorithm to minimize the nonlinear least squares is actually

$$\tilde{\alpha}_1 = \alpha_1^0 - \left.\frac{\dfrac{\partial L(\alpha_1)}{\partial \alpha_1}}{\dfrac{\partial^2 L(\alpha_1)}{\partial \alpha_1^2}}\right|_{\alpha_1=\alpha_1^0}. \tag{12.13}$$

To expand the estimation process to more than one parameter, we return to the problem in Equation 12.5. In addition, we need to introduce the concept of a gradient vector, which is essentially a vector of scalar derivatives. The gradient of Equation 12.5 with respect to the three parameters ($\{\alpha_0, \alpha_1, \alpha_2\}$) is a 3×1 vector

$$\nabla_\alpha L(\alpha) = \begin{bmatrix} \dfrac{\partial L(\alpha)}{\partial \alpha_0} \\ \dfrac{\partial L(\alpha)}{\partial \alpha_1} \\ \dfrac{\partial L(\alpha)}{\partial \alpha_2} \end{bmatrix} = \begin{bmatrix} -2\sum_{i=1}^{40} x_{1i}^{\alpha_1}x_{2i}^{\alpha_2}\left(y-\alpha_0 x_{1i}^{\alpha_1}x_{2i}^{\alpha_2}\right) \\ -2\alpha_0\sum_{i=1}^{40} x_{1i}^{\alpha_1}x_{2i}^{\alpha_2}\ln(x_{1i})\left(y-\alpha_0 x_{1i}^{\alpha_1}x_{2i}^{\alpha_2}\right) \\ -2\alpha_0\sum_{i=1}^{40} x_{1i}^{\alpha_1}x_{2i}^{\alpha_2}\ln(x_{2i})\left(y-\alpha_0 x_{1i}^{\alpha_1}x_{2i}^{\alpha_2}\right) \end{bmatrix}. \tag{12.14}$$

The Hessian matrix, which is the 3×3 matrix equivalent to the second derivative, is defined as

$$\nabla^2_{\alpha\alpha} L(\alpha) = \begin{bmatrix} \frac{\partial^2 L(\alpha)}{\partial \alpha_0^2} & \frac{\partial^2 L(\alpha)}{\partial \alpha_0 \partial \alpha_1} & \frac{\partial^2 L(\alpha)}{\partial \alpha_0 \partial \alpha_2} \\ \frac{\partial^2 L(\alpha)}{\partial \alpha_1 \partial \alpha_0} & \frac{\partial^2 L(\alpha)}{\partial \alpha_1^2} & \frac{\partial^2 L(\alpha)}{\partial \alpha_1 \partial \alpha_2} \\ \frac{\partial^2 L(\alpha)}{\partial \alpha_2 \partial \alpha_0} & \frac{\partial^2 L(\alpha)}{\partial \alpha_2 \partial \alpha_2} & \frac{\partial^2 L(\alpha)}{\partial \alpha_2^2} \end{bmatrix}. \tag{12.15}$$

The matrix form of Equation 12.13 can then be expressed as

$$\tilde{\alpha} = \alpha^0 - \left[\nabla^2_{\alpha\alpha} L(\alpha)\right]^{-1} \nabla_\alpha L(\alpha)\Big|_{\alpha=\alpha^0}. \tag{12.16}$$

The numerical solution to the three parameter Cobb–Douglas is presented in Appendix A.

To develop the distribution of the nonlinear least squares estimator, consider a slight reformulation of Equation 12.5:

$$L(x, y|\,\alpha) = [y - f(x,\alpha)]'\,[y - f(x,\alpha)]. \tag{12.17}$$

So we are separating the dependent variable y from the predicted component $f(x,\alpha)$. Given this formulation, we can then define the overall squared error of the estimate $(s^*(\alpha))$:

$$s^*(\alpha) = \sigma^2 + \int_X [f(x,\alpha^*) - f(x,\alpha)]^2\, d\mu(x) \tag{12.18}$$

where α^* is the level of the parameters that minimize the overall squared error and α is a general value of the parameters. Also notice that Gallant [14] uses measure theory. Without a great loss in generality, we rewrite Equation 12.18 as

$$s^*(\alpha) = \sigma^2 + \int_{-\infty}^{\infty} [f(x,\alpha^*) - f(x,\alpha)]^2\, dG(x) \tag{12.19}$$

where $dG(x) = g(x)\,dx$ is the probability density function of x. Next, we substitute a nonoptimal value (α^0) for α^*, yielding an error term $f(x,\alpha^0) + e = f(x,\alpha^*)$. Substituting this result into Equation 12.20,

$$s\left(e,\alpha^0,\alpha\right) = \sigma^2 + \int_{-\infty}^{\infty} \left[e + f\left(x,\alpha^0\right) - f(x,\alpha)\right]^2 dG(x). \tag{12.20}$$

Looking forward, if $\alpha^0 \to \alpha^*$ then $e + f\left(x,\alpha^0\right) \to f(x,\alpha^*)$ and $s\left(e,\alpha^0,\alpha\right) \to s^*(\alpha)$. Next, we take the derivative (gradient) of Equation 12.20,

$$\nabla_\alpha s\left(e,\alpha^0,\alpha\right) = -2\int_{-\infty}^{\infty} \left[e + f\left(x,\alpha^0\right) - f(x,\alpha)\right] \nabla_\alpha f(x,\alpha)\, dG(x). \tag{12.21}$$

Notice that in this formulation $(e + f(x,\alpha^0) - f(x,\alpha))$ is a scalar number while $\nabla_\alpha f(x,\alpha)$ is a vector with the same number of rows as the number of parameters in the nonlinear expression. Taking the second derivative of Equation 12.20 with respect to the parameter vector (or the gradient of Equation 12.21) yields

$$\begin{aligned}\nabla^2_{\alpha\alpha} s\left(e,\alpha^0,\alpha\right) = 2\int_{-\infty}^{\infty} & \left[\nabla_\alpha f(x,\alpha)\,\nabla_\alpha f(x,\alpha)'\right. \\ & \left. -2\left[e + f\left(x,\alpha^0\right) - f(x,\alpha)\right]\nabla^2_{\alpha\alpha} f(x,\alpha)\right] dG(x).\end{aligned} \tag{12.22}$$

A couple of things about Equations 12.21 and 12.22 are worth noting. First, we can derive the sample equivalents of Equations 12.21 and 12.22 as

$$\begin{aligned}\nabla_\alpha s\left(e,\alpha^0,\alpha\right) &\Rightarrow 2\left[e + F\left(x,\alpha^0\right) - F(x,\alpha)\right]'\nabla_\alpha F(x,\alpha) \\ \nabla^2_{\alpha\alpha} s\left(e,\alpha^0,\alpha\right) &\Rightarrow 2\nabla_\alpha F(x,\alpha)'\,\nabla_\alpha F(x,\alpha) \\ & -2\sum_{i=1}^{N}\left[e_i + f\left(x_i,\alpha^0\right) - f(x_i,\alpha)\right]\nabla^2_{\alpha\alpha} f(x_i,\alpha).\end{aligned} \tag{12.23}$$

Second, following the concept of a limiting distribution,

$$\frac{1}{\sqrt{N}}\nabla_\alpha F\left(\alpha^0\right) \overset{LD}{\to} \mathrm{N}_p\left(0, \sigma^2\frac{1}{N}\nabla_\alpha F(\alpha)'\,\nabla_\alpha F(\alpha)\right) \tag{12.24}$$

where p is the number of parameters. This is basically the result of the central limit theorem given that $1/N\left[e + F\left(x,\alpha^0\right) - F(x,\alpha)\right]'1 \to 0$ (where 1 is a conformable column vector of ones). By a similar conjecture,

$$\sqrt{N}\left(\hat{\alpha} - \alpha^0\right) \overset{LD}{\to} \mathrm{N}_p\left(0, \sigma^2\frac{1}{N}\left[\nabla_\alpha F(\alpha)'\,\nabla_\alpha F(\alpha)\right]^{-1}\right). \tag{12.25}$$

Finally, the standard error can be estimated as

$$s^2 = \frac{1}{N-p}\left[y - F(x,\alpha)\right]'\left[y - F(x,\alpha)\right]. \tag{12.26}$$

Of course the derivation of the limiting distributions of the nonlinear least squares estimator is typically superfluous given the assumption of normality. Specifically, assuming that the residual is normally distributed, we could rewrite the log-likelihood function for the three parameter Cobb–Douglas problem as

$$\ln\left(L\left(x,y|\,\alpha,\sigma^2\right)\right) \propto -\frac{N}{2}\ln\left(\sigma^2\right) - \frac{1}{2\sigma^2}\sum_{i=1}^{N}\left(y_i - \alpha_0 x_{1i}^{\alpha_1} x_{2i}^{\alpha_2}\right)^2. \tag{12.27}$$

Based on our standard results for maximum likelihood,

$$\hat{\alpha} \sim N\left(\alpha, -\frac{1}{N}\left[\sum_{i=1}^{N}\nabla^2_{\alpha\alpha}\left(y_i - \alpha_0 x_{1i}^{\alpha_1} x_{2i}^{\alpha_2}\right)^2\right]^{-1}\right). \tag{12.28}$$

Comparing the results of Equation 12.24 with Equation 12.28, any real differences in the distribution derive from $\nabla_{\alpha} F(\alpha)' \nabla_{\alpha} F(\alpha) \approx \nabla^2_{\alpha\alpha} F(\alpha)$.

12.2 Bayesian Estimation

Historically, Bayesian applications were limited by the existence of well formed conjugate families. In Section 8.2 we derived the Bayesian estimator of a Bernoulli parameter with a beta prior. This combination led to a closed-form posterior distribution – one that we could write out. Since the late 1990s advances in both numerical procedures and computer power have opened the door to more general Bayesian applications based on simulation.

12.2.1 Basic Model

An empirical or statistical model used for research purposes typically assumes that our observations are functions of unobservable parameters. Mathematically,

$$p(y|\theta) \tag{12.29}$$

where $p(.)$ is the probability of y – a set of observable outcomes (i.e., crop yields) and θ is a vector of unobservable parameters (or latent variables). Under normality $p(y|\mu, \sigma^2)$ means that the probability of an observed outcome $(y = 50.0)$ is a function of unobserved parameters such as the mean of the normal (μ) and its variance (σ^2). The formulation in Equation 12.29 is for any general outcome (i.e., any potential level of y). It is important to distinguish between general outcome – y — and a specific outcome that we observe – y^0. In general, we typically refer to a specific outcome as "data."

If we do not know what the value of θ is, we can depict the density function for θ as $p(\theta)$. We can then combine the density function for θ $(p(\theta))$ with our formulation of the probability of the observables $(p(y|\theta))$ to produce information about the observables that are not conditioned on knowing the value of θ.

$$p(y) = \int p(\theta) p(y|\theta) d\theta. \tag{12.30}$$

Next, we index our relation between observables and unobservables as model A. Hence, $p(y|\theta)$ becomes $p(y|\theta_A, A)$, and $p(\theta)$ becomes $p(\theta_A|A)$. We shall denote the object of interest on which decision making depends, and which all models relevant to the decision have something to say, by the vector ω. We shall denote the implications of the model A for ω by $p(\omega|y, \theta_A, A)$. In summary, we have identified three components of a complete model, A, involving unobservables (often parameters) θ_A, observables y, and a vector of interest ω.

12.2.2 Conditioning and Updating

Given this setup, we can sketch out the way to build on information from our prior beliefs about the unobservable parameters using sample information. As a starting point, we denote the prior distribution as $p(\theta_A | A)$. This captures our initial intuition about the probability of the unobservable parameters (i.e., the mean and variance of the distribution). These expectations are conditioned on a model – A (i.e., normality). In part, these prior beliefs specify a distribution for the observable variables $p(y | \theta_A, A)$. Given these two pieces of information, we can derive the probability of the unobservable parameters based on an observed sample (y^0).

$$p(\theta_A | y^0, A) = \frac{p(\theta_A, y^0 | A)}{p(y^0 | A)} = \frac{p(\theta_A | A) p(y^0 | \theta_A, A)}{p(y^0 | A)} \tag{12.31}$$

where $p(\theta_A | y^0, A)$ is referred to as the posterior distribution.

As a simple example, assume that we are interested in estimating a simple mean for a normal distribution. As a starting point, assume that our prior distribution for the mean is a normal distribution with a mean of $\tilde{\mu}$ and a known variance of 1. Our prior distribution then becomes

$$f_1(\mu | \tilde{\mu}, 1) = \frac{1}{\sqrt{2\pi}} \exp\left[-\frac{(\mu - \tilde{\mu})^2}{2}\right]. \tag{12.32}$$

Next, assume that the outcome is normally distributed with a mean of μ from Equation 12.32.

$$f_2(y | \mu, k) = \frac{1}{\sqrt{2\pi k}} \exp\left[-\frac{(y - \mu)^2}{2k}\right] \tag{12.33}$$

where k is some fixed variance. The posterior distribution then becomes

$$\begin{aligned} f(\mu | y^0, k) &\propto f_1(\mu | \tilde{\mu}, 1) f_2(y^0 | \mu, k) \\ &= \frac{1}{2\pi\sqrt{k}} \exp\left[-\frac{(\mu - \tilde{\mu})^2}{2} - \frac{(y^0 - \mu)^2}{2k}\right] \end{aligned} \tag{12.34}$$

(ignoring for the moment the denominator of Equation 12.32). One application of this formulation is to assume that we are interested in the share of cost associated with one input (say the constant in the share equation in a Translog cost function). If there are four inputs, we could assume *a priori* (our prior guess) that the average share for any one input would be 0.25. Hence, we set $\tilde{\mu} = 0.25$ in Equation 12.34. The prior distribution for any y^0 would then become

$$f(\mu | y^0, k) \propto \frac{1}{2\pi\sqrt{k}} \exp\left[-\frac{(\mu - 0.25)^2}{2} - \frac{(y^0 - \mu)^2}{2k}\right]. \tag{12.35}$$

Next, assume that we have a single draw from the sample $y^0 = 0.1532$ – the empirical posterior distribution would then become

$$f\left(\mu \left| y^0 = 0.1532, k\right.\right) \propto \frac{1}{2\pi\sqrt{k}} \exp\left[-\frac{(\mu - 0.25)^2}{2} - \frac{(0.1532 - \mu)^2}{2k}\right]. \tag{12.36}$$

Returning to the denominator from Equation 12.32, $p\left(y^0 \left| A\right.\right)$ is the probability of drawing the observed sample unconditional on the value of θ, in this case, unconditioned on the value of μ. Mathematically,

$$f(0.1532) = \int_{-\infty}^{\infty}\left(\frac{1}{2\pi\sqrt{k}} \exp\left[-\frac{(\mu - 0.25)^2}{2} - \frac{(0.1532 - \mu)^2}{2k}\right]\right) d\mu = 0.28144 \tag{12.37}$$

(computed using *Mathematica*). The complete posterior distribution can then be written as

$$f\left(\mu \left| y^0 = 0.1532, k\right.\right) = \frac{\frac{1}{2\pi\sqrt{k}} \exp\left[-\frac{(\mu - 0.25)^2}{2} - \frac{(0.1532 - \mu)^2}{2k}\right]}{0.28144}. \tag{12.38}$$

One use of the posterior distribution in Equation 12.38 is to compute a Bayesian estimate of μ. This is typically accomplished by minimizing the loss function

$$\begin{aligned} &\mathrm{Min}_{\mu_B} L(\mu, \mu_B) \equiv \mathrm{E}\left[\mu - \mu_B\right]^2 \Rightarrow \mu_B = \mathrm{E}[\mu] \\ &\Rightarrow \mathrm{E}[\mu] = \int_{-\infty}^{\infty} f\left(\mu \left| y^0 = 0.1532, k\right.\right) \mu\, d\mu = 0.2016 \end{aligned} \tag{12.39}$$

again relying on *Mathematica* for the numeric integral.

Next, consider the scenario where we have two sample points $y^0 = \{0.1532, 0.1620\}$. In this case the sample distribution becomes

$$f_2\left(y^0 \left| \mu, k\right.\right) = \frac{1}{\sqrt{2\pi k^2}} \exp\left[-\frac{(0.1532 - \mu)^2 + (0.1620 - \mu)^2}{2k}\right]. \tag{12.40}$$

The posterior distribution can then be derived based on

$$f\left(\mu \left| y^0, k\right.\right) \propto \frac{1}{2\pi\sqrt{k^2}} \exp\left[-\frac{(\mu - 0.25)^2}{2} - \frac{(0.1532 - \mu)^2 + (0.1620 - \mu)^2}{2k}\right]. \tag{12.41}$$

Integrating the denominator numerically yields 0.2300. Hence, the posterior distribution becomes

$$f\left(\mu\left|y^0,k\right.\right)=\frac{\frac{1}{2\pi\sqrt{k^2}}\exp\left[-\frac{(\mu-0.25)^2}{2}-\frac{(0.1532-\mu)^2+(0.1620-\mu)^2}{2k}\right]}{0.2300}\tag{12.42}$$

which yields a Bayesian estimate for μ of 0.1884. Note that this estimator is much higher than the standard mean of 0.1576 – the estimate is biased upward by the prior.

In the foregoing development we have been playing a little fast and loose with the variance of the normal. Let us return to the prior distribution for the mean.

$$f_1\left(\mu\left|\tilde{\mu},\sigma_\mu^2\right.\right)=\frac{1}{\sqrt{2\pi\sigma_\mu^2}}\exp\left[-\frac{(\mu-\tilde{\mu})^2}{2\sigma_\mu^2}\right].\tag{12.43}$$

Here we explicitly recognize the fact that our prior for μ (based on the normal distribution) has a variance parameter σ_μ^2. For the next wrinkle, we want to recognize the variance of the observed variable (y):

$$f_2\left(y\left|\mu,\sigma_y^2\right.\right)=\frac{1}{\sqrt{2\pi\sigma_y^2}}\exp\left[-\frac{(y-\mu)^2}{2\sigma_y^2}\right]\tag{12.44}$$

where σ_y^2 is the variance of the normal distribution for our observed variable.

Based on the formulation in Equation 12.44 we have acquired a new unknown parameter – σ_y^2. Like the mean, we need to formulate a prior for this variable. In formulating this variable we need to consider the characteristics of the parameter – it needs to be always positive (say $V\equiv\sigma_y^2\gg 0$). Following Section 3.5, the gamma distribution can be written as

$$f\left(V\left|\alpha,\beta\right.\right)=\begin{cases}\frac{1}{\Gamma(\alpha)\beta^\alpha}V^{\alpha-1}\exp\left[\frac{-V}{\beta}\right] & \text{such that } 0<v<\infty\\ & 0 \text{ otherwise.}\end{cases}\tag{12.45}$$

Letting $\alpha=2.5$ and $\beta=2$, the graph of the gamma function is depicted in Figure 12.2.

Folding Equation 12.43 and Equation 12.45 together gives the complete prior (i.e., with the variance terms)

$$\begin{aligned}f_1\left(\mu,V\left|\tilde{\mu},\sigma_\mu^2,\alpha,\beta\right.\right)&=f_{11}\left(\mu\left|\tilde{\mu},\sigma_\mu^2\right.\right)\times f_{12}\left(V\left|\alpha,\beta\right.\right)\\&=\frac{1}{\sqrt{2\pi\sigma_\mu^2}}\exp\left[-\frac{(\mu-\tilde{\mu})^2}{2\sigma_\mu^2}\right]\\&\quad\times\frac{1}{\Gamma(\alpha)\beta^\alpha}V^{\alpha-1}\exp\left[\frac{-V}{\beta}\right].\end{aligned}\tag{12.46}$$

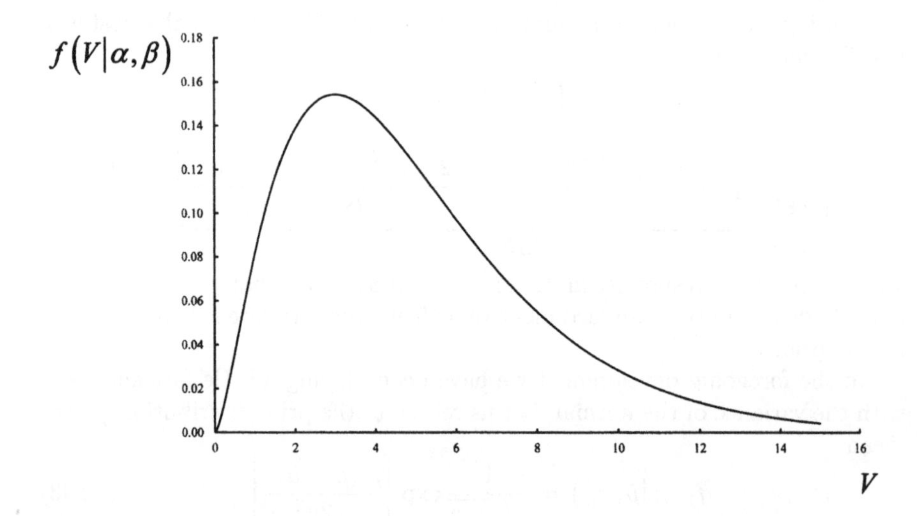

FIGURE 12.2
Gamma Distribution Function.

The sample distribution then becomes

$$f_2\left(y\left|\mu,\sigma_y^2\right.\right)=\frac{1}{V\sqrt{2\pi}}\exp\left[-\frac{(y-\mu)^2}{2V}\right]. \tag{12.47}$$

The posterior distribution then becomes

$$\begin{aligned} f\left(\tilde{\mu},\sigma_\mu^2,\alpha,\beta\left|y\right.\right) &= \frac{1}{\sqrt{2\pi\sigma_\mu^2}\left(\sqrt{2\pi V}\right)\Gamma(\alpha)\,\beta^\alpha}V^{\alpha-1} \\ &\quad\times\exp\left[-\frac{(\mu-\tilde{\mu})^2}{2\sigma_\mu^2}-\frac{(y-\mu)^2}{2V}-\frac{V}{\beta}\right] \\ &= \frac{1}{2\pi\sigma_\mu V^{1/2}\Gamma(\alpha)\,\beta^\alpha}V^{\alpha-1} \\ &\quad\times\exp\left[-\frac{(\mu-\tilde{\mu})^2}{2\sigma_\mu^2}-\frac{(y-\mu)^2}{2V}-\frac{V}{\beta}\right]. \end{aligned} \tag{12.48}$$

For simplification purposes, the gamma distribution is typically replaced with the inverse gamma.

Focusing on the exponential term in Equation 12.48, we can derive a form of a mixed estimator. The first two terms in the exponent of Equation 12.48 can be combined.

$$-\frac{(\mu-\tilde{\mu})^2}{2\sigma_\mu^2}-\frac{(y-\mu)^2}{2V}=-\frac{(\mu-\tilde{\mu})^2V+(y-\mu)^2\sigma_\mu^2}{2\sigma_\mu^2V}. \tag{12.49}$$

With a little bit of flare, we are going to divide the numerator by σ_μ^2 and let $\phi = V/\sigma_\mu^2$, yielding

$$-\frac{(\mu-\tilde{\mu})^2}{2\sigma_\mu^2} - \frac{(y-\mu)^2}{2V} = -\frac{\phi(\mu-\tilde{\mu})^2 + (y-\mu)^2}{2V}. \tag{12.50}$$

Maximizing Equation 12.50 with respect to μ gives an estimator of μ conditional on $\tilde{\mu}$ and ϕ:

$$\hat{\mu} = \frac{\phi\tilde{\mu} + y}{1+\phi}. \tag{12.51}$$

Next, let us extend the sample distribution to include two observations,

$$\begin{aligned} \tilde{f}_2\left(y_1, y_2 \left| \mu, \sigma_y^2\right.\right) &= \prod_{i=1}^{2} f_2\left(y_i \left| \mu, \sigma_y^2\right.\right) \\ &= \left(\frac{1}{\sqrt{2\pi V}}\right)^2 \exp\left[-\frac{(y_1-\mu)^2 + (y_2-\mu)^2}{2V}\right]. \end{aligned} \tag{12.52}$$

Folding the results for Equation 12.52 into Equation 12.48 implies a two observation form of Equation 12.53.

$$-\frac{(\mu-\tilde{\mu})^2}{2\sigma_\mu^2} - \frac{(y_1-\mu)^2 + (y_2-\mu)^2}{2V} = -\frac{\phi(\mu-\tilde{\mu})^2 + (y_1-\mu)^2 + (y_2-\mu)^2}{2V}. \tag{12.53}$$

With a little bit of effort, the value of μ that maximizes the exponent can be derived as

$$\hat{\mu} = \frac{\frac{\phi}{2}\tilde{\mu} + \frac{y_1+y_2}{2}}{1+\frac{\phi}{2}}. \tag{12.54}$$

Defining $\bar{y} = (y_1+y_2)/2$, this estimator becomes

$$\hat{\mu} = \frac{\frac{\phi}{2}\tilde{\mu} + \bar{y}}{1+\frac{\phi}{2}}. \tag{12.55}$$

This formulation clearly combines (or mixes) sample information with prior information.

Next, consider extending the sample to n observations. The sample distribution becomes

$$\begin{aligned} \tilde{f}_2\left(y_1, y_2, \cdots y_n \left| \mu, \sigma_y^2\right.\right) &= \prod_{i=1}^{n} f_2\left(y_i \left| \mu, \sigma_y^2\right.\right) \\ &= \left(\frac{1}{\sqrt{2\pi V}}\right)^n \exp\left[-\frac{\sum_{i=1}^{n}(y_i-\mu)^2}{2V}\right]. \end{aligned} \tag{12.56}$$

TABLE 12.3
Capital Share in KLEM Data

Year	Share
1960	0.1532
1961	0.1620
1962	0.1548
1963	0.1551
1964	0.1343
1965	0.1534
1966	0.1587
1967	0.1501
1968	0.1459
1969	0.1818

Combining the exponent term for posterior distribution yields

$$-\frac{(\mu-\tilde{\mu})^2}{2\sigma_\mu^2}-\frac{\sum_{i=1}^{n}(y_i-\mu)^2}{2V}=-\frac{\phi(\mu-\tilde{\mu})^2+\sum_{i=1}^{n}(y_i-\mu)^2}{2V}. \tag{12.57}$$

Following our standard approach, the value of μ that maximizes the exponent term is

$$\hat{\mu}=\frac{\frac{\phi}{n}\tilde{\mu}+\bar{y}}{1+\frac{\phi}{n}} \text{ s.t. } \bar{y}=\frac{\sum_{i=1}^{n}y_i}{n}. \tag{12.58}$$

Consider the data on the share of cost spent on capital inputs for agriculture from Jorgenson's KLEM data presented in Table 12.3. Let us assume a prior for this share of 0.25 (i.e., 1/4 of overall cost). First, let us use only the first observation and assume $\phi = 2.0$. The mixed estimator of the average share would be

$$\hat{\mu}=\frac{2.0\times 0.25+0.1532}{1+2.0}=0.2177. \tag{12.59}$$

Next, consider an increase in ϕ to 3.0.

$$\hat{\mu}=\frac{3.0\times 0.25+0.1532}{1+3.0}=0.2258. \tag{12.60}$$

Finally, consider reducing ϕ to 1.0.

$$\hat{\mu}=\frac{1.0\times 0.25+0.1532}{1+1.0}=0.2016. \tag{12.61}$$

As ϕ increases, more weight is put on the prior. The smaller the ϕ, the closer the estimate is to the sample value.

Next, consider using the first two data points so that $\bar{y} = 0.1576$. Assuming $\phi = 2.0$,

$$\hat{\mu} = \frac{\frac{2.0}{2} \times 0.25 + 0.1576}{1 + \frac{2.0}{2}} = 0.2038. \tag{12.62}$$

Notice that this estimate is closer to the sample average than for the same ϕ with one observation. Expanding to the full sample, $\phi = 2.0$ and $\bar{y} = 0.1549$ yields

$$\hat{\mu} = \frac{\frac{2.0}{10} \times 0.25 + 0.1549}{1 + \frac{2.0}{10}} = 0.1708. \tag{12.63}$$

Finally, notice that as $n \to \infty$ the weight on the prior vanishes so that the estimate converges to the sample average.

12.2.3 Simple Estimation by Simulation

Consider the small sample on the share of capital cost for production agriculture from the KLEM data. Suppose we assume a prior normal distribution with a mean of $\tilde{\mu} = 0.25$ and a coefficient of variation of 1.25. The variance would then be $(0.25 \times 1.25)^2 = 0.097656$. The posterior distribution would then be

$$f\left[\alpha \left| y^0 \right.\right] = \frac{1}{\sqrt{2\pi \times 0.096757}} \exp\left[-\frac{(\alpha - 0.25)^2}{2 \times 0.096757}\right]$$
$$\times \left(\frac{1}{\sqrt{2\pi\sigma_y^2}}\right)^{10} \exp\left[-\frac{(0.1532 - \alpha)^2 + (0.1620 - \alpha)^2 + \cdots (0.1818 - \alpha)^2}{2\sigma_y^2}\right]. \tag{12.64}$$

The concept is then to estimate α by integrating the probability density function by simulation. Specifically,

$$\hat{\alpha}_{Bayes} = \int_{-\infty}^{\infty} \alpha f\left[\alpha \left| y^0 \right.\right] d\alpha. \tag{12.65}$$

As a starting point, assume that we draw a value of α based on the prior distribution. For example, in the first row of Table 12.4 we draw $\alpha = 0.8881$. Given this draw, we compute the likelihood function of the sample.

$$\begin{aligned} L\left(\alpha \left| y^0 \right.\right) &\propto \exp \\ &\times \left[-\frac{(0.1532 - 0.8881)^2 + (0.1620 - 0.8881)^2 + \cdots (0.1818 - 0.8881)^2}{2 \times 0.096757}\right] \\ &= 8.5594E - 13. \end{aligned} \tag{12.66}$$

TABLE 12.4
Simulation Share Estimator

Draw	α	$L(\alpha \mid y^0)$	$\alpha \times L(\alpha \mid y^0)$
1	0.8881	8.5594E-13	7.6018E-13
2	0.1610	9.9130E-01	1.5960E-01
3	0.1159	9.1806E-01	1.0642E-01
4	0.3177	2.5264E-01	8.0261E-02
5	0.0058	3.1454E-01	1.8129E-03
6	−0.0174	2.1401E-01	−3.7269E-03
7	0.2817	4.3307E-01	1.2198E-01
8	0.9582	3.2764E-15	3.1395E-15
9	0.3789	7.4286E-02	2.8149E-02
10	−0.1751	3.5680E-03	−6.2479E-04
11	−0.0477	1.1906E-01	−5.6766E-03
12	0.2405	6.8021E-01	1.6360E-01
13	0.2461	6.4644E-01	1.5908E-01
14	0.2991	3.3906E-01	1.0143E-01
15	0.7665	3.9991E-09	3.0655E-09
16	0.0952	8.2591E-01	7.8617E-02
17	−0.0060	2.6062E-01	−1.5565E-03
18	0.2745	4.7428E-01	1.3020E-01
19	0.6151	1.7537E-05	1.0787E-05
20	0.3079	2.9626E-01	9.1228E-02

Summing over the 20 observations,

$$\hat{\alpha}_{Bayes} = \frac{\sum_{i=1}^{20} \alpha_i L\left[\alpha_i \mid y^0\right]}{\sum_{i=1}^{20} L\left[\alpha_i \mid y^0\right]} = 0.1769. \tag{12.67}$$

Expanding the sample to 200 draws, $\hat{\alpha}_{Bayes} = 0.1601$.

12.3 Least Absolute Deviation and Related Estimators

With the exception of some of the maximum likelihood formulations, we have typically applied a squared error weighting throughout this textbook.

$$\begin{aligned} \epsilon &= y_i - X_i\beta \\ \rho(\epsilon_i) &= \left(\epsilon_i^2\right) \\ L(\beta) &= \frac{1}{N}\sum_{i=1}^{N} \rho(\epsilon_i). \end{aligned} \tag{12.68}$$

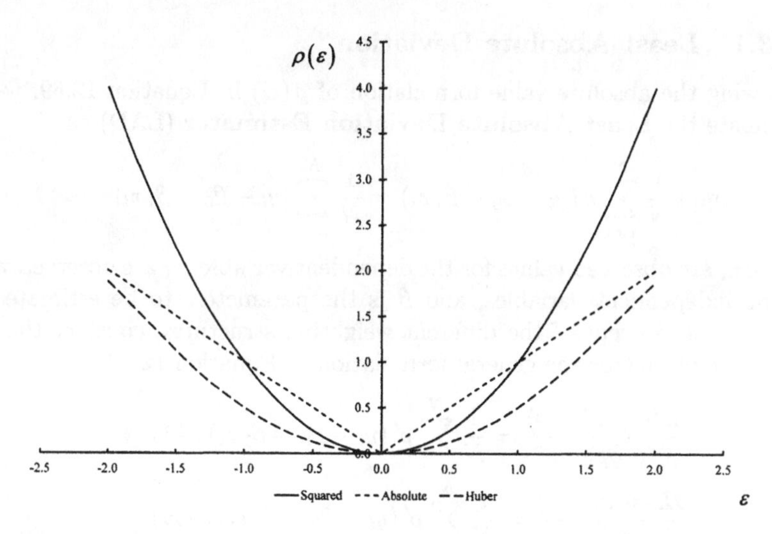

FIGURE 12.3
Alternative Residual Functions.

However, the squared error choice is to some extent arbitrary. Figure 12.3 presents three different residual functions: the standard residual squared function, the absolute value function, and a Huber weighting function [19]. The absolute value function assumes

$$\rho(\epsilon_i) = |\epsilon_i| \tag{12.69}$$

while the Huber function is

$$\rho(\epsilon_i) = \begin{cases} \frac{1}{2}\epsilon_i^2 \text{ for } |\epsilon_i| \leq k \\ k|\epsilon_i| - \frac{1}{2}k^2 \text{ for } |\epsilon_i| > k \end{cases} \tag{12.70}$$

with $k = 1.5$ (taken from Fox [12]). Each of these functions has implications for the weight given observations farther away from the middle of the distribution. In addition, each estimator has slightly different consequences for the "middle" of the distribution. As developed throughout this textbook, the "middle" for Equation 12.68 is the mean of the distribution while the "middle" for Equation 12.69 yields the median (δ).

$$\delta \text{ s.t. } F(\epsilon_i) = 1 - F(\epsilon_i) = 0.50 \tag{12.71}$$

where $F(\epsilon_i)$ is the cumulative density function of ϵ_i. The "middle" of the Huber function is somewhat flexible, leading to a "robust" estimator.

12.3.1 Least Absolute Deviation

Following the absolute value formulation of $\rho(\epsilon_i)$ in Equation 12.69, we can formulate the **Least Absolute Deviation Estimator** (LAD) as

$$\min_{\beta} \frac{1}{N} \sum_{i=1}^{N} \rho(y_i - \beta_0 - \beta_1 x_i) = \frac{1}{N} \sum_{i=1}^{N} |y_i - \beta_0 - \beta_1 x_i| \tag{12.72}$$

where y_i are observed values for the dependent variable, x_i are observed values of the independent variables, and β is the parameters to be estimated. To develop the concept of the different weighting structures, consider the first-order conditions for the general formulation in Equation 12.72.

$$\begin{aligned} \frac{\partial L(y, x|\beta, \rho)}{\partial \beta_0} &= \frac{1}{N} \sum_{i=1}^{N} \rho'(y_i - \beta_0 - \beta_1 x_i)(-1) \\ \frac{\partial L(y, x|\beta, \rho)}{\partial \beta_1} &= \frac{1}{N} \sum_{i=1}^{N} \rho'(y_i - \beta_0 - \beta_1 x_i)(-x_i). \end{aligned} \tag{12.73}$$

If $\rho(\epsilon_i)$ is the standard squared error, Equation 12.73 yields the standard set of normal equations.

$$\frac{\partial \rho(y_i - \beta_0 - \beta_1 x_i)}{\partial \beta_1} = 2(y_i - \beta_0 - \beta_1 x_i) x_i. \tag{12.74}$$

However, if $\rho(\epsilon_i)$ is the absolute value, the derivative becomes

$$\frac{\partial \rho(y_i - \beta_0 - \beta_1 x_i)}{\partial \beta_1} = \begin{cases} x_i \text{ for } y_i < \beta_0 + \beta_1 x_i \\ -x_i \text{ for } y_i > \beta_0 + \beta_1 x_i \end{cases} \tag{12.75}$$

which cannot be solved using the standard calculus (i.e., assuming a smooth derivative).

Bassett and Koenker [3] develop the asymptotic distribution of the parameters as

$$\sqrt{N}(\beta_n^* - \beta) \overset{\text{LD}}{\to} \text{N}\left(0, \frac{1}{N}\omega^2 [X_n' X_n]^{-1}\right) \tag{12.76}$$

where ω^2 is an asymptotic estimator of the variance (i.e., $\omega^2 = 1/N \sum_{i=1}^{N} (y_i - X_i\beta)^2$).

In order to demonstrate the applications of the LAD estimator, consider the effect of gasoline and corn prices on ethanol.

$$p_{et} = \beta_0 + \beta_1 p_{gt} + \beta_2 p_{ct} + \epsilon_t \tag{12.77}$$

where p_{et} is the price of ethanol at time t, p_{gt} is the price of gasoline at time t, p_{ct} is the price of corn at time t, ϵ_t is the residual, and β_0, β_1, and β_2 are the parameters we want to estimate. Essentially, the question is whether gasoline or corn prices determine the price of ethanol. The data for 1982 through 2013 are presented in Table 12.5. The parameter estimates using OLS, LAD, quantile regression with $\tau = 0.50$ discussed in the next section, and two different Huber weighting functions are presented in Table 12.6.

TABLE 12.5
Ethanol, Gasoline, and Corn Prices 1982–2013

	Nominal Prices				Real Prices		
Year	Ethanol	Gasoline	Corn	PCE	Ethanol	Gasoline	Corn
1982	1.71	1.00	2.55	50.479	3.631	2.123	5.415
1983	1.68	0.91	3.21	52.653	3.420	1.853	6.535
1984	1.55	0.85	2.63	54.645	3.040	1.667	5.159
1985	1.60	0.85	2.23	56.581	3.031	1.610	4.225
1986	1.07	0.51	1.50	57.805	1.984	0.946	2.781
1987	1.21	0.57	1.94	59.649	2.174	1.024	3.486
1988	1.13	0.54	2.54	61.973	1.954	0.934	4.393
1989	1.23	0.61	2.36	64.640	2.040	1.012	3.913
1990	1.35	0.75	2.28	67.439	2.146	1.192	3.624
1991	1.27	0.69	2.37	69.651	1.954	1.062	3.647
1992	1.33	0.64	2.07	71.493	1.994	0.960	3.103
1993	1.16	0.59	2.50	73.278	1.697	0.863	3.657
1994	1.19	0.56	2.26	74.802	1.705	0.802	3.238
1995	1.15	0.59	3.24	76.354	1.614	0.828	4.548
1996	1.35	0.69	2.71	77.980	1.856	0.948	3.725
1997	1.15	0.55	2.43	79.326	1.554	0.743	3.283
1998	1.05	0.43	1.94	79.934	1.408	0.577	2.601
1999	0.98	0.59	1.82	81.109	1.295	0.780	2.405
2000	1.35	0.93	1.85	83.128	1.741	1.199	2.385
2001	1.48	0.88	1.97	84.731	1.872	1.113	2.492
2002	1.12	0.81	2.32	85.872	1.398	1.011	2.896
2003	1.35	0.98	2.42	87.573	1.652	1.199	2.962
2004	1.69	1.25	2.06	89.703	2.019	1.494	2.462
2005	1.80	1.66	2.00	92.260	2.091	1.929	2.324
2006	2.58	1.94	3.04	94.728	2.919	2.195	3.440
2007	2.24	2.23	4.20	97.099	2.473	2.462	4.636
2008	2.47	2.57	4.06	100.063	2.646	2.753	4.349
2009	1.79	1.76	3.55	100.000	1.919	1.886	3.805
2010	1.93	2.17	5.18	101.654	2.035	2.288	5.462
2011	2.70	2.90	6.22	104.086	2.780	2.986	6.405
2012	2.37	2.95	6.89	106.009	2.396	2.983	6.967
2013	2.47	2.90	4.50	107.187	2.470	2.900	4.500

12.3.2 Quantile Regression

The least absolute deviation estimator in Equation 12.72 provides a transition to the **Quantile Regression** estimator. Specifically, following Koenker and Bassett [27] we can rewrite the estimator in Equation 12.72 as

$$\min_{\beta}\left[\sum_{i\in i:y_i\geq x_i\beta}\theta\left|y_i-x_i\beta\right|+\sum_{i\in i:y_i<x_i\beta}(1-\theta)\left|y_i-x_i\beta\right|\right] \tag{12.78}$$

TABLE 12.6
Least Absolute Deviation Estimates of Ethanol Price

	OLS	LAD	$Q(\tau = 0.50)$	M-Robust Estimator 1.25	1.5
β_0	0.952	1.254	1.254	1.045	1.028
	(0.245)	(0.245)		(0.220)	(0.216)
β_1	0.308	0.325	0.325	0.338	0.335
	(0.136)	(0.135)		(0.114)	(0.114)
β_2	0.189	0.090	0.090	0.137	0.145
	(0.079)	(0.079)		(0.073)	(0.072)

TABLE 12.7
Quantile Regression Estimates for Ethanol Prices

Parameter	Quantile Regression (τ) 0.2	0.4	0.6	0.8	OLS	LAD
β_0	1.1085	1.1729	1.1481	0.4874	0.9519	1.2537
Lower Bound	0.6278	0.5184	0.1691	−0.0264	0.4504	0.7526
Upper Bound	1.2898	1.5508	1.5706	0.8447	1.4535	1.7548
β_1	0.3248	0.3962	0.3112	0.6733	0.3084	0.3254
Lower Bound	−0.1656	0.2543	0.0840	0.1721	0.0313	0.0485
Upper Bound	0.5374	0.5180	0.8321	1.3640	0.5855	0.6023
β_1	0.0520	0.0662	0.1475	0.2772	0.1886	0.0902
Lower Bound	−0.0469	−0.0523	0.0435	0.2413	0.0266	−0.0717
Upper Bound	0.2925	0.4022	0.3903	0.4312	0.3506	0.2521

where $\theta = 0.5$. Intuitively, the first sum in Equation 12.78 corresponds to the observations where $y_i \geq x_i\beta$ or the observation is above the regression relationship, while the second sum corresponds to the observations where $y_i < x_i\beta$ or the observation is below the regression relationship.

Generalizing this relationship slightly,

$$Q(\tau) = \min_{\beta} \left[\sum_{i \in i: y_i \geq x_i\beta} \tau \left| y_i - x_i\beta \right| + \sum_{i \in i: y_i < x_i\beta} (1-\tau) \left| y_i - x_i\beta \right| \right] \Rightarrow \beta(\tau) \tag{12.79}$$

for any $\tau \in (0,1)$.

Turning to the effect of the price of gasoline and corn on ethanol prices reported in Table 12.7, the effect of gasoline on ethanol prices appears to be the same for quantiles 0.2 through 0.6. However, the effect of gasoline on ethanol prices increases significantly at the 0.8 quantile. On the other hand, the regression coefficients for the price of corn on ethanol prices increase rather steadily throughout the entire sample range.

TABLE 12.8
Quantile Regression on Factors Affecting Farmland Values

	Selected Quantiles				
Coefficient	20	40	50	70	90
Constant	411.71***	1172.27***	1395.03***	1868.91***	3164.32***
	(64.78)[a]	(56.44)	(56.57)	(139.98)	(258.88)
Cash Income/	0.004	0.07**	0.10***	0.25***	0.64***
Owned Acre	(0.010)	(0.03)	(0.027)	(0.043)	(0.126)
CRP/	0.08	−0.78	−1.06*	−1.59***	−2.89***
Owned Acre	(0.325)	(0.492)	(0.603)	(0.355)	(1.119)
Direct Payment/	−0.06	−.34*	0.46***	1.00***	1.85***
Owned Acre	(0.084)	(0.212)	(0.167)	(0.24)	(0.94)
Indirect Payment/	0.01	0.006	0.07**	0.28**	0.74**
Owned Acre	(0.058)	(0.162)	(0.030)	(0.090)	(0.34)
Off-Farm Income/	0.10**	0.24***	0.39***	0.86***	2.12***
Owned Acre	(0.021)	(0.033)	(0.042)	(0.080)	(0.190)

***, **, and * denote statistical significance at the 0.01, 0.05, and 0.10 levels of confidence, respectively. [a] Numbers in parenthesis denote standard errors.

Quantile regressors are sometimes used to develop the distributional effect of an economic policy. Mishra and Moss [31] estimate the effect of off-farm income on farmland values using a quantile regression approach. Table 12.8 presents some of the regression estimates for selected quantiles. These results indicate that the amount that each household is willing to pay for farmland increases with the quantile of the regression. In addition, the effect of government payments on farmland values increases with the quantile.

12.4 Chapter Summary

- One of the factors contributing to the popularity of linear econometric models such as those presented in Chapter 11 is their simplicity. As long as the independent variables are not linearly dependent, $\beta = (X'X)\, X'y$ exists and is relatively simple to compute.
- The advances in computer technology and algorithms have increased the use of nonlinear estimation techniques. These techniques largely involve iterative optimization algorithms.
- Nonlinear least squares allows for flexible specifications of functions and distributions (i.e., the Cobb–Douglas production function can be estimated without assuming that the residuals are log-normal).
- There are similarities between nonlinear maximum likelihood and nonlin-

ear least squares. One of the differences is the derivation of the variance matrix for the parameters.

- The development of the Gibbs sampler and other simulation techniques provides for the estimation of a variety of priors and sampling probabilities. Specifically, we are no longer bound to conjugate families with simple closed form solutions.
- Least absolute deviation models provide one alternative to the traditional concept of minimizing the squared error of the residual.
- The least absolute deviation is a special case of the quantile regression formulation.
- Another weighting of the residual is the M-robust estimator proposed by Huber [19].

12.5 Review Questions

12-1R. What information is required to estimate either nonlinear least squares or maximum likelihood using Newton–Raphson?

12-2R. How do the least absolute deviation and M-robust estimators reduce the effect of outliers?

12.6 Numerical Exercises

12-1E. Estimate the Cobb–Douglas production function for three inputs given the data in Table 12.1 using nonlinear least squares. Compare the results to the linear transformation using the same data. Are the results close?

12-2E. Using the 1960 through 1965 data for the interest rate paid by Alabama farmers in Appendix A, construct a Bayesian estimator for the average interest rate given the prior distribution for the mean is $N(0.05, 0.005)$.

12-3E. Given the setup in Exercise 12-2E, use simulation to derive an empirical Bayesian estimate of the average interest rate.

12-4E. Estimate the effect of the market interest rate (R_t) and changes in the debt to asset level ($\Delta\left(D/A\right)_t$) on the interest rate paid by farmers in South Carolina (r_t),

$$r_t = \alpha_0 + \alpha_1 R_t + \alpha_2 \Delta\left(D/A\right)_t + \epsilon_t \tag{12.80}$$

using least absolute deviation and ordinary least squares. How different are the results?

13

Conclusions

As I stated in Chapter 1, one of the biggest problems in econometrics is that students have a tendency to learn statistics and econometrics as a set of tools. Typically, they do not see the unifying themes involved in quantifying sample information to make inferences. To introduce this concept, Chapter 1 starts by reviewing how we think of science in general and economics as a science. In addition to the standard concepts of economics as a science, Chapter 1 also introduces econometrics as a policy tool and framework for economic decisions.

Chapter 2 develops the concept of probabilities, starting with a brief discussion of the history of probability. I have attempted to maintain these debates in the text, introducing not only the dichotomy between classical and Bayesian ideas about probability, but also introducing the development of Huygens, Savage and de Finetti, and Kolmogorov.

Chapter 3 then builds on the concepts of probability introduced in Chapter 2. Chapter 3 develops probability with a brief introduction to the concept of measure theory. I do not dwell on it, but I have seen significant discussions of measure theory in econometric literature – typically in the time series literature. After this somewhat abstract introduction, Chapter 3 proceeds with a fairly standard development of probability density functions. The chapter includes a fairly rigorous development of the normal distribution – including the development of trigonometric transformations in Appendix C.

Chapter 4 presents the implications of the probability measures developed in Chapter 3 on the moments of the distribution. Most students identify with the first and second moments of distribution because of their prominence in introductory statistical classes. Beginning with the anomaly that the first moment (i.e., the mean) need not exist for some distributions such as the Cauchy distribution, this chapter develops the notion of boundedness of an expectation. Chapter 4 also introduces the notion of sample, population, and theoretical moments. In our discussion of the sample versus population moments, we then develop the cross-moment of covariance and its normalized version, the correlation coefficient. The covariance coefficient then provides our initial development of the least squares estimator.

The development of the binomial and normal distributions are then presented in Chapter 5. In this chapter, we demonstrate how the binomial distribution converges to the normal as the sample size increases. After linking these distributions, Chapter 5 generalizes the univariate normal distribution to first the bivariate and then the more general multivariate normal. The extension of the univariate normal to the bivariate normal is used to develop the role of the correlation coefficient.

Given these foundations, the next part of the textbook develops the concepts of estimation. Chapter 6 presents the concept of large samples based on the notion of convergence. In its simplest form, the various modes of convergence give us a basis for saying that sample parameters will approach (or converge to) the true population parameters. Convergence is important for a variety or reasons, but in econometrics we are often interested in the convergence properties of ordinary least squares estimates. In general, Chapter 6 demonstrates that ordinary least squares coefficients converge to their true values as the sample size becomes large. In addition, Chapter 6 develops the central limit theorem, which states that the linear estimators such as ordinary least squares estimators are asymptotically distributed normal.

Chapter 7 focuses on the development of point estimators – the estimation of parameters of distributions or functions of parameters of distribution. As a starting point, the chapter introduces the concept of the sample as an image of the population. Given that the sample is the image of the population, we can then use the estimates of the sample parameters to make inferences about the population parameters. As a starting point, Chapter 7 considers some familiar estimators such as the sample mean and variance. Given these estimators, we introduce the concept of a measure of the goodness of an estimator. Specifically, Chapter 7 highlights that an estimator is a random variable that provides a value for some statistic (i.e., the value of some parameter of the distribution function). Given that the estimator is a random variable with a distribution, we need to construct a measure of closeness to describe how well the estimator represents the sample. Chapter 7 makes extensive use of two such measures – mean squared error of the estimator and the likelihood of the sample. Using these criteria, the chapter develops the point estimators.

Given that most estimators are continuous variables, the probability that any estimate is correct is actually zero. As a result, most empirical studies present confidence intervals (i.e., ranges of values that are hypothesized to contain the true parameter value with some level of statistical confidence). Chapter 8 develops the mathematics of confidence intervals.

Chapter 9 presents the mathematical basis for testing statistical hypotheses. The dominant themes in this chapter include the development of Type I and Type II errors and the relative power of hypothesis tests. In addition to the traditional frequentist approach to hypothesis testing, Chapter 9 presents an overview of the Neyman–Pearson Lemma, which provides for the selection of an interval of rejection with different combinations of Type I and Type II error.

The third part of the textbook shifts from a general development of mathematical statistics to focus on applications particularly popular in economics. To facilitate our development of these statistical methodologies, Chapter 10 provides a brief primer on matrix analysis. First, the chapter reviews the basic matrix operations such as matrix addition, multiplication, and inversion. Using these basic notions, Chapter 10 turns to the definition of a vector space spanned by a linear system of equations. This development allows for the general development of a projection matrix which maps points from a more general space into a subspace. Basically, ordinary least squares is simply one formulation for such a projection.

With this background, Chapter 11 develops the typical linear models applied in econometrics. As a starting point, we prove that ordinary least squares is a best linear unbiased estimator of the typical linear relationship under homoscedasticity (i.e., the assumption that all the errors are independently and identically distributed). After this basic development, we generalize the model by first allowing the residuals to be heteroscedastic (i.e., have different variances or allowing the variances to be correlated across observations). Given the possibility of heteroscedasticity, we develop the generalized least squares estimator. Next, we allow for the possibility that some of the regressors are endogenous (i.e., one of the "independent variables" is simultaneously determined by the level of a dependent variable). To overcome endogeneity, we develop Theil's two stage least squares and the instrumental variable approach. Finally, Chapter 11 develops the linear form of the generalized method of moments estimator.

The textbook presents three nonlinear econometric techniques in Chapter 12. First, nonlinear least squares and numerical applications of maximum likelihood are developed. These extensions allow for more general formulations with fewer restrictions on the distribution of the residual. Next, the chapter develops a simple Bayesian estimation procedure using simulation to integrate the prior distribution times the sampling distribution. Simulation has significantly expanded the variety of models that can be practically estimated using Bayesian techniques. Finally, Chapter 12 examines the possibility of nonsmooth error functions such as the least absolute deviation estimator and quantile regression. The least absolute deviation estimator is a robust estimator – less sensitive to extreme outliers. On the other hand, quantile regression allows the researcher to examine the distributional differences in the results.

In summation, this textbook has attempted to provide "bread crumbs" for the student to follow in developing a theoretical understanding of econometric applications. Basically, why are the estimated parameters from a regression distributed Student's t or normal? What does it mean to develop a confidence interval? And, why is this the rejection region for a test? It is my hope that after reading this textbook, students will be at least partially liberated from the cookbook approach to econometrics. If they do not know the why for a specific result, at least they will have a general reason.

There are some things that this book is not. It is not a stand-alone book on econometrics. Chapters 11 and 12 barely touch on the vast array of econometric techniques in current use. However, the basic concepts in this book form a foundation for serious study of this array of techniques. For example, maximum likelihood is the basis for most models of limited dependent variables where only two choices are possible. Similarly, the section on matrix algebra and the mathematics of projection matrices under normality form the basis for many time series technique such as Johansen's [21] error correction model. This book is intended as an introductory text to provide a foundation for further econometric adventures.

Appendix A

Symbolic Computer Programs

CONTENTS

Beginning in the late 1980s, several computer programs emerged to solve symbolic and numeric representations of algebraic expressions. In this appendix, we briefly introduce two such computer programs – Maxima and *Mathematica*.

A.1 Maxima

Maxima is a an open-source code for symbolic analysis. For example, suppose we are interested in solving for the zeros of

$$f(x) = 8 - \frac{(x-4)^2}{2} \tag{A.1}$$

and then plotting the function. To accomplish this we write a batch (ASCII) file:

```
/*************************************************************/
/* Setup the simple quadratic function                       */

f(x):=8-(x-4)^2/2;

/* Solve for those points where f(x) = 0                     */

solve(f(x)=0,x);

/* Plot the simple function                                   */

plot2d(f(x),[x,0,8]);

/*************************************************************/
```

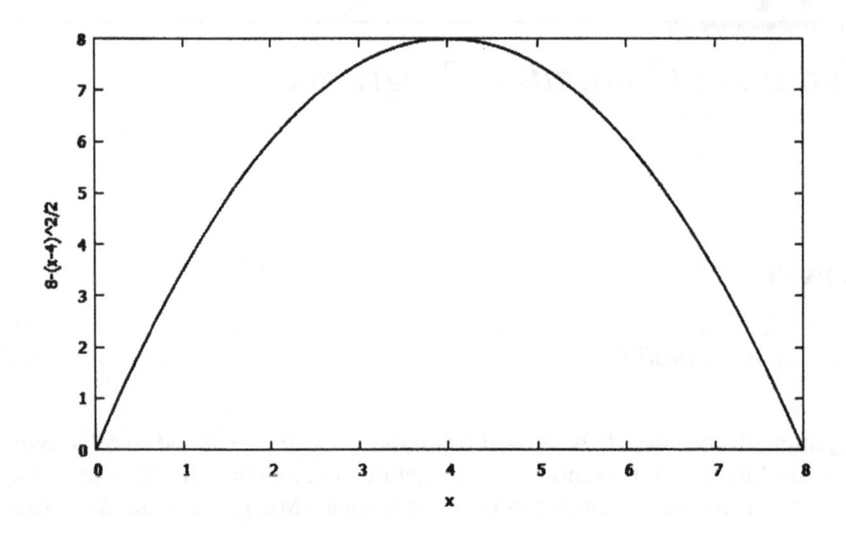

FIGURE A.1
Maxima Plot of Simple Quadratic Function.

Executing this batch file yields:

```
(%i2) f(x):=8-(x-4)^2/2
(%o2) f(x):=8-(x-4)^2/2
(%i3) solve(f(x) = 0,x)
(%o3) [x=0,x=8]
(%i4) plot2d(f(x),[x,0,8])
```

and the graphic output presented in Figure A.1. The output (%o3) presents the solutions to $f(x) = 0$ as $x = 0$ and $x = 8$.

Next, consider using Maxima to compute the mean, variance, skewness, and kurtosis of the simple quadratic distribution. The Maxima code for these operations becomes:

```
/*************************************************************/
/* Integrating f(x) for 0 to 8                              */

f(x):=8-(x-4)^2/2;

k: integrate(f(x),x,0,8);

/* Setup a valid probability density function               */

g(x):= f(x)/k;
```

```
/* Verify that the new function integrates to one                */

integrate(g(x),x,0,8);

/* Compute the expected value of the distribution                */

avg: integrate(x*g(x),x,0,8);

/* Compute the variance of the distribution                      */

var: integrate((x-avg)^2*g(x),x,0,8);

/* Compute the skewness of the distribution                      */

skw: integrate((x-avg)^3*g(x),x,0,8);

/* Compute the kurtosis of the distribution                      */

krt: integrate((x-avg)^4*g(x),x,0,8);

/****************************************************************/
```

The output for these computations is then:

```
(%i2) f(x):=8-(x-4)^2/2
(%o2) f(x):=8-(x-4)^2/2
(%i3) k:integrate(f(x),x,0,8)
(%o3) 128/3
(%i4) g(x):=f(x)/k
(%o4) g(x):=f(x)/k
(%i5) integrate(g(x),x,0,8)
(%o5) 1
(%i6) avg:integrate(x*g(x),x,0,8)
(%o6) 4
(%i7) var:integrate((x-avg)^2*g(x),x,0,8)
(%o7) 16/5
(%i8) skw:integrate((x-avg)^3*g(x),x,0,8)
(%o8) 0
(%i9) krt:integrate((x-avg)^4*g(x),x,0,8)
(%o9) 768/35
```

From this output we see that the constant of integration is $k = 128/3$. The valid probability density function for the quadratic is then

$$f(x) = \frac{3\left(8 - \frac{(x-4)^2}{2}\right)}{128}. \tag{A.2}$$

As demonstrated in the Maxima output $\int_0^8 f(x)\,dx = 1$ so $f(x)$ is a valid probability density function. Next, we develop the Maxima code to derive the cumulative density function:

```
/*************************************************************/
/* Starting with the valid probability density function      */

f(x):=3/128*(8-(x-4)^2/2);

/* We derive the general form of the cumulative density      */
/*  function                                                 */

r1: integrate(f(z),z,0,x);

/* We can then plot the cumulative density function          */

plot2d(r1,[x,0,8]);

/* Next, assume that we want to compute the value of the     */
/* cumulative density function at x = 3.5                    */

subst(x=3.5,r1);

/*************************************************************/
```

The result of this code is then:

```
(%i2) f(x):=3*(8-(x-4)^2/2)/128
(%o2) f(x):=(3*(8-(x-4)^2/2))/128
(%i3) r1:integrate(f(z),z,0,x)
(%o3) -(x^3-12*x^2)/256
(%i4) plot2d(r1,[x,0,8])
(%o4)
(%i5) subst(x = 3.5,r1)
(%o5) 0.40673828125
```

with the plot depicted in Figure A.2.

A.2 *Mathematica*™

Mathematica is a proprietary program from Wolfram Research. To use *Mathematica* efficiently, the user writes a **Notebook** file using *Mathematica*'s front-end program. In this section, we will present some of the commands and

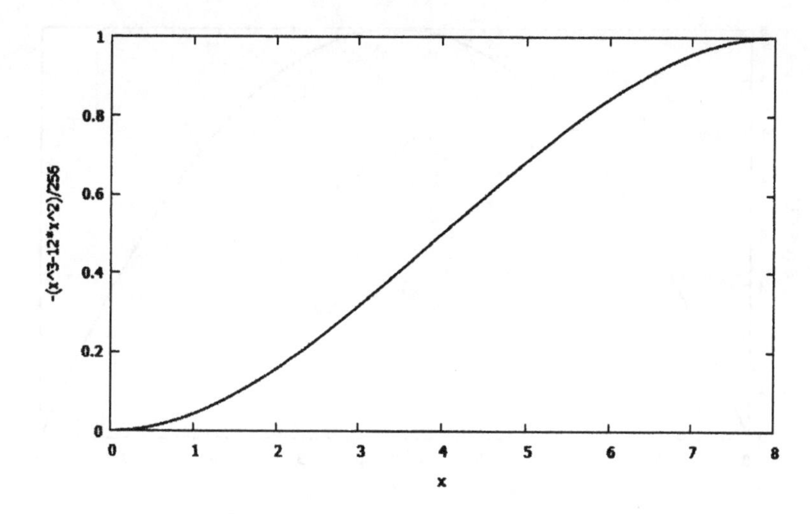

FIGURE A.2
Maxima Cumulative Distribution Function for Quadratic Distribution.

responses from *Mathematica.* We will not discuss the frontend program in detail.

As a starting point, consider the same set of operations from our Maxima example. We start by defining the quadratic function

```
f[x_] := 8 - (x - 4)^2/2;
Print[f[x]];
```

Mathematica responds with the general form of the function (i.e., from the Print command):

```
8-1/2 (-4+x)^2
```

The input command and resulting output for solving for the zeros of the quadratic function is then:

```
sol1 = Solve[f[x] == 0, x];
Print[sol1];

{{x->0},{x->8}}
```

We can then generate the plot of the quadratic function:

```
Print[Plot[f[x],{x,0,8}]];
```

The plot for this input is presented in Figure A.3. Following the Maxima example, we integrate the quadratic function over its entire range to derive the normalization factor:

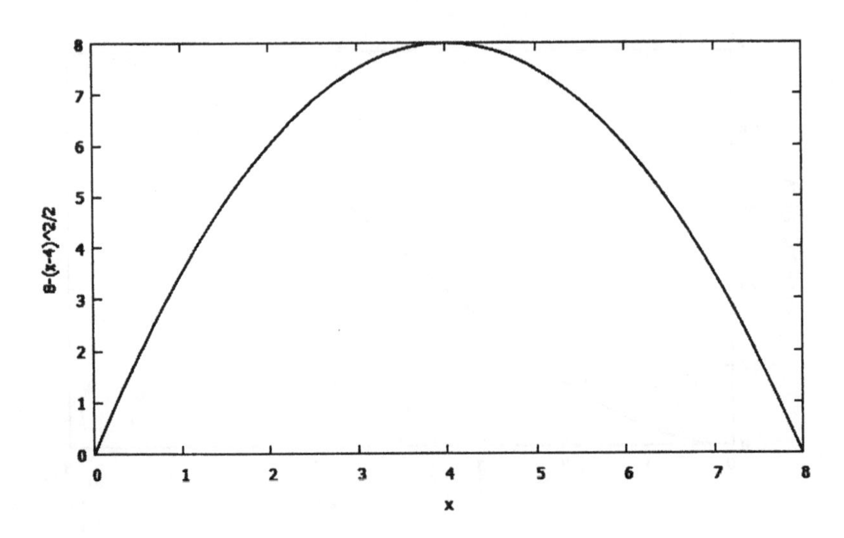

FIGURE A.3
Mathematica Plot of Simple Quadratic Function.

```
k = Integrate[f[x], {x, 0, 8}];
Print[k];
g[x_] := f[x]/k;
Print[g[x]];

128/3
3/128 (8-1/2 (-4+x)^2)
```

Next we test the constant of integration and compute the average, variance, skewness, and kurtosis:

```
tst = Integrate[g[x], {x, 0, 8}];
Print[tst];
1

avg = Integrate[x g[x], {x, 0, 8}];
Print[avg];
4

var = Integrate[(x - avg)^2 g[x], {x, 0, 8}];
Print[var];
16/5

skw = Integrate[(x - avg)^3 g[x], {x, 0, 8}];
Print[skw];
0
```

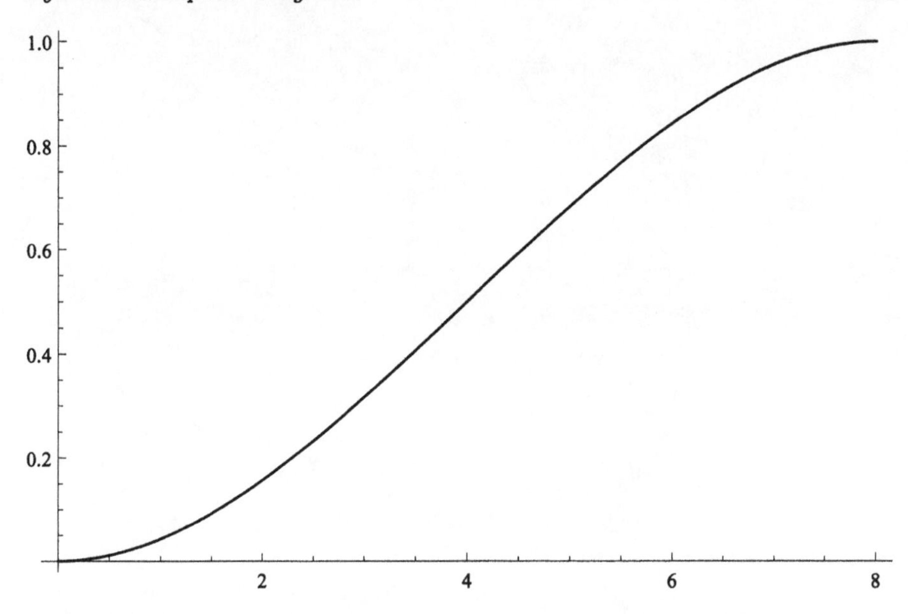

FIGURE A.4
Mathematica Cumulative Distribution Function for Quadratic Distribution.

```
krt = Integrate[(x - avg)^4 g[x], {x, 0, 8}];
Print[krt];
768/35
```

To finish the example, we derive the cumulative density function by integrating the probability density function. The graph of the cumulative density function is shown in Figure A.4.

```
h[x_] := Integrate[g[z], {z, 0, x}];
Print[h[x]];

(3 x^2)/64-x^3/256
```

Appendix B

Change of Variables for Simultaneous Equations

CONTENTS

One of the more vexing problems in econometrics involves the estimation of simultaneous equations. Developing a simple model of supply and demand,

$$\begin{aligned} q_S &= \alpha_{10} + \alpha_{11} p + \alpha_{12} p^A + \alpha_{13} w + \epsilon_1 \\ q_D &= \alpha_{20} + \alpha_{21} p + \alpha_{24} p^B + \alpha_{25} Y + \epsilon_2 \end{aligned} \tag{B.1}$$

where q_S is the quantity supplied, q_D is the quantity demanded, p is the price, p^A is the price of an alternative good to be consided in the production (i.e., another good that the firm could produce using its resources), p^B is the price of an alternative good in the consumption equation (i.e., a complement or substitute in consumption), w is the price of an input, Y is consumer income, ϵ_1 is the error in the supply relationship, and ϵ_2 is the error in the demand relationship. In estimating these relationships, we impose the market clearing condition that $q_S = q_D$. As developed in Section 11.4.2, a primary problem with this formulation is the correlation between the price in each equation and the respective residual. To demonstrate this correlation we solve the supply and demand relationships in Equation B.1 to yield

$$p = \frac{-\alpha_{10} - \alpha_{12} p^A - \alpha_{13} w + \alpha_{20} + \alpha_{24} p^B + \alpha_{25} Y - \epsilon_1 + \epsilon_2}{\alpha_{11} - \alpha_{21}}. \tag{B.2}$$

Given the results in Equation B.2, it is apparent that $\mathrm{E}\left[p\epsilon_1\right], \mathrm{E}\left[p\epsilon_2\right] \nrightarrow 0$. Thus, the traditional least squares estimator cannot be unbiased or asymptotically consistent. The simultaneous equations bias will not go away.

Chapter 11 presents two least squares remedies for this problem – two stage least squares and instrumental variables. However, in this appendix we develop the likelihood for this simultaneous problem to develop the **Full Information Maximum Likelihood** (FIML) estimator.

B.1 Linear Change in Variables

To develop the FIML estimator we return to the linear change in variables notion developed in Theorem 3.36. The transformation envisioned in Chapter 3 is

$$\left.\begin{array}{l} Y_1 = a_{11}X_1 + a_{12}X_2 \\ Y_2 = a_{21}X_1 + a_{22}X_2 \end{array}\right\} \Rightarrow \left\{\begin{array}{l} X_1 = b_{11}Y_1 + b_{12}Y_2 \\ X_2 = b_{21}Y_1 + b_{22}Y_2. \end{array}\right. \tag{B.3}$$

Given this formulation, the contention was Equation 3.94:

$$g(y_1, y_2) = \frac{f(b_{11}y_1 + b_{12}y_2, b_{21}y_1 + b_{22}y_2)}{|a_{11}a_{22} - a_{12}a_{21}|}.$$

The real question is then, what does this mean? The overall concept is that I know (or at least hypothesize that I know) the distribution of X_1 and X_2 and want to use it to derive the distribution of a function of Y_1 and Y_2.

Consider a simple example of the mapping in Equation B.3:

$$\begin{aligned} Y_1 &= X_2 + \frac{1}{2}X_2 \\ Y_2 &= \frac{3}{4}X_2 \end{aligned} \tag{B.4}$$

such that $f(X_1, X_2) = 1$. Applying Equation 3.94, the new probability density function (i.e., in terms of Y_1 and Y_2) is then

$$g(Y_1, Y_2) = \frac{1}{\left|1 \times \frac{3}{4}\right|}. \tag{B.5}$$

The problem is that the range of Y_1 and Y_2 (depicted in Figure B.1) is not $Y_1, Y_2 \in [0,1]$. Specifically $Y_2 \in \left[0, \frac{4}{3}\right]$, but the range of Y_1 is determined by the linear relationship $Y_1 = X_1 - \frac{3}{2}X_2$. This range is a line with an intercept at $Y_1 = 0$ and a slope of $-\frac{2}{3}$. In order to develop the ranges, consider the solution of Equation B.4 in terms of X_1 and X_2:

$$\begin{aligned} X_1 &= Y_1 - \frac{2}{3}Y_2 \\ X_2 &= \frac{4}{3}Y_2. \end{aligned} \tag{B.6}$$

First we solve for the bounds of Y_1:

$$Y_1 = X_1 - \frac{2}{3}Y_2 \text{ and } X_1 \in [0,1] \Rightarrow Y_1 \in \left[-\frac{2}{3}Y_2, 1 - \frac{2}{3}Y_2\right]. \tag{B.7}$$

If you want to concentrate on the "inside integral,"

$$\int_{-\frac{2}{3}Y_2}^{1-\frac{2}{3}Y_2} dY_1 = \frac{3}{4}\left[\left(1 - \frac{2}{3}Y_2\right) - \left(-\frac{2}{3}Y_2\right)\right] = 1. \tag{B.8}$$

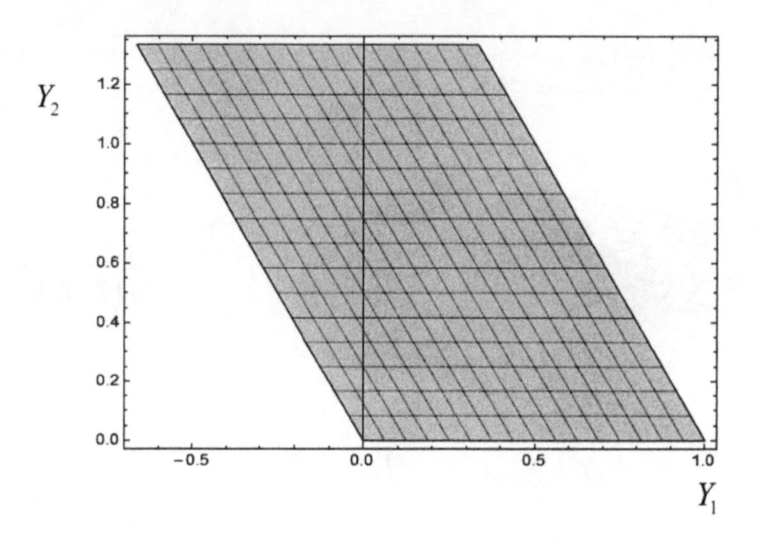

FIGURE B.1
Transformed Range of Y_1 and Y_2.

The "outside integral" is then determined by

$$Y_2 = \frac{4}{3}X_2 \text{ and } X_2 \in [0,1] \Rightarrow Y_2 \in \left[0, \frac{3}{4}\right]. \tag{B.9}$$

Integrating over the result from Equation B.8 then gives

$$\frac{4}{3}\int_0^{\frac{3}{4}} dY_2 = \frac{4}{3}\left[\frac{3}{4} - 0\right] = 1. \tag{B.10}$$

The complete form of the integral is then

$$\frac{4}{3}\int_0^{\frac{3}{4}}\int_{-\frac{2}{3}Y_2}^{1-\frac{2}{3}Y_2} dY_1 dY_2 = 1. \tag{B.11}$$

The implication of this discussion is simply that the probability function is valid.

Next, consider the scenario where the $f(X_1, X_2)$ is a correlated normal distribution:

$$f(X_1, X_2 | \mu, \Sigma) = (2\pi)^{-1} |\Sigma|^{-\frac{1}{2}} \exp\left[-\frac{1}{2}\left(\left[\begin{array}{c} X_1 \\ X_2 \end{array}\right] - \mu\right)' \Sigma^{-1}\left(\left[\begin{array}{c} X_1 \\ X_2 \end{array}\right] - \mu\right)\right]. \tag{B.12}$$

In order to simplify our discussion, let $\mu = 0$ and $\Sigma = I_2$. This simplifies the distribution function in Equation B.12 to

$$f(X_1, X_2) = (2\pi)^{-1} \exp\left[-\frac{1}{2}\left(\begin{array}{c} X_1 \\ X_2 \end{array}\right)'\left(\begin{array}{c} X_1 \\ X_2 \end{array}\right)\right]. \tag{B.13}$$

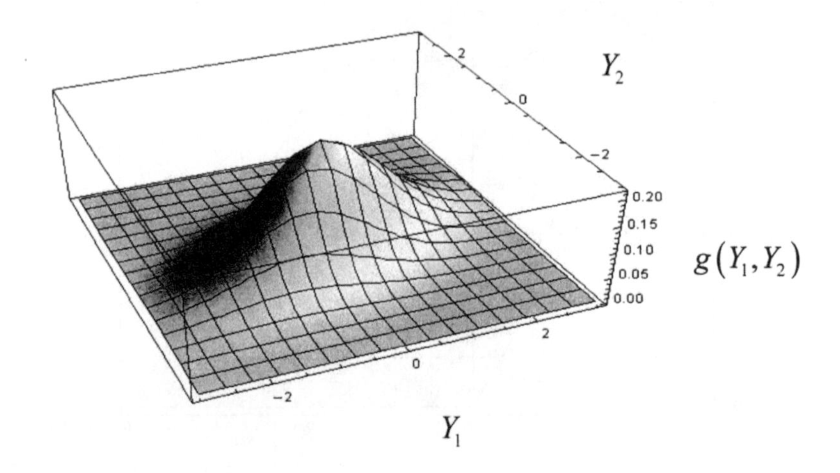

FIGURE B.2
Transformed Normal Distribution.

Next, consider rewriting $f(X_1, X_2)$ (in Equation B.13) as $g(Y_1, Y_2)$. Following Equation 3.94 yields

$$g(Y_1, Y_2) = \frac{1}{\left|1 \times \frac{3}{4}\right|} (2\pi)^{-1} \exp\left[-\frac{1}{2}\begin{pmatrix} Y_1 - \frac{2}{3}Y_2 \\ \frac{3}{4}Y_2 \end{pmatrix}'\begin{pmatrix} Y_1 - \frac{2}{3}Y_2 \\ \frac{3}{4}Y_2 \end{pmatrix}\right] \quad \text{(B.14)}$$

which is presented graphically in Figure B.2.

Next, we redevelop the results in matrix terms.

$$\begin{bmatrix} X_1 \\ X_2 \end{bmatrix} = \begin{bmatrix} 1 & -\frac{2}{3} \\ 0 & \frac{3}{4} \end{bmatrix}\begin{bmatrix} Y_1 \\ Y_2 \end{bmatrix} \Leftrightarrow X = \Gamma Y. \quad \text{(B.15)}$$

Substituting Equation B.15 into Equation B.14 yields

$$\begin{aligned} g(Y_1, Y_2) &= \frac{1}{\left|1 \times \frac{3}{4}\right|} (2\pi)^{-1} \\ &\times \exp\left[-\frac{1}{2}\left(\begin{bmatrix} 1 & -\frac{2}{3} \\ 0 & \frac{3}{4} \end{bmatrix}\begin{bmatrix} Y_1 \\ Y_2 \end{bmatrix}\right)'\left(\begin{bmatrix} 1 & -\frac{2}{3} \\ 0 & \frac{3}{4} \end{bmatrix}\begin{bmatrix} Y_1 \\ Y_2 \end{bmatrix}\right)\right] \\ &\Rightarrow \frac{1}{\left|1 \times \frac{3}{4}\right|} (2\pi)^{-1} \\ &\times \exp\left[-\frac{1}{2}\begin{bmatrix} Y_1 \\ Y_2 \end{bmatrix}'\begin{bmatrix} 1 & -\frac{2}{3} \\ 0 & \frac{3}{4} \end{bmatrix}'\begin{bmatrix} 1 & -\frac{2}{3} \\ 0 & \frac{3}{4} \end{bmatrix}\begin{bmatrix} Y_1 \\ Y_2 \end{bmatrix}\right]. \end{aligned} \quad \text{(B.16)}$$

Using the $X = \Gamma Y$ form in Equation B.15, we can rewrite Equation B.16 as

$$g(Y_1, Y_2) = (2\pi)^{-1} \exp\left[-\frac{1}{2} Y'\Gamma'\Gamma Y\right] |\Gamma|^{-1}. \tag{B.17}$$

Going back to the slightly more general formulation,

$$g(Y_1, Y_2) = (2\pi)^{-\frac{p}{2}} |\Sigma|^{-\frac{1}{2}} \exp\left[-\frac{1}{2} Y'\Gamma'\Sigma^{-1}\Gamma Y\right] |\Gamma|^{-1} \tag{B.18}$$

where p is the number of variables in the multivariate distribution.

B.2 Estimating a System of Equations

Next, we return to an empirical version of Equation B.1.

$$\begin{aligned} q_s &= 115.0 + 6.0p - 2.5p^A - 12.0w + \epsilon_1 \\ q_d &= 115.5 - 10.0p + 3.3333p^B + 0.00095Y + \epsilon_2. \end{aligned} \tag{B.19}$$

We can write Equation B.19 in the general form

$$\Gamma Y = AX + \epsilon \Rightarrow Y - \Gamma^{-1}AX = \Gamma^{-1}\epsilon \tag{B.20}$$

as

$$\begin{bmatrix} 1 & -6.0 \\ 1 & 10.0 \end{bmatrix} \begin{bmatrix} q \\ p \end{bmatrix} = \begin{bmatrix} 115.0 & -2.5 & -12.0 & 0.0 & 0.0 \\ 115.5 & 0.0 & 0.0 & 3.3333 & 0.00095 \end{bmatrix} \begin{bmatrix} 1 \\ p^A \\ w \\ p^B \\ Y \end{bmatrix} + \begin{bmatrix} \epsilon_1 \\ \epsilon_2 \end{bmatrix}. \tag{B.21}$$

Using this specification, we return to the general likelihood function implied by Equation B.18.

$$\begin{aligned} f(Y) &= (2\pi)^{-\frac{p}{2}} |\Sigma|^{-\frac{1}{2}} \\ &\times \exp\left[-\frac{1}{2}\left[Y - \Gamma^{-1}AX\right]' (\Gamma')^{-1} \Sigma^{-1} \left(\Gamma^{-1}\right) \left[Y - \Gamma^{-1}AX\right]\right] \left|\Gamma^{-1}\right|^{-1}. \end{aligned} \tag{B.22}$$

Since $\left|\Gamma^{-1}\right|^{-1} = |\Gamma|$, Equation B.22 can be rewritten as

$$\begin{aligned} f(Y) &= (2\pi)^{-\frac{p}{2}} |\Sigma|^{-\frac{1}{2}} \\ &\times \exp\left[-\frac{1}{2}\left[Y - \Gamma^{-1}AX\right]' (\Gamma')^{-1} \Sigma^{-1} \left(\Gamma^{-1}\right) \left[Y - \Gamma^{-1}AX\right]\right] |\Gamma|. \end{aligned} \tag{B.23}$$

Consider the simulated data for Equation B.21 presented in Table B.1. We estimate this system of equations using an iterative technique similar to those presented in Appendix D. Specifically, we use a computer algorithm that starts with an estimated set of parameters and then attempts to compute a set of parameters closer to the optimal. As an additional point of complexity, Equation B.23 includes a set of variance parameters (Σ). In our case the Σ matrix involves three additional parameters (σ_{11}, σ_{12}, and σ_{22}). While we could search over these parameters, it is simpler to concentrate these parameters out of the distribution function. Notice that these parameters are the variance coefficients for the untransformed model (i.e., the model as presented in Equation B.19). Hence, we can derive the estimated variance matrix for each value of parameters as

$$\epsilon(\alpha) = \begin{bmatrix} q - \alpha_{10} - \alpha_{11}p - \alpha_{12}p^A - \alpha_{13}w \\ q - \alpha_{20} - \alpha_{21}p - \alpha_{24}p^B - \alpha_{25}Y \end{bmatrix}$$
$$\hat{\Sigma}(\alpha) = \frac{1}{N}\epsilon(\alpha)'\epsilon(\alpha). \tag{B.24}$$

Further simplifying the optimization problem by taking the natural logarithm and discarding the constant yields

$$\alpha = \{\alpha_{10}, \alpha_{11}, \alpha_{12}, \alpha_{13}, \alpha_{20}, \alpha_{21}, \alpha_{24}, \alpha_{25}\}$$

$$\Gamma = \begin{bmatrix} 1 & -\alpha_{11} \\ 1 & -\alpha_{21} \end{bmatrix}$$

$$A = \begin{bmatrix} \alpha_{10} & \alpha_{12} & \alpha_{13} & 0 & 0 \\ \alpha_{20} & 0 & 0 & \alpha_{24} & \alpha_{25} \end{bmatrix}$$

$$Y_i = \begin{bmatrix} q_i \\ p_i \end{bmatrix} X_i = \begin{bmatrix} 1 & p_i^A & w_i & p_i^B & Y_i \end{bmatrix}$$

$$\nu_i(\alpha) = Y_i - \Gamma A X_i$$

$$\hat{\Sigma}(\alpha) = \frac{1}{N}\sum_{i=1}^{N} \nu_i(\alpha)\nu_i(\alpha)'$$

$$l(\alpha) = \sum_{i=1}^{N}\left[-\frac{1}{2}\ln\left|\hat{\Sigma}(\alpha)\right| - \frac{1}{2}\nu_i(\alpha)'(\Gamma')^{-1}\hat{\Sigma}(\alpha)^{-1}\Gamma^{-1}\nu_i(\alpha) + |\Gamma|\right]. \tag{B.25}$$

TABLE B.1
Data Set for Full Information Maximum Likelihood

Obs.	Q	p	p^A	w	p^b	y
1	101.013	5.3181	5.7843	2.6074	8.7388	1007.20
2	91.385	5.6342	6.1437	3.5244	6.8113	1005.17
3	103.315	4.6487	6.4616	1.9669	7.4628	1001.18
4	105.344	4.1317	6.1897	1.6157	6.5012	1002.51
5	98.677	4.8446	6.2839	2.3121	6.5719	1006.60
6	106.015	4.6119	5.9365	1.7852	8.1629	990.01
7	100.507	4.8681	5.2442	2.6392	7.2785	1001.22
8	97.746	5.1389	5.5195	2.8337	7.2162	1003.83
9	104.609	4.5759	4.9144	2.1443	7.5860	1011.73
10	100.274	5.2478	6.4346	2.4840	8.3292	999.61
11	105.011	4.3364	4.8738	1.9210	7.0101	997.99
12	99.311	5.0234	5.9073	2.5028	7.3492	998.36
13	100.133	4.9524	5.6071	2.5546	7.4004	1001.29
14	99.163	4.8527	7.0464	2.3681	6.8731	986.34
15	100.986	4.9196	6.0364	2.3506	7.5375	1001.05
16	97.317	5.5198	6.9735	2.8430	8.2804	999.71
17	101.919	4.5613	5.2070	2.3069	6.7528	999.88
18	98.571	5.2390	6.1843	2.7258	7.7439	1014.08
19	97.835	5.8442	5.3252	3.2269	9.4427	979.63
20	93.062	5.3741	4.5755	3.4441	6.4937	1004.44
21	96.886	5.5546	5.5188	3.1334	8.1915	1004.27
22	102.866	4.7970	6.0927	2.0457	7.7434	992.51
23	101.721	4.7528	5.1983	2.4157	7.2560	1007.71
24	98.433	5.3295	6.4392	2.5720	8.0082	992.39
25	95.441	5.3400	6.1446	2.9980	7.1166	1021.44
26	101.268	4.7685	5.5490	2.3263	7.1841	996.66
27	100.197	5.1127	6.9770	2.2818	7.8262	1016.34
28	96.613	4.6348	6.9509	2.3346	5.3826	991.89
29	97.385	4.9461	6.7019	2.5715	6.5967	986.52
30	103.915	4.6871	6.3564	1.9521	7.7442	1001.89
31	95.602	5.1170	6.4526	2.8207	6.5441	990.76
32	102.136	4.7981	7.0438	2.0735	7.5421	1005.92
33	93.416	5.4138	6.6783	3.1374	6.7559	1001.24
34	95.716	5.4084	6.3358	2.9980	7.4294	998.78
35	94.905	5.0263	6.0993	3.0018	6.0674	1004.53
36	99.439	4.7399	6.2888	2.3126	6.5385	1005.63
37	110.629	3.9795	5.7451	1.1964	7.6438	1003.48
38	100.805	4.5947	5.6510	2.3125	6.4995	1008.28
39	100.951	5.1130	5.0394	2.6601	8.1279	1006.28
40	104.540	4.4065	5.7951	1.8974	7.0803	994.11
41	103.274	4.8168	6.5948	2.1816	7.9596	1006.27

Continued on Next Page

Continued from the Previous Page

Obs.	Q	p	p^A	w	p^b	y
42	94.083	5.2631	5.1958	3.3148	6.5127	996.84
43	99.577	5.3378	6.0798	2.7206	8.4150	992.78
44	96.338	4.9589	5.9935	2.8168	6.2990	990.47
45	104.351	4.7474	5.3707	2.1632	8.0089	1011.81
46	100.210	4.9229	6.6945	2.3060	7.3371	994.40
47	94.072	5.7720	5.5735	3.4364	8.0373	995.79
48	100.460	5.1441	5.3778	2.7249	8.1262	989.14
49	99.129	5.3715	6.4842	2.6675	8.3774	997.82
50	102.551	4.8758	6.2919	2.1630	7.8877	997.69
51	97.505	4.7343	6.3408	2.5109	5.9771	989.90
52	97.932	5.2163	6.2582	2.8204	7.5332	1010.31
53	100.669	5.2442	6.6273	2.4942	8.4434	1012.06
54	102.016	5.1210	4.8340	2.6142	8.4602	999.90
55	102.327	4.6689	6.6954	1.9703	7.1832	1007.31
56	96.758	5.3178	6.8145	2.7339	7.4476	1002.68
57	100.947	5.1529	5.8795	2.5656	8.2317	1007.66
58	93.441	5.7598	5.1249	3.6586	7.8207	1006.79
59	99.241	5.1567	6.9333	2.4682	7.7528	996.58
60	98.562	4.6966	6.7658	2.2488	6.1362	1008.28
61	109.635	4.3500	5.6216	1.4074	8.4442	998.47
62	97.033	5.2723	6.8122	2.7122	7.3950	1014.60
63	98.975	4.7467	6.1962	2.3677	6.4070	1001.00
64	101.968	4.8730	5.3569	2.5012	7.7700	995.71
65	104.448	4.5595	6.2224	1.8166	7.4865	998.85
66	101.025	5.1205	6.8176	2.3072	8.1799	993.81
67	104.936	4.3629	5.5058	1.9391	7.0954	1007.82
68	106.865	4.0540	5.5562	1.5484	6.7380	993.36
69	97.024	4.9168	5.4087	2.8418	6.3121	1007.82
70	97.513	5.1881	6.3041	2.7227	7.2752	1010.94
71	89.772	5.9486	7.7699	3.4693	7.2697	1003.00
72	105.279	4.3413	5.2095	1.9837	7.1083	1000.92
73	106.535	4.1015	6.0280	1.5198	6.7768	994.72
74	97.998	5.2607	5.7413	2.8433	7.6496	1006.89
75	101.167	4.8004	6.1460	2.2469	7.2899	983.29
76	97.666	4.9058	5.7852	2.6019	6.5298	991.44
77	102.534	4.6082	5.0741	2.3579	7.1192	995.70
78	101.353	4.5451	6.0791	2.1908	6.5430	1001.58
79	95.475	5.8495	5.4118	3.4785	8.7392	988.47
80	95.387	5.0495	5.9921	2.9921	6.2858	1000.46
81	102.550	4.9486	7.3503	2.1588	8.1570	998.02
82	99.617	5.1257	4.7759	2.9729	7.7619	1003.14
83	98.207	4.8788	5.9363	2.5928	6.5777	1004.16

Continued on Next Page

Continued from the Previous Page

Obs.	Q	p	p^A	w	p^b	y
84	107.700	4.3114	6.3258	1.4821	7.7802	1001.08
85	94.616	5.1705	6.5016	2.8974	6.4460	981.95
86	96.412	5.3390	5.8664	2.9854	7.5201	980.00
87	104.131	4.5881	5.6296	2.0724	7.5209	998.92
88	98.171	4.8662	7.1715	2.3646	6.5489	1003.12
89	100.330	4.7532	4.3557	2.7301	6.8322	1010.36
90	102.046	4.5079	6.2594	1.9468	6.6121	997.16
91	102.058	4.8524	7.2229	2.0469	7.6635	1008.76
92	101.538	5.1086	6.6352	2.2712	8.2555	1007.75
93	106.127	4.3825	5.9567	1.6748	7.4946	999.99
94	96.395	4.9926	6.7225	2.6088	6.4035	988.35
95	100.459	4.9249	5.8115	2.4600	7.4220	991.86
96	97.746	5.1175	5.9726	2.5973	7.1797	986.47
97	96.813	5.0942	6.7325	2.6480	6.8318	995.47
98	98.187	5.0592	4.8647	2.9029	7.1851	982.71
99	106.491	4.6461	6.2315	1.7326	8.3240	1014.81
100	95.846	5.2445	5.8738	2.9156	6.9860	1003.44

The R Code for Full Information Maximum Likelihood

```
##########################################################################
# Read the data from a comma separated variables file                    #

dta <- read.csv("AppendixD.csv")

# Setup the X matrix for each operation                                  #

xx <- as.matrix(cbind(dta[,1],matrix(1,nrow=nrow(dta),1),dta[,2:6]))

# Defining the maximum likelihood function                               #

ml <- function(b) {
   g  <- cbind(rbind(1,1),rbind(-b[2],-b[6]))
   a  <- cbind(rbind(b[1],b[5]),rbind(b[3],0),rbind(b[4],0),
          rbind(0,b[7]),rbind(0,b[8]))
   aa <- cbind(rbind(1,1),-rbind(b[1],b[5]),-rbind(b[2],b[6]),
            -rbind(b[3],0),-rbind(b[4],0),-rbind(0,b[7]),
            -rbind(0,b[8]))
   vv <- var(xx%*%t(aa))
   for (i in 1:nrow(dta)) {
     err <- rbind(dta[i,1],dta[i,2]) - solve(g)%*%a%*%rbind(1,
        dta[i,3],dta[i,4],dta[i,5],dta[i,6])
     li <- 1/2*log(det(vv))+1/2*t(err)%*%solve(t(g))%*%solve(vv)
        %*%solve(g)%*%err - det(g) }
   return(li) }
```

TABLE B.2
Full Information Maximum Likelihood Estimates

Parameter	Estimates			
	α	α^*	$\hat{\alpha}$	OLS
α_{10}	115.000	115.677	116.495	116.460
		(12.379)	(12.350)	(1.321)
α_{11}	6.000	5.320	5.222	5.327
		(4.561)	(4.582)	(0.469)
α_{12}	−2.500	−2.414	−2.442	−2.392
		(0.980)	(0.992)	(0.125)
α_{13}	−12.000	−11.507	−11.530	−11.489
		(3.655)	(3.670)	(0.378)
α_{20}	115.500	115.732	115.602	116.600
		(7.363)	(7.358)	(0.793)
α_{21}	−10.000	−9.991	−9.981	−9.979
		(0.178)	(0.178)	(0.018)
α_{24}	−3.333	3.331	3.330	3.308
		(0.101)	(0.101)	(0.010)
α_{25}	0.00095	0.00920	0.00918	0.00848
		(0.00535)	(0.00535)	(0.00079)

```
###########################################################################
# Call for the values used in the simulation as the initial values       #

b0 <- rbind(115,6,-2.5,-12,115.5,-10,3.333,0.00095)
res <- optim(b0,ml,control=list(maxit=3000))
print(res)
h   <- optimHess(res$par,ml)

###########################################################################
# Estimate ordinary least squares to use as initial values               #

b1 <- lm(dta[,1] ~ dta[,2] + dta[,3] + dta[,4])
b2 <- lm(dta[,1] ~ dta[,2] + dta[,5] + dta[,6])
b0 <- rbind(as.matrix(b1$coefficients),as.matrix(b2$coefficients))
res <- optim(b0,ml,control=list(maxit=3000))
print(res)
h   <- optimHess(res$par,ml)
###########################################################################
```

Table B.2 presents the true values of the parameters (i.e., those values used to simulate the data) along with three different sets of estimates. The first set of estimates (presented in column 3) uses the true values as the initial parameter values for the iterative techniques. The second set of estimates in column 4 uses the ordinary least squares estimates in the fifth column as the initial parameter values.

Each set of paramters is fairly close to the true values – even the ordinary least squares results. The primary difference between the ordinary least squares results is the relatively small standard deviation of the parameter estimates. For example, the maximum likelihood estimate for the standard deviation of α_{13} is 3.670 in the fourth column compared to a standard deviation of 0.378 under ordinary least squares. To compare the possible consequences of this difference, we construct the z values for the difference between the estimated value and the true value for the FIML and OLS estimates.

$$z_{FIML} = \frac{|-11.000 + 11.530|}{3.670} = 0.128 \ll 1.354 = \frac{|-11.000 + 11.489|}{0.378} = z_{OLS}. \tag{B.26}$$

Hence, the confidence interval under ordinary least squares is smaller. However, in this case that may not be a good thing since the estimated interval is less likely to include the true value because of the simultaneous equations bias.

Appendix C

Fourier Transformations

CONTENTS

Section 3.7 integrates the normal distribution by transforming from the standard Cartesian space ($\{x, y\}$) into polar space ($\{r, \theta\}$). In this appendix, we develop this transformation in a little more detail.

C.1 Continuing the Example

We continue with the formulation from Equation 3.103:

$$y = f(x) = 5 - \frac{x^2}{5} \; x \in [-5, 5] .$$

Table C.1 presents the value of x in column 1, the value of $f(x)$ in column 2, and the radius (r) in column 3. Given these results, there are two ways to compute the inscribed angle (θ). The way described in the text involves solving for the inverse of the tangent function:

$$\theta = \tan^{-1}\left(\frac{f(x)}{x}\right). \tag{C.1}$$

Column four of Table C.1 gives the value of $f(x)/x$ while column 5 presents the inverse tangent value (note that the values are in radians – typically radians are given in ratios of π, so for $x = 4.0$ the value of the angle is $\theta = 0.4229 = 0.1346\pi$, as discussed in the text). The other method is implied by the cosine result in Equation 3.110:

$$y = r\cos(\theta) \Rightarrow \theta = \cos^{-1}\left(\frac{f(x)}{r}\right). \tag{C.2}$$

Columns 6 and 7 of Table C.1 demonstrate this approach.

Examining the results for θ in columns 5 and 7 of Table C.1, we see that the computed values of the inscribed angle are the same up until $x = 0$.

TABLE C.1
Transformation to Polar Space

x	$f(x)$	r	$\frac{f(x)}{x}$	$\tan^{-1}\left(\frac{f(x)}{x}\right)$	$\frac{f(x)}{r}$	$\cos^{-1}\left(\frac{f(x)}{r}\right)$	$\pi + \tan^{1}\left(\frac{f(x)}{x}\right)$	$\hat{r}(\theta)$	$\hat{r}(\theta) \times \sin(\theta)$
5.00	0.0000	5.0000	0.0000	0.0000	1.0000	0.0000		4.5161	0.0000
4.75	0.4875	4.7750	0.1026	0.1023	0.9948	0.1023		4.5397	0.4635
4.50	0.9500	4.5992	0.2111	0.2081	0.9784	0.2081		4.5625	0.9424
4.25	1.3875	4.4708	0.3265	0.3156	0.9506	0.3156		4.5840	1.4226
4.00	1.8000	4.3863	0.4500	0.4229	0.9119	0.4229		4.6036	1.8892
3.75	2.1875	4.3414	0.5833	0.5281	0.8638	0.5281		4.6212	2.3285
3.50	2.5500	4.3304	0.7286	0.6296	0.8082	0.6296		4.6366	2.7303
3.25	2.8875	4.3474	0.8885	0.7264	0.7476	0.7264		4.6498	3.0884
3.00	3.2000	4.3863	1.0667	0.8176	0.6839	0.8176		4.6610	3.4004
2.75	3.4875	4.4413	1.2682	0.9031	0.6192	0.9031		4.6704	3.6674
2.50	3.7500	4.5069	1.5000	0.9828	0.5547	0.9828		4.6782	3.8925
2.25	3.9875	4.5785	1.7722	1.0571	0.4914	1.0571		4.6845	4.0799
2.00	4.2000	4.6519	2.1000	1.1264	0.4299	1.1264		4.6898	4.2342
1.75	4.3875	4.7236	2.5071	1.1913	0.3705	1.1913		4.6940	4.3600
1.50	4.5500	4.7909	3.0333	1.2523	0.3131	1.2523		4.6974	4.4612
1.25	4.6875	4.8513	3.7500	1.3102	0.2577	1.3102		4.7001	4.5414
1.00	4.8000	4.9031	4.8000	1.3654	0.2040	1.3654		4.7023	4.6035
0.75	4.8875	4.9447	6.5167	1.4185	0.1517	1.4185		4.7039	4.6495
0.50	4.9500	4.9752	9.9000	1.4701	0.1005	1.4701		4.7051	4.6813
0.25	4.9875	4.9938	19.9500	1.5207	0.0501	1.5207		4.7059	4.7000
0.00	5.0000	5.0000			0.0000	1.5708		4.7064	4.7064
−0.25	4.9875	4.9938	−19.9500	−1.5207	−0.0501	1.6209	1.6209	4.7064	4.7005
−0.50	4.9500	4.9752	−9.9000	−1.4701	−0.1005	1.6715	1.6715	4.7061	4.6823
−0.75	4.8875	4.9447	−6.5167	−1.4185	−0.1517	1.7231	1.7231	4.7054	4.6510
−1.00	4.8000	4.9031	−4.8000	−1.3654	−0.2040	1.7762	1.7762	4.7043	4.6054
−1.25	4.6875	4.8513	−3.7500	−1.3102	−0.2577	1.8314	1.8314	4.7027	4.5439
−1.50	4.5500	4.7909	−3.0333	−1.2523	−0.3131	1.8892	1.8892	4.7005	4.4642
−1.75	4.3875	4.7236	−2.5071	−1.1913	−0.3705	1.9503	1.9503	4.6977	4.3634
−2.00	4.2000	4.6519	−2.1000	−1.1264	−0.4299	2.0152	2.0152	4.6941	4.2381
−2.25	3.9875	4.5785	−1.7722	−1.0571	−0.4914	2.0845	2.0845	4.6896	4.0843
−2.50	3.7500	4.5069	−1.5000	−0.9828	−0.5547	2.1588	2.1588	4.6840	3.8973
−2.75	3.4875	4.4413	−1.2682	−0.9031	−0.6192	2.2385	2.2385	4.6771	3.6727
−3.00	3.2000	4.3863	−1.0667	−0.8176	−0.6839	2.3239	2.3239	4.6688	3.4060
−3.25	2.8875	4.3474	−0.8885	−0.7264	−0.7476	2.4152	2.4152	4.6587	3.0942
−3.50	2.5500	4.3304	−0.7286	−0.6296	−0.8082	2.5119	2.5119	4.6466	2.7362
−3.75	2.1875	4.3414	−0.5833	−0.5281	−0.8638	2.6135	2.6135	4.6326	2.3342
−4.00	1.8000	4.3863	−0.4500	−0.4229	−0.9119	2.7187	2.7187	4.6165	1.8944
−4.25	1.3875	4.4708	−0.3265	−0.3156	−0.9506	2.8260	2.8260	4.5984	1.4271
−4.50	0.9500	4.5992	−0.2111	−0.2081	−0.9784	2.9335	2.9335	4.5787	0.9458
−4.75	0.4875	4.7750	−0.1026	−0.1023	−0.9948	3.0393	3.0393	4.5577	0.4653
−5.00	0.0000	5.0000	0.0000	0.0000	−1.0000	3.1416	3.1416	4.5360	0.0000

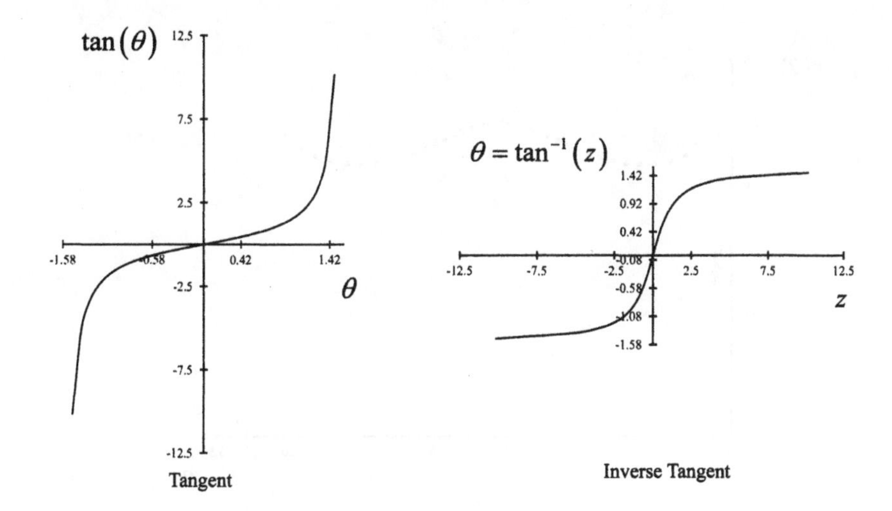

FIGURE C.1
Standard Table for Tangent.

Consider a couple of things about that point. First, $x = 0$ represents the 90° angle – the tangent at this point is positive infinity from the right and negative infinity from the left. Second, the inverse functions for the tangent are typically defined on the interval $\theta \in \left(-\frac{\pi}{2}, \frac{\pi}{2}\right)$, as depicted in Figure C.1. Given that we are interested in the range $\theta \in [0, \pi]$, we transform the inverse tangent in column 8 of Table B.1 by adding π to each value. Given this adjustment, the θs based on the tangents are the same as the θs based on the cosines. Figure C.2 presents the transformed relationship between $r(\theta)$ and θ.

Based on the graphical results in Figure C.2, we can make a variety of approximations to the original function. First, we could use the average value of the radius ($\bar{r} = 4.6548$). Alternatively, we can regress

$$r(\theta) = \gamma_0 + \gamma_1 \theta + \gamma_2 \theta^2 + \gamma_3 \theta^3. \tag{C.3}$$

Figure C.3 presents each approximation in the original Cartesian space. The graph demonstrates that the linear approximation lies slightly beneath the cubic approximation.

Given that we can approximate the function in a polar space, we demonstrate the integration of the transformed function. Using the result from Equation 3.111,

$$dydx = rdrd\theta$$

we can write the integral of the transformed problem as

$$\int_0^{\pi} \int_0^{4.6548} rdrd\theta = 34.0347. \tag{C.4}$$

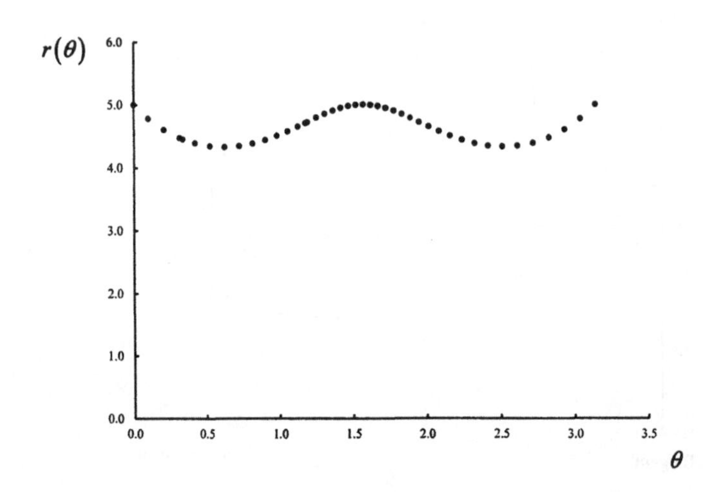

FIGURE C.2
Transformed Function.

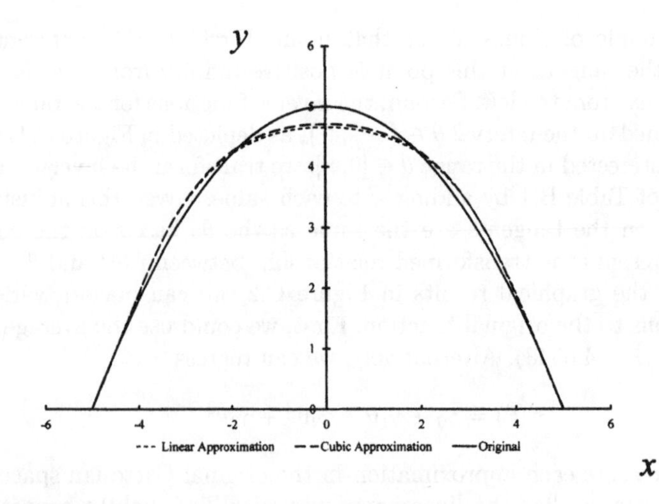

FIGURE C.3
Approximation in (x, y) Space.

This result can be compared with the integral of the original function of

$$\int_{-5}^{5}\left(5-\frac{x^2}{5}\right)dx = \frac{100}{3} \approx 33.3333. \tag{C.5}$$

Hence, the approximation is close to the true value of the integral.

Obviously the polar integral is much more difficult for the quadratic function in Equation 3.103. The point of the exercise is that some functions such as the normal distribution function are more easily integrated in polar space. If the original function is not integrable, the polar space may yield a close approximation. The most viable candidates for transformation involve the sum of squared terms – such as the exponent in the normal distribution.

C.2 Fourier Approximation

A related but slightly different formulation involving trigonometric forms is the Fourier approximation. To develop this approximation, consider the second order Taylor series expansion that we use to approximate a nonlinear function.

$$\begin{aligned} f(x) = f(\tilde{x}) + & \begin{bmatrix} \frac{\partial f(x)}{\partial x_1} \\ \frac{\partial f(x)}{\partial x_2} \end{bmatrix}' \begin{bmatrix} x_1 - \tilde{x}_1 \\ x_2 - \tilde{x}_2 \end{bmatrix} + \\ & \frac{1}{2} \begin{bmatrix} x_1 - \tilde{x}_1 \\ x_2 - \tilde{x}_2 \end{bmatrix}' \begin{bmatrix} \frac{\partial^2 f(x)}{\partial x_1^2} & \frac{\partial^2 f(x)}{\partial x_2 \partial x_1} \\ \frac{\partial^2 f(x)}{\partial x_1 \partial x_2} & \frac{\partial^2 f(x)}{\partial x_2^2} \end{bmatrix} \begin{bmatrix} x_1 - \tilde{x}_1 \\ x_2 - \tilde{x}_2 \end{bmatrix}. \end{aligned} \tag{C.6}$$

Letting the first and second derivatives be constants and approximating around a fixed point $(\tilde{x})$, we derive a standard quadratic approximation to an unknown function.

$$y = a_1 + \begin{bmatrix} a_1 \\ a_2 \end{bmatrix}' \begin{bmatrix} x_1 \\ x_2 \end{bmatrix} + \begin{bmatrix} x_1 \\ x_2 \end{bmatrix}' \begin{bmatrix} a_{11} & a_{12} \\ a_{12} & a_{22} \end{bmatrix} \begin{bmatrix} x_1 \\ x_2 \end{bmatrix}. \tag{C.7}$$

We can solve for the coefficients in Equation C.7 that minimize the approximation error to produce an approximation to any nonlinear function. However, the simple second order Taylor series expansion may not approximate some functions – the true function may be a third or higher order function.

An alternative approximation is the Fourier approximation

$$f(x) = \alpha_0 + \sum_{j=1}^{N} \left[\alpha_1 \cos\left(\frac{x \times \pi}{k_i}\right) + \alpha_2 \sin\left(\frac{x \times \pi}{k_i}\right) \right] \tag{C.8}$$

where k_i is a periodicity. Returning to our example, we hypothesize two periodicities $k_1 = 5/\pi$ and $k_2 = 5/2\pi$. The approximations for the two periodicities are estimated by minimizing the squared difference between the function and

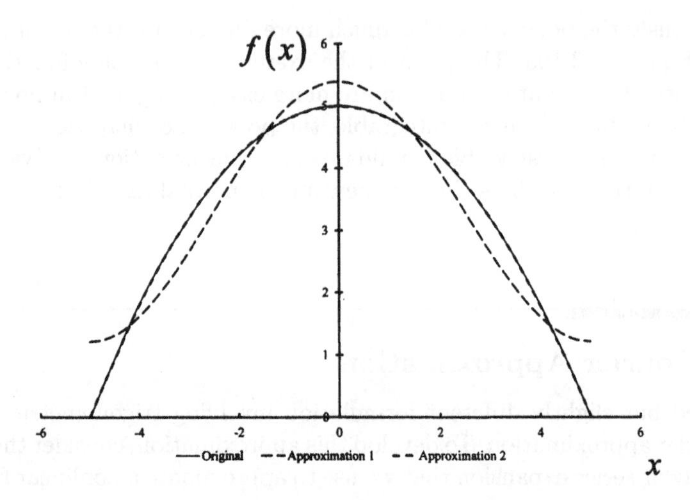

FIGURE C.4
Fourier Approximations.

Equation C.8 and are presented in Figure C.4. Given that the original function is a quadratic, the second order Taylor series expansion is exact. However, the Fourier approximation is extremely flexible. A variant of this approximation is used in Theorem 6.24.

Appendix D

Farm Interest Rate Data

TABLE D.1
Farm Interest Rates for Southeastern United States

	Alabama		Florida		Georgia		South Carolina		Baa
Year	Int Rate	$\Delta (D/A)$	Int Rate	$\Delta (D/A)$	Int Rate	$\Delta (D/A)$	Int Rate	$\Delta (D/A)$	Rate
1960	0.0553	0.1127	0.0511	−0.0588	0.0507	0.0789	0.0595	0.1232	0.0506
1961	0.0547	0.0501	0.0500	0.0771	0.0514	0.0008	0.0584	0.0427	0.0495
1962	0.0551	0.0580	0.0512	0.0864	0.0516	0.1405	0.0584	0.0793	0.0490
1963	0.0546	0.0278	0.0558	0.0651	0.0521	0.0286	0.0589	0.0561	0.0474
1964	0.0537	0.0269	0.0586	0.0826	0.0516	0.0462	0.0602	0.0105	0.0472
1965	0.0543	0.0290	0.0509	0.0612	0.0541	0.0324	0.0547	0.0378	0.0475
1966	0.0545	−0.0024	0.0517	0.1083	0.0550	0.0339	0.0549	0.0462	0.0551
1967	0.0551	0.0059	0.0542	0.0093	0.0580	0.0113	0.0585	−0.0232	0.0604
1968	0.0573	−0.0299	0.0556	0.0063	0.0603	−0.0362	0.0595	−0.0277	0.0671
1969	0.0601	0.0247	0.0575	−0.0277	0.0631	0.0443	0.0628	0.1077	0.0752
1970	0.0609	−0.0486	0.0627	−0.0138	0.0669	−0.0079	0.0681	0.0292	0.0871
1971	0.0614	0.0166	0.0603	−0.0001	0.0651	−0.0185	0.0650	−0.0145	0.0822
1972	0.0588	0.0791	0.0592	−0.0273	0.0624	0.0208	0.0647	0.0449	0.0784
1973	0.0630	−0.1264	0.0622	−0.0522	0.0653	−0.0443	0.0631	0.0483	0.0792
1974	0.0695	0.0566	0.0682	0.0987	0.0731	0.0577	0.0753	0.0392	0.0907
1975	0.0700	0.0404	0.0716	−0.0057	0.0760	0.1529	0.0794	0.1296	0.1008
1976	0.0718	−0.0079	0.0720	−0.0459	0.0757	−0.0232	0.0750	−0.0006	0.0930
1977	0.0724	0.1041	0.0675	0.0208	0.0716	0.0539	0.0708	0.1041	0.0859
1978	0.0743	−0.0040	0.0706	−0.0417	0.0754	−0.0371	0.0861	−0.0805	0.0906
1979	0.0825	−0.0585	0.0768	−0.0094	0.0817	0.0485	0.0829	0.0075	0.1016
1980	0.0912	0.0273	0.0830	−0.0070	0.0889	0.0720	0.0907	0.0674	0.1281
1981	0.1014	0.1143	0.0877	0.1584	0.1028	0.1242	0.1018	0.1199	0.1488
1982	0.1081	0.0499	0.0946	0.0059	0.1133	0.0158	0.1129	0.0757	0.1494
1983	0.1059	0.0065	0.0950	0.0002	0.1134	0.0103	0.1139	0.0424	0.1271
1984	0.1017	0.0121	0.0900	0.0729	0.1122	−0.0143	0.1113	−0.0145	0.1327
1985	0.0903	−0.0327	0.0942	0.0100	0.1010	−0.0017	0.1030	0.0099	0.1197
1986	0.0928	−0.1189	0.0916	−0.0917	0.1044	−0.1526	0.1008	0.0116	0.0989
1987	0.0943	−0.1231	0.0982	−0.1512	0.0926	−0.0528	0.0961	−0.1878	0.1005
1988	0.0963	−0.1058	0.0971	−0.0111	0.0951	−0.1124	0.0978	−0.2189	0.1028
1989	0.0993	−0.0577	0.0990	−0.0733	0.0987	−0.1397	0.0978	−0.1068	0.0969
1990	0.0962	−0.0414	0.0973	0.0214	0.0965	−0.0423	0.0908	−0.0842	0.0985
1991	0.0867	−0.0321	0.0915	0.0014	0.0887	0.0421	0.0848	−0.0705	0.0935
1992	0.0816	−0.0744	0.0842	−0.0003	0.0836	−0.0455	0.0800	−0.0210	0.0860
1993	0.0831	−0.0680	0.0757	0.0470	0.0772	−0.0241	0.0769	−0.0640	0.0763
1994	0.0835	−0.0427	0.0787	0.0366	0.0785	−0.0113	0.0784	−0.0468	0.0827
1995	0.0814	0.0472	0.0810	−0.0303	0.0768	0.0475	0.0775	−0.0131	0.0788
1996	0.0849	0.0292	0.0836	0.0409	0.0836	−0.0147	0.0815	0.0408	0.0775
1997	0.0830	0.0212	0.0789	0.0466	0.0801	0.0029	0.0791	−0.0102	0.0757
1998	0.0805	0.0518	0.0766	0.0512	0.0789	−0.0212	0.0770	0.0020	0.0697
1999	0.0794	0.0000	0.0752	0.0004	0.0790	−0.0909	0.0779	0.0521	0.0758
2000	0.0794	0.0287	0.0756	−0.0053	0.0777	−0.0278	0.0673	0.1619	0.0803
2001	0.0682	0.0169	0.0668	−0.0176	0.0671	−0.0100	0.0603	0.0272	0.0765
2002	0.0629	0.0097	0.0628	−0.0151	0.0619	−0.0265	0.0611	−0.0019	0.0751
2003	0.0509	−0.0229	0.0528	−0.0251	0.0506	−0.0319	0.0498	−0.0248	0.0655

Appendix E

Nonlinear Optimization

CONTENTS

As an introduction to numerical procedures for nonlinear least squares and maximum likelihood, this appendix presents the solution for the three parameter Cobb–Douglas production function presented in Section 12.1. Then we present R code that solves the maximum likelihood formulation of the same problem. After the maximum likelihood example, the appendix then presents a simple application of Bayesian estimation using simulation. Finally, the appendix presents the iterative solution to the Least Absolute Deviation problem using a conjugate gradient approach.

E.1 Hessian Matrix of Three-Parameter Cobb–Douglas

As a starting point, we derive the Hessian matrix (Equation 12.15) for the three parameter Cobb–Douglas production function. Notice that by Young's theorem (i.e., $\partial^2 f/\partial x_1 \partial x_2 = \partial^2 f/\partial x_2 \partial x_1$) we only have to derive the upper triangle part of the Hessian.

$$
\begin{aligned}
\frac{\partial^2 L(\alpha)}{\partial \alpha_0^2} &= 2\sum_{i=1}^{40} x_{1i}^{2\alpha_1} x_{2i}^{2\alpha_2} \\
\frac{\partial^2 L(\alpha)}{\partial \alpha_0 \partial \alpha_1} &= 2\sum_{i=1}^{40} \left[\alpha_0 x_{1i}^{2\alpha_1} x_{2i}^{2\alpha_2} - x_{1i}^{\alpha_1} x_{2i}^{\alpha_2} \left(y_i - \alpha_0 x_{1i}^{\alpha_1} x_{2i}^{\alpha_2}\right)\right] \ln(x_{1i}) \\
\frac{\partial^2 L(\alpha)}{\partial \alpha_0 \partial \alpha_2} &= 2\sum_{i=1}^{40} \left[\alpha_0 x_{1i}^{2\alpha_1} x_{2i}^{2\alpha_2} - x_{1i}^{\alpha_1} x_{2i}^{\alpha_2} \left(y_i - \alpha_0 x_{1i}^{\alpha_1} x_{2i}^{\alpha_2}\right)\right] \ln(x_{2i}) \\
\frac{\partial^2 L(\alpha)}{\partial \alpha_1^2} &= 2\alpha_0 \sum_{i=1}^{40} \left[\alpha_0 x_{1i}^{2\alpha_1} x_{2i}^{2\alpha_2} \ln(x_{1i}) - x_{1i}^{\alpha_1} x_{2i}^{\alpha_2} \left(y_i - \alpha_0 x_{1i}^{\alpha_1} x_{2i}^{\alpha_2}\right)\right] \ln(x_{1i}) \\
\frac{\partial^2 L(\alpha)}{\partial \alpha_1 \partial \alpha_2} &= 2\alpha_0 \sum_{i=1}^{40} \left[\alpha_0 x_{1i}^{2\alpha_1} x_{2i}^{2\alpha_2} - x_{1i}^{\alpha_1} x_{2i}^{\alpha_2} \left(y_i - \alpha_0 x_{1i}^{\alpha_1} x_{2i}^{\alpha_2}\right)\right] \ln(x_{1i}) \ln(x_{2i}) \\
\frac{\partial^2 L(\alpha)}{\partial \alpha_2^2} &= 2\alpha_0 \sum_{i=1}^{40} \left[\alpha_0 x_{1i}^{2\alpha_1} x_{2i}^{2\alpha_2} \ln(x_{2i}) - x_{1i}^{\alpha_1} x_{2i}^{\alpha_2} \left(y_i - \alpha_0 x_{1i}^{\alpha_1} x_{2i}^{\alpha_2}\right)\right] \ln(x_{2i}) .
\end{aligned}
\tag{E.1}
$$

Using $\alpha_0 = \{60.00, 0.25, 0.15\}$ as the starting value, the first gradient vector becomes

$$\frac{\nabla_\alpha L(\alpha)}{100{,}000} = \begin{bmatrix} 1.408 \\ 459.680 \\ 383.507 \end{bmatrix} \tag{E.2}$$

where we have normalized by 100,000 as a matter of convenience. The Hessian matrix for the same point becomes

$$\frac{\nabla^2_{\alpha\alpha} L(\alpha)}{100{,}000} = \begin{bmatrix} 0.037 & 18.955 & 15.791 \\ 18.955 & 5860.560 & 4874.010 \\ 15.791 & 4874.010 & 4092.800 \end{bmatrix} . \tag{E.3}$$

The next point in our sequence is then

$$\begin{bmatrix} \alpha_0 \\ \alpha_1 \\ \alpha_2 \end{bmatrix} = \begin{bmatrix} 60.00 \\ 0.25 \\ 0.15 \end{bmatrix} - \begin{bmatrix} -0.130 \\ 0.053 \\ 0.031 \end{bmatrix} = \begin{bmatrix} 60.130 \\ 0.197 \\ 0.119 \end{bmatrix} . \tag{E.4}$$

The first six iterations of this problem are presented in Table E.1. A full treatment of numeric Newton–Raphson is beyond the scope of this appendix, but if we accept that the last set of gradients is "close enough to zero," the estimated Cobb–Douglas production function becomes

$$f(x_1, x_2) = 62.946 x_1^{0.160} x_2^{-0.019} . \tag{E.5}$$

TABLE E.1
Newton–Raphson Iterations

Iteration	Parameters α_0	Parameters α_1	Parameters α_2	Function Value	Gradient α_0	Gradient α_1	Gradient α_2
1	60.000	0.250	0.150	3,058,700	1.491	459.680	383.507
2	60.130	0.197	0.119	816,604	0.508	156.642	130.767
3	60.259	0.161	0.082	180,003	0.160	49.278	41.197
4	60.410	0.149	0.038	41,179	0.041	12.733	10.684
5	60.747	0.157	−0.001	24,492	0.006	1.942	1.645
6	62.946	0.160	−0.019	24,003	0.000	0.032	0.030

E.2 Bayesian Estimation

Applications of applied Bayesian econometric techniques have increased significantly over the past twenty years. As stated in the text, most of this expansion is due to the development of empirical integration techniques. These technical advancements include the development of the Gibbs sampler. While a complete development of these techniques is beyond the scope of the current text, we develop a small R code to implement the simple example from the textbook.

The code presented below inputs the data (i.e., the dta < − rbind(...)) command and then draws 200 random draws from the prior distribution (i.e., a < − rnorm(2000,mean=0.25,sd=sqrt(0.096757))). Given each of these draws, the code then computes the likelihood function for the sample and saves the simulated value, the likelihood value, and the likelihood value times the random draw (i.e., a[i]*exp(l)). The last line of the program then computes the Bayesian estimation

$$\hat{a} = \frac{\sum_{i=1}^{200} \mathrm{a}\,[\mathrm{i}] \times L\,[x|\,\mathrm{a}\,[\mathrm{i}]]}{\sum_{i=1}^{200} L\,[x|\,\mathrm{a}\,[\mathrm{i}]]} \tag{E.6}$$

R Code for Simple Bayesian Estimation

```
dta <- rbind(0.1532,0.1620,0.1548,0.1551,0.1343,0.1534,0.1587,
             0.1501,0.1459,0.1818)

a <- rnorm(200,mean=0.25,sd=sqrt(0.096757))

for (i in 1:200) {
  for (t in 1:10) {
```

```
    if (t == 1) l <- 0
    l <- l -((dta[t]-a[i])^2)/(2*0.096757) }
  if (i == 1) res <- cbind(a[i],exp(l),a[i]*exp(l)) else \$
       res <- rbind(res,cbind(a[i],exp(l),a[i]*exp(l))) }

ahat <- sum(res[,3])/sum(res[,2])

print(res)
print(ahat)
```

E.3 Least Absolute Deviation Estimator

Section E.1 presented the overall mechanics of the Newton–Raphson algorithm. This algorithm is efficient if the Hessian matrix is well behaved. However, because of the difficulty in computing the analytical Hessian and possible instability of the Hessian for some points in the domain of some functions, many applications use an approximation to the analytical Hessian matrix. Two of these approximations are the Davidon–Fletcher–Powel (DFP) and the Broyden–Fletcher–Goldfarb–Shanno (BFGS). These algorithms are typically referred to as conjugate gradient routines – they conjugate or build information from the gradients to produce an estimate of the Hessian.

These updates start from the first-order Taylor series expansion of the gradient

$$\nabla_x f(x_t + s_t) = \nabla_x f(x_t) + \nabla^2_{xx} f(x_t)\, s_t \tag{E.7}$$

where x_t is the current point of approximation and s_t is the step or change in x_t produced by the algorithm. Hence, the change in the gradient gives information about the Hessian

$$\nabla_x f(x_t + s_t) - \nabla_x f(x_t) = \nabla^2_{xx} f(x_t)\, s_t. \tag{E.8}$$

The information in the Hessian can be derived from

$$s_t' \left(\nabla_x f(x_t + s_t) - \nabla_x f(x_t)\right) = s_t' \nabla^2_{xx} f(x_t)\, s_t. \tag{E.9}$$

Many variants of this code are implemented in numerical codes. In R, these codes can be used in the function *optim.* The code presented below provides an example of this code based on the estimation of the Least Absolute Deviation estimator.

R Code for Least Absolute Deviation Estimator

```
dta <- read.csv("EthanolGasoline.csv")

bols <- lm(dta[,6]~dta[,7] + dta[,8])
print(summary.lm(bols))

x <- as.matrix(cbind(matrix(1,nrow=nrow(dta),ncol=1),dta[,7:8]))

lad <- function(b) {
    bvec <- rbind(b[1],b[2],b[3])
    ll <- abs(dta[,6] - x%*%bvec)
    return(1/nrow(dta)*sum(ll)) }

b0 <- cbind(0.95,0.31,0.19)

blad <- optim(b0,lad)
```

Glossary

actuarial value a fair market value for a risky payoff – typically the expected value.

aleatoric involving games of chance.

asymptotic the behavior of a function $f(x)$ as x becomes very large (i.e., $x \to \infty$).

Bayes' theorem the theorem defining the probability of a conditional event based on the joint distribution and the marginal distribution of the conditioning event.

best linear unbiased estimator the linear unbiased estimator (i.e., an estimator that is a linear function of sample observations) that produces the smallest variance for the estimated parameter.

binomial probability the probability function of repeated Bernoulli draws.

Borel set an element of a σ-algebra.

composite event an event that is not a simple event.

conditional density the probability density function – relative probability – for one variable conditioned or such that the outcome of another random variable is known.

conditional mean the mean of a random variable or a function of random variables given that you know the value of another random variable.

continuous random variable a random variable such that the probability of any one outcome approaches zero.

convergence for our purposes, the tendency of sample statistics to approach population statistics or the tendency of unknown distributions to be arbitrarily close to known distributions.

convergence in distribution when the distribution function F of sequence of random variables $\{X_n\}$ becomes infinitely close to the distribution function of the random variable X.

convergence in mean square a sequence of random variables $\{X_n\}$ converges in mean square if $\lim_{n\to\infty} E[X_n - X]^2 = 0$.

convergence in probability a sequence of random variables $\{X_n\}$ converges in probability to a random variable X if $\mathrm{P}(|X_n - X| < \epsilon) 1 - \delta$.

correlation coefficient normalized covariance between two random variables.

discrete random variable a random variable that can result in a finite number of values such as a coin toss or the number of dots visible on the roll of a die.

endogeneity the scenario where one of the regressors is correlated with the error of the regression.

Euclidean Borel field a Borel field defined on real number space for multiple variables.

event a subset of the sample space.

expected utility theory the economic theory that suggests that economic agents choose the outcome that maximizes the expected utility of the outcomes.

frequency approach probability as defined by the relative frequency or relative count of outcomes.

Generalized Instrumental Variables using instruments that are correlated with the residuals and imperfectly correlated with X, to remove endogeneity.

gradient vector a vector of scalar derivatives for a multivariate function.

heteroscedasticity the scenario where either the variances for the residuals are different and/or the variances are correlated across observations.

homoscedastic the scenario where the errors are independently and identically distributed $(\sigma^2 I_{T\times T})$.

joint probability the probability that two or more random variables occur at the same time.

kurtosis the normalized fourth moment of the distribution. The kurtosis is 3.0 for the standard normal.

Lebesgue integral a broader class of integrals covering a broader group of functions than the Riemann sum.

marginal distribution the function depicting the relative probability of one variable in a multivariate distribution regardless or independent of the value of the other random variables.

nonparametric statistics that do not assume a specific distribution.

ordinary least squares typically refers to the linear estimator that minimizes the least squares of the residual $\hat{\beta} = (X'X)^{-1}(X'Y)$.

probability density function a function that gives the relative probability for a continuous random variable.

quantile the random variable such that k percent of the other random variables in the sample are less than $x^*(k) \Rightarrow \int_{\infty}^{x} f(z)\,dz$.

quartiles the random variable at the sample such that 25%, 50%, and 75% of the random variables are smaller.

Reimann sum the value of the integral of a function – typically the antiderivative of the function.

Saint Petersburg paradox basically the concept of a gamble with an infinite value that economic agents are only willing to pay a finite amount for.

sample of convenience a sample that was not designed by the researcher – typically a sample that was created for a different reason.

sample space the set of all possible outcomes.

sigma-algebra a collection F of subsets of a nonempty set Ω satisfying closure under complementarity and infinite union.

simple event an event which cannot be expressed as a union of other events.

skewness the third central moment of the distribution. Symmetric distributions such as the normal and Cauchy have a zero skewness. Thus, a simple description of the skewness is that skewness is a measure of nonsymmetry.

statistical independence where the value of one random variable does not affect the probability of the value of another random variable.

Two Stage Least Squares a two stage procedure to eliminate the effect of endogeneity. In the first stage the endogenous regressor is regressed on the truly exogenous variables. Then the estimated values of the endogenous regressors are used in a standard regression equation.

Type I error the possibility of rejecting the null hypothesis when it is correct.

Type II error the possibility of failing to reject the null hypothesis when it is incorrect.

Uniform distribution the uniform distribution has a constant probability for each continuous outcome of a random variable over a range $x \in [a, b]$. The standard uniform distribution is the $U[0,1]$ such that $f(x) = 1$ for $x \in [0,1]$.

Working's law of demand the conjecture that the share of the consumer's expenditure on food declines as the natural logarithm of income increases.

References

[1] T. Amemiya. *Introduction to Statistics and Econometrics.* Harvard University Press, Cambridge, MA, 1994.

[2] Q. Ashraf and O. Galor. The "out of Africa" hypothesis, human genetic diversity, and comparative economic development. *American Economic Review*, 103(1):1–46, 2013.

[3] G. Bassett and R. Koenker. Asymptotic theory of least absolute error regression. *Journal of the American Statistical Association*, 73(363):618–622, September 1978.

[4] G. S. Becker. *Economic Theory.* Aldine Transaction, New Brunswick, NJ, 2008.

[5] H. J. Bierens. *Introduction to the Mathematical and Statistical Foundations of Econometrics.* Camb, New York, 2004.

[6] C. Borell. Lecture notes in measure theory. Webpage, January 2006.

[7] G. Casella and R. L. Berger. *Statistical Inference.* Duxbury, Pacific Grove, CA, second edition, 2002.

[8] A. DasGupta. *Asymptotic Theory of Statistics and Probability.* Springer, New York, NY, 2008.

[9] P. de Fermat. Letter from Fermat to Pascal. Webpage: http://science.larouchepac.com/fermat/, 1654.

[10] B. de Finetti. *Theory of Probability: A Critical Introductory Treatment*, volume 1. John Wiley & Sons, New York, NY, 1970.

[11] J. L. Doob. *Stochastic Processes.* John Wiley & Sons, New York, NY, 1953.

[12] J. Fox. Robust regression, CRAN.R-project.org/doc/contrib/Fox-companion/appendix-robust-regression.pdf. January 2002.

[13] M. Friedman. *Essays in Positive Economics*, chapter The Methodology of Positive Economics, pages 3–43. Chicago University Press, Chicago, IL, 1953.

[14] A. R. Gallant. *Nonlinear Statistical Models.* John Wiley & Sons, New York, NY, 1987.

[15] T. Haavelmo. Prize lecture: Econometrics and the welfare state. Webpage http://www.nobelprize.org/nobel_prizes/economics/laureates/1989/, April 2013.

[16] I. Hacking. *The Emergence of Probability: A Philosophical Study of Early Ideas about Probability, Induction and Statistical Inference.* Cambridge University Press, New York, NY, second edition, 2006.

[17] A. R. Hall. *Generalized Method of Moments.* Oxford University, New York, NY, 2005.

[18] R. V. Hogg, J. W. McKean, and A. T. Craig. *Introduction to Mathematical Statistics.* Pearson Prentice Hall, Upper Saddle River, NJ, sixth edition, 2005.

[19] P. J. Huber. *Robust Statistics.* John Wiley & Sons, New York, NY, 1981.

[20] C. Huygens. *The Value of All Chances in Games of Fortune; Cards, Dice, Wagers, Lotteries etc.* S. Keimer, 1714.

[21] S. Johansen. *Likelihood Based Inference on Cointegration in the Vector Autoregressive Model.* Oxford University Press, second edition, 1996.

[22] D. W. Jorgenson. Dataverse – 35 sector klem. Webpage, July 2008. Webpage http://scholar.harvard.edu/jorgenson/data.

[23] G. G. Judge, R. C. Hill, W. E. Griffiths, H. Lutkepohl, and T.-C. Lee. *Introduction to the Theory and Practice of Econometrics.* John Wiley & Sons, New York, NY, 1982.

[24] L. R. Klein. *A Textbook of Econometrics.* Row, Peterson, and Co., Evanston, IL, 1953.

[25] J. R. Kling, S. Mullainathan, E. Shafir, L. C. Vermeulen, and M. V. Wrobel. Comparison friction: Experimental evidence from Medicare drug plans. *Quarterly Journal of Economics*, 127(1):199–235, 2012.

[26] J. Kmenta. *Elements of Econometrics.* University of Michigan Press, Ann Arbor, second edition, 1986.

[27] R. Koenker and G. Bassett. Regression quantiles. *Econometrica*, 46(1):33–50, 1978.

[28] S. Kullback. *Information Theory and Statistics.* Dover Publications, Mineola, NY, 1968.

[29] J. R. Magnus and H. Neudecker. *Matrix Differential Calculus with Applications in Statistics and Econometrics.* Wiley & Sons, New York, NY, revised edition, 1999.

[30] A. G. Malliaris. *Stochastic Methods in Economics and Finance.* North-Holland, New York, NY, 1982.

[31] A. K. Mishra and C. B. Moss. Modeling the effect of off-farm income on farmland values: A quantile regression approach. *Economic Modelling*, 32:361–368, 2013.

[32] C. B. Moss. *Risk, Uncertainty, and the Agricultural Firm.* World Scientific, Hackensack, NJ, 2010.

[33] C. B. Moss and J. S. Shonkwiler. Estimating yield distributions using a stochastic trend model and nonnormal errors. *American Journal of Agricultural Economics*, 75(5):1056–1062, November 1993.

[34] C. B. Moss and T. G. Schmitz. A semiparametric estimator of the Zellner production function for corn: Fitting the univariate primal. *Applied Economics*, pages 863–7, 2006.

[35] S. Naidu and N. Yuchtman. Coercive contract enforcement: Law and the labor market in nineteenth century Britain. *American Economic Review*, 103(1):107–144, 2013.

[36] C. H. Nelson. The influence of distributional assumptions on the calculation of crop insurance premia. *North Central Journal of Agricultural Economics*, 12(1):71–78, January 1990.

[37] B. Norwood, M. C. Roberts, and J. L. Lusk. Ranking crop yield models using out-of-sample likelihood functions. *American Journal of Agricultural Economics*, 86(4):1032–1043, November 2004.

[38] K. Popper. *The Logic of Scientific Discovery.* Routledge Classics, New York, NY, 2010.

[39] D. Salsburg. *The Lady Tasting Tea: How Statistics Revolutionized Science in the Twentieth Century.* Henry Hold and Company, New York, NY, 2001.

[40] A. Sandmo. On the theory of the competitive firm under price uncertainty. *American Economic Review*, 61(1):65–73, March 1971.

[41] L. J. Savage. *The Foundations of Statistics.* Dover Publications, New York, NY, 1972.

[42] A. Schmitz, C. B. Moss, T. G. Schmitz, H. W. Furtan, and H. C. Schmitz. *Agricultural Policy, Agribusiness, and Rent-Seeking Behaviour.* University of Toronto Press, Toronto, Canada, 2010.

[43] G. Schmoller. *The Mercantile System and Its Historical Significance.* Evergeen Review, Inc., New York, NY, 2008.

[44] A. Smith. *The Wealth of Nations.* Penguin Books, New York, NY, 1982.

[45] A. Spanos. *Statistical Foundations of Economic Modelling.* Cambridge University Press, New York, NY, 1986.

[46] E. Staley. *World Economy in Transition: Technology vs. Politics, Laissez Faire vs. Planning, and Power vs. Welfare.* Kennikat Press, Port Washington, NY, 1939.

[47] S. M. Stigler. *The History of Statistics: The Measurement of Uncertainty before 1900.* Belknap Press of Harvard University Press, Cambridge, MA, 1986.

[48] A. E. Taylor and W. R. Mann. *Advanced Calculus.* John Wiley & Sons, New York, NY, third edition, 1983.

[49] H. Theil. *Principles of Econometrics.* John Wiley, New York, 1971.

[50] H. Theil, C. F. Chung, and Seale, J. L., Jr. *International Evidence on Consumption Patterns.* JAI Press, Greenwich, CT, 1989.

[51] J. Tinbergen. *Econometrics.* The Blakiston Company, New York, 1951.

[52] J. Tinbergen. *On the Theory of Economic Policy.* North-Holland Publishing Co., New York, NY, 1952.

[53] H. White. *Asymptotic Theory for Econometricians.* Academic Press, San Diego, CA, revised edition, 2001.

[54] A. Zellner. Statistical theory and econometrics. In K. J. Arrow and M. D. Intriligator, editors, *Handbook of Econometrics*, volume 1, chapter 2, pages 67–178. North-Holland Publishing Co., New York, NY, 1983.

[55] H. Theil. Estimation and Simultaneous Correlation in Complete Equation Systems in Henri Theil's *Contributions to Economics and Econometric: Econometric Theory and Methodology*, Volume 1. Edited by B. Raj and J. Koerts. Klewer Academic Publishers, Norwell, MA, 1992.

Index